GEOMETRY, GEODESICS, AND THE UNIVERSE

GEOMETRY, GEODESICS, AND THE UNIVERSE

THE LINES THAT LED FROM
EUCLID TO EINSTEIN

ROBERT G. BILL

Library of Congress Control Number:		2022919890
ISBN:	Hardcover	978-1-6698-5306-0
	Softcover	978-1-6698-5305-3
	eBook	978-1-6698-5304-6

Cover with photo of the Orion Nebula and interior figures by the author.

Print information available on the last page.

Rev. date: 07/11/2023

To order additional copies of this book, contact:
Xlibris
844-714-8691
www.Xlibris.com
Orders@Xlibris.com
848109

CONTENTS

PART II
Following the Path of Number

PART III
The Geometric Universe

LIST OF FIGURES

LIST OF TABLES

Foreword

The subject of my book is the development of geometry from the concepts of the ancient Greeks, familiar from high school, to the four-dimensional space-time that is central to our modern vision of the universe. For the specialist, there are many technical books that will delve into the subjects of modern geometry and physics at an advanced mathematical level. However, for many who are not specialists, this approach will be impossible to follow. Alternatively, there are books without any, or virtually without any, mathematics whatsoever. Such books can qualitatively describe geometrical concepts including those leading to Special and General Relativity, but I believe that many readers could obtain a more exact understanding of these subjects from a mathematical presentation that takes advantage of their typical math background.

Mathematics is, of course, the natural language to express the concepts of geometry and Relativity and brings with it a precision in meaning which I believe is otherwise impossible. Moreover, the experience of following a mathematical description rather than a qualitative verbal description is akin to the difference between playing music and just listening to it. The experiences are intellectually and aesthetically quite different.[i] Along the way you will encounter concepts that are crucial to mathematics, science, the meaning of truth, space and time, the infinite, and the origin of the universe - not a bad list!

My ideal readers would be those who enjoyed encountering new concepts in math and physics in high school or introductory college courses and

---

[i]   Of course, to carry my analogy further, composing is a very different creative experience than playing. It is a privilege to be able to follow the discoveries of great mathematicians and physicists bringing some part of their work to your attention.

puzzling out their distinctive problems, although they did not pursue them further in their vocation. I only assume in my presentation familiarity with high school geometry, algebra, and Newton's laws of motion. These staples of school will be presented sufficiently to reacquaint you with the necessary concepts. Also helpful would be some familiarity with calculus as it would be obtained in a first introductory course; however even in the absence of such a background, concepts from the calculus will be developed as needed. In particular, the basic meaning of the differential and integral calculus will be geometrically motivated by appealing to their relation to the slopes of curves and to the area under a curve. In both these cases, key results will be illustrated using numerical methods that readers could perform for themselves with a spread sheet such as EXCEL. It seems to me that a numerical approach has an advantage over traditional analytic results in that it can make clearer the meaning of the various calculus operations.

A major theme of the book is the expansion of our understanding of geometry beyond that of Euclid. Two thousand years after Euclid, a new geometry was discovered that changed the understanding of geometry and mathematics, while paving the way for Einstein's General Relativity. In order to follow the revolutionary path to the first new geometry, you will need no other additional skills than those developed in high school to understand Euclid and, unlike many of the followers of Euclid, an open mind. The next steps leading to General Relativity will introduce concepts necessary to characterize the new geometries, including notably the metric tensor and the concept of curvature. These are unquestionable sophisticated mathematical entities, but fortunately you already have at least an intuitive understanding of a non-Euclidean geometry that can be used to introduce the new mathematics, that is, the spherical geometry familiar from the globe found in classrooms and homes. Spherical geometry is therefore used in detailed example calculations so that you can gain a comfort level with all newly introduced mathematical concepts.

As should be clear, my presentation is not meant for specialists in mathematics and physics. With this in mind, I have made use of traditional older tensor component notation rather than its modern abstract coordinate independent counterpart. I believe that results are more easily visualized in the component notation. In this way the presentation should be accessible to all readers from the continuing math enthusiast to those with a science and engineering background who would not have been introduced in their studies to the concepts told here.

With the new mathematical concepts in place, the principles behind Einstein's General Relativity will be described with the object of providing an introduction to the formulation and the meaning of Einstein's Field Equations. The solutions to these equations are the source of our understanding of such phenomena as the Big Bang and black holes. Although the full details of the solutions of the Field Equations are beyond the scope of my book, the level of understanding achieved will allow you to see how models of the universe arise from General Relativity, explaining the interconnectedness of the geometry of space-time with the experimentally observed expansion of the universe.

Throughout my book, I have cited references where more details can be obtained. The references have been selected to help readers to continue their journey in mathematics. Thus, they support an equally important goal of my book: opening readers to some of the pleasures of mathematics. In this regard, I dedicate my book to those who long ago put their mathematics aside, but also to those involved in science or engineering who received much training in the practical use of advanced mathematics, but with little regard to its foundations. The concepts addressed here are implicit in much of mathematics taught in high school and beyond, but hidden by an emphasis of math as a tool. My role here is to be a tour guide along a deductive journey pointing out some of the wonders discovered by mathematicians. As such, I have endeavored to include everything needed for your understanding. Thus, I hope you will enjoy the journey and be inspired to continue to explore mathematics as one of humanity's great expressive and creative achievements.

R. G. B. May 2021

Prologue

Our current understanding of the origin and structure of the universe is based on General Relativity published by Albert Einstein in 1915. More than two hundred years before, Isaac Newton had inaugurated modern physics with his Laws of Motion and Universal Law of Gravitation. These laws mathematically described the motions of the planets and the earth's moon, the motion of objects on the earth under the force gravity, and the tides. They remain central to much of engineering. However, Einstein, in a revolutionary theory, replaced Newton's description, in which gravity acted as a force, with a geometric description in which the planets simply followed geometric paths, called geodesics. The paths are dictated by the geometry of space-time, in itself a revolutionary development, which had its origins in Einstein's Special Relativity. In Newton's description, motion occurred on a stage of absolute three-dimensional space with objects following Newton's Laws, moving to the flow of an absolute time. Newton's Laws also provided a description for other special observers moving with uniform velocity relative to absolute space.

Einstein's description banished absolute space and is valid for any observer. Time is no longer absolute and is inescapably intertwined with space to form a four-dimensional geometry. In addition to the dismissal of the familiar gravitational forces holding the planets in their orbits, the new theory, as a generalization of Einstein's earlier Special Relativity, retained such extraordinary effects as the dependence of length, time, and simultaneity of events on the observer's motion.

One might imagine that with the introduction of so many counterintuitive phenomena, Einstein's theories burst forth with no connection to the past. However, the mathematics that was needed by Einstein was made possible by developments in the understanding of geometry that had occurred throughout the nineteenth century and that followed from two thousand years of questions about the geometric concepts developed by the Greeks of the first millennium BC. This book follows that development.

The geometry of the ancient Greeks holds a special place in mathematics as attested to by its continued use, virtually unchanged in secondary math education. The achievements of the Greeks are therefore a natural place to begin our story. However, their revolution in mathematics may be better appreciated by a brief review of the sources and content of the mathematics that the Greeks inherited from the past. Thus, our story begins tens of thousands of years earlier with the first communities of *Homo sapiens* of the Upper Paleolithic Period. Among the many revolutionary activities that would characterize our ancient ancestors was an awareness of form and number. For example, the discoveries of numerous cave paintings from over 30,000 years ago attest to an early delight in form and its visual expression that is vital not only to the beginnings of art, but also to geometry.[1][ii] Occurring about the same time period, wolf bones discovered in 1937 excavations, showed markings with notches in groupings of five.[2] These artifacts suggest the idea of counting and development of the concept of number with a rudimentary base 5 system. As we shall see, the themes of form and number evolving over millennia become one of the most important modes of human expression - mathematics.

Evidence of the development of the concepts of form and number goes well beyond these first steps with the burst of creativity associated with the dawn of civilization and the advent of history. Of the importance of form, one has only to think of the pyramids of Egypt or the ziggurats of Mesopotamia to see the hold of form on the ancient imagination. The practical requirements of construction and agriculture would naturally lead these ancient civilizations to the development of insights into geometric forms and quantitative aspects that could be expressed with

---

[ii]   Numbered notes refer to citations given at the back of the book (Notes). Full details of the cited references are given in the Bibliography.

2

their number systems. In geometry, the Egyptians were concerned with areas and volumes of geometric figures.[3] Methods for computing the volume of a pyramid and a truncated pyramid were known, but we do not know how they were derived. These methods were provided through instructions for specific examples, although not always strictly correct. For example, the area of an arbitrary quadrilateral was calculated as the product of half the sum of two opposite sides and half the sum of the other two sides.[4] Their method for determining the area of a circle implies a reasonable approximation for π (256/81 or 3.16...). They were aware of the right triangle, and of specific lengths of sides which produced them, for example: 3, 4, 5. Similar, comments could be used to summarize the achievements of the Babylonians in geometry, although they seemed to know some general results such as the Pythagorean Theorem, but without a proof.[5]

In regard to the concept of number, written notation for numbers occurred as documented through the discovery of ancient Egyptian papyri and the discovery of clay tablets from the civilizations of Mesopotamia. For example, about 4,000 years ago in Egypt, symbols were introduced for the powers of 10 up to 1,000,000. The system generated numbers through repetition, for example, a number such as 50 was represented using the symbol for 10 five times. When the repetition increased to the next power of ten, a new symbol was introduced. Thus, the position of the symbols was not critical, and no symbol for zero existed.[6] Using their symbols, the Egyptians developed techniques for addition, subtraction, multiplication, use of fractions in which the numerator was unity (except for the fraction 2/3), and division. However, the methodologies for employing these arithmetic operations were, like geometric calculations, supplied through specific instructions for practical calculations without the development of a more general or symbolic approach. Other types of examples demonstrated methods for determining unknowns in specific simple algebraic problems, again using problem specific instructions[7].

The Babylonian number system from about the same period was a sexagesimal (base 60) positional system, conceptually similar to our decimal system (base 10) with the first fifty nine numbers using a symbolic system like that of the Egyptians with symbols for units and tens. Like our decimal system the number of units of a given power of 60 depended on the placement of the symbols; however, it was not always clear to which power of 60 the symbols applied. A method of separating the numbers in situations analogous to those in which our zero is used in the decimal system was eventually developed about 300 BC; however, it was not conceived of as an actual number like our zero.[8] Mathematical techniques were developed by the Babylonians with some steps towards greater generalization than provided by Egyptian approaches. These techniques included: the arithmetic operations, approximations of square roots (even for some cases that we know of as irrational numbers – numbers that cannot be expressed as the ratio of two integers), and solutions to algebraic problems including forms of the quadratic equation with positive solution.[9] Remarkably, in the absence of a system of symbolic representation, the Babylonians were able to solve cubic equations (as expressed here in our modern algebraic notation) of the form

$$ax^3 + bx^2 = c$$

with a, b, and c being specific positive numbers.[10] Solutions of cubic equations were unknown to the Egyptians.

By the dawn of Greek civilization in the first millennium BC, the ideas of numeration, the operations of arithmetic, simple solutions of algebraic type problems (however, without a general symbolic approach), and some elements of geometry, including quantification of area and volume had appeared. It is worth repeating to underscore the achievements of the Greeks that a key aspect of this knowledge was its practical character emphasizing the transmission of knowledge through specific examples. For the development of general approaches with results based upon proofs, progress would await new and revolutionary insights introduced by the Greeks. Over a period of a few hundred years, they developed a method to uncover general abstract truths of geometry. Over a period of two thousand years, through frequent examinations of the foundations of the Greeks' discovery, a new geometry would be found eventually leading to a new understanding of mathematics and a method to describe the structure of the universe. These new discoveries follow what I call the path of form as they follow the long line of inspiration from the paleolithic cave paintings and the pyramids. The mathematics of the ancient Greeks would not provide a systematic understanding of numbers; however, their geometry would lead to fundamental questions about numbers, in particular irrational numbers, that would also be crucial to the development of mathematics. But for now, let us take a closer look at the details of the Greeks' geometric discoveries.

PART I

FOLLOWING THE PATH OF FORM

1 Lessons From School: Euclid's Legacy

The system of geometry developed by the ancient Greeks was organized by Euclid in his textbook the *Elements* in the fourth century BC. One of the remarkable aspects of the Greeks' development of geometry with its deductive method of proof is that the subject continued to be a core subject in mathematics with its key principles intact for over two thousand years, even to the present day in high school textbooks. During that two-thousand-year period, there were many attempts to improve on Euclid's system, as will be discussed in Chapter 2, but it was still seen as embodying absolute truths just as imagined by the ancient Greeks. Nevertheless, the search for improvements in Euclid's system undoubtedly added depth to the understanding of Euclid's geometry and facilitated the discovery of new geometries (Chapter 3). With this in mind, key elements and proofs in Euclid's geometry are given in the following sections. These elements should be familiar from high school geometry, but a review will reacquaint you with the methods of the Greek mathematicians and prepare you by providing a contrast with the revolutionary new geometry discovered in the nineteenth century. Before beginning that review, however, I shall summarize some of the contributions of the Greeks prior to Euclid that underscore the innovative nature of changes made to the mathematics of the Egyptians and Babylonians.

The pyramids of ancient Egypt are monuments attesting to the ancient world's knowledge of practical aspects of geometry long before the Greeks of the first millennium BC began their study of geometric principles. In the busy commercial world of the eastern Mediterranean, the Greeks would certainly have become aware of the geometric knowledge of neighboring communities. Indeed, the first Greek generally recognized as a mathematician is Thales (ca. 640-550 BC) of Miletus on the west coast of Asia Minor. Thales was a merchant who while travelling became familiar with the astronomy and geometry of the Egyptians and Babylonians.[11] As with many of the Greeks of the classic period, little is directly known of his life except through references to him from a work by Proclus (410-485), *Commentary on the First Book of Euclid's Elements*,[iii] written about a thousand years after Thales' death. Proclus refers to an earlier work, now lost, summarizing Thales' propositions.[12] The following geometric propositions, attributed to Thales, are given below:[13]

1. *The angles at the base of an isosceles triangle are equal.*
2. *If two straight lines cut one another, the vertically opposite angles are equal.*
3. *A triangle is determined if its base and base angles are given* [that is to say, triangles with equal bases and base angles are congruent].
4. *A circle is bisected by any diameter.*
5. *The angle subtended by any diameter is a right angle.*[iv]

All of the propositions listed above would eventually find their way into Euclid's *Elements*. Significantly, in contrast to the measurement based, empirical approach to geometry of the Egyptians and the Babylonians, the propositions are of a general abstract nature rather than prescriptions for how to calculate some geometric property such as length in a specific case. This desire by the Greeks to understand the general nature of geometric properties, in distinction to the Egyptian and Babylonian approach is what I have designated as the path of form.

While it is not clear how Thales proved his propositions (or even whether he proved all of them), his interest in general properties of geometric figures was an enormous advance in mathematics. Proclus

---

[iii] The work by the geometer Proclus also supplies some of the few known details on the life of Euclid.; Heath, Vol. 1, p. 1.

[iv] In addition to these propositions, Ball, pp.15-16, attributes an additional one to Thales: The sides of equiangular triangles are proportional.

noted specifically that Thales proved Proposition 4.[14] Although it seems to be a particularly simple conclusion, the effort to find a logical basis for a conclusion of such a general nature was the beginning of the deductive approach to geometry and mathematics. Thales did not develop a complete ordered system of geometry - that would take several hundred more years. He was, however, the first known to use deductive methods.[15] Thales' approach to geometry should be seen in a wider context as consistent with his efforts to understand the world around him through rational explanations.[16] In essence, Thales asked, What are all things? This is not a question one encounters in a world explained in mythic terms. Thales' answer, as reported by Aristotle, was that "all things are water,"[17] that is, all things resulted from various transformations of water as liquid, solid, or gas. This is not a very promising answer from our perspective, but it is perhaps the first attempted unified theory of the universe. In the world of the ancient Greeks, Thales' implicit question, What is the world made of? would eventually lead to Democritus' (ca. 400 BC) statement that," Nothing exists but atoms and the void"[v] with its echoes in modern physics.[18] That Thales explained the immense variety of nature by transformations from a single substance is not surprising given his evident interest in deductive proofs in which accepted statements lead to a succession of connected statements and a conclusion. Thales' explanation for all things can be seen as the first concept of matter and with his concept of transformation, arguably, Thales is not only the first mathematician, but the first physicist. For his pioneering attempt to explain reality without reference to the gods, he is the father of philosophy.[19] From the time of Thales, the subjects of geometry and physics would take many intricately winding paths that would eventually join in General Relativity.

Among the most familiar names from mathematics is Pythagoras (569-500 BC).,[20] known primarily for the famous theorem named for him. Pythagoras probably taught at his school the essentials of what would become the first two books of Euclid's *Elements*. Pythagoras' lectures

---

[v]   Democritus' detailed atomic theory had earlier roots, notably Empedocles (ca 440 BC) - see Brumbaugh, pp. 68-73.

apparently included proofs of the properties of parallels, triangles, and parallelograms, although in some cases his proofs were faulty. For example, some proofs assumed that if a statement had been proved to be true, than its converse was also true.[21] Like many of the proofs of the propositions ascribed to Pythagoras, we cannot be certain whether the Pythagorean Theorem was original with him or was developed by the followers of his school. However, the eminent scholar Sir Thomas L. Heath[vi] saw no reason to question the tradition of Pythagoras being the originator of the first proof of that famous theorem among the Greeks.[22] Whether the proofs were discovered by Pythagoras or his followers, Pythagoras and his school developed geometry as a deductive and ordered structure in which each new proposition followed from those previously proved.[23] Geometry continued to be developed vigorously after Pythagoras with its main lines visible long before Euclid. For example, according to Proclus, Hippocrates of Chios (ca. 430 BC) wrote a textbook comparable to that of Euclid over one hundred years earlier. Hippocrates' textbook like all such documents prior to Euclid have been lost.[24] No contemporary manuscript of the *Elements* exists; however, it was copied countless times and transferred to Western Europe in the Middle Ages by the Islamic civilization, becoming one of the most wide-spread books in the world.[25] [vii]

One more development of the school of Pythagoras will be mentioned as it would also have an important impact on mathematics, that is, the Pythagoreans devotion to whole numbers. According to Boethius, a sixth century Roman, the relationship between different lengths of vibrating strings and their musical tone was discovered by Pythagoras.[26] For example, if a string was halved in length, it would produce the same tone an octave higher. As an example, on a modern instrument, if the A string of a violin is touched in the middle, that is the length of the string is reduced by a a ratio of 1 to 2, the A tone is produced an octave higher, ie., at twice the frequency. Similarly Pythagoras noticed that harmonious

---

[vi]   Sir Thomas L. Heath wrote the definitive work translating and commenting on the Euclid's *Elements* - see Bibliography.

[vii]   The earliest existing manuscript of Euclid's *Elements* is from a commentary made 700 years after Euclid - Eves, p. 29.

tones were formed when string lengths of instruments such as the lyre were in a ratio of whole numbers. For example, the fifth is formed when the string lengths have a ratio of two to three - E in the case of the A string when bowed (or plucked) on the violin. This may have been the first example of a quantitative physical law, beginning our expectation that the physical universe could be described by mathematics. The Pythagoreans saw in this result evidence of the specific power of whole numbers to control all phenomena. They would associate all manner of things with whole numbers: odd numbers were masculine, even number were feminine, six was the number of the soul....[27] More relevant for us is the investigation of the properties of classes of whole numbers such as whether numbers were odd, even, prime, or composite. For example, they knew that the sum of two even numbers was even, the product of two odd numbers was odd, and when and odd number divides and even number, it also divides its half.[28] Thus, their worship of whole numbers became a major contribution to what we call number theory.

One discovery about numbers that the Pythagoreans found disconcerting was that all geometric lengths could not be given as the ratio of two whole numbers, that is, as a rational number. From the Pythagorean Theorem, they knew the relationship between the side of a square and its diagonal. In the specific case of a square with unit sides, high school students can show that the length of the diagonal is what we designate as $\sqrt{2}$, the square root of two. The Pythagoreans proved that this length could not be be written as the ratio of two whole number by an indirect proof, that is the assumption that the length was rational led to a contradiction. (The proof is given in Appendix A). As a result, the Greeks avoided numerical geometric descriptions which probably delayed some advances in mathematics. The importance of linking the points of a line with numbers would be vital to the expansion of our understanding of geometry. However, this did not effectively occur until the seventeenth century when René Descartes (1596-1650) and Pierre Fermat (1601-1665)[29] used coordinate descriptions of curves aided by advances in symbolic representation in algebra (Chapter 4).

Although the Greeks were fascinated with the properties of numbers, their system of numeration was a non-positional system with letters of the alphabet as symbols for numbers, e.g., the first nine letters stood for 1 through 9 and continuing on with letter symbols for 10, 20...,100, 200,...,1000, 2000,...,9000. Larger numbers were expressed through multiplying symbols.[30] The system was not a conceptual advance over that of the Egyptians nor well devised to advance the theory of numbers.

1.1   Euclid's self-evident truths

The revolutionary contribution of the Greeks to geometry and mathematics in general was the shift in focus from the search for solutions to specific problems to an approach that would develop general geometric conclusions. The deduction of propositions from previously proven propositions, in an unbroken chain of reasoning connecting back to definitions and accepted assumptions became the basis for deductive mathematics, the axiomatic method. This method is named for its dependence on the assumptions which the ancient Greeks presumed to be self–evident truths called axioms.[viii] For two thousand years, the Greeks and later mathematicians thought this approach provided the only description of spatial relations in the universe. This turning point in mathematics in Greek communities of the first millenium BC was consistent with the broad, tenacious search within these communities of that period for explanations for everything that was encountered in the world about them. This search has influenced art, drama, literature, philosophy, and social organization to this day.

Euclid's *Elements* consists of thirteen books which cover plane geometry (Books I to VI), number theory (Books VII to IX), and solid geometry (Books X to XIII).[31] My purpose here is to present the critical role in Euclid's geometry of his fifth postulate (called the

---

[viii]   Axioms in modern terminology refer to all the abstract assumptions that form the basis of a mathematical system. In contrast, Aristotle, however, saw them as accepted truths that were broken up into two categories: common notions and postulates with axiom being an alternative name for common notion. The distinction between the two categories is not very clear.

Parallel Postulate) to the subsequent developments. For that purpose, only a discussion of aspects of Book I and II are necessary. Euclid's development of plane geometry began with twenty-three definitions, five postulates, and five common notions. Much of our understanding of the significance of these elements of Euclid's system to the Greeks comes from discussions of them by Aristotle and Proclus.[32] Ultimately, these elements, while remaining in the modern axiomatic system, would have to be reinterpreted, as we shall see in Chapter 7. However, for the moment let us look at them through Aristotle's eyes.

Aristotle explained that one begins with certain definitions and what we would call assumptions which he variously discussed as postulates and common things. He noted that

> *"Now the things peculiar to the science, the existence of which must be assumed, are the things with reference to which the science investigates the essential attributes, e.g. arithmetic to units and geometry with reference to points and lines. With these things it is assumed they exist and that they are of such and such a nature. But, with regard to their essential properties, what is assumed is only the meaning of each term employed...."*[33]

Aristotle distinguished between definitions and assumptions noting *"Now definitions are not hypotheses, for they do not assert the existence or non-existence of anything.... Definitions only require to be understood...."*[34] With this background, it is time to list some of Euclid's definitions from Book I to illustrate their character and as a reminder of some of the geometry we encountered in school.

Definitions[ix]

D1. *A point is that which has no part.*
D2. *A line is breadthless length.*

[ix]   Definitions and subsequent Postulates, Common Notions, and Propositions of Euclid's *Elements* as translated by Heath (Vol.1).

D3. *The extremities of a line are points.*

D4. *A straight line is a line which lies evenly with the points on itself.*

D5. *A plane surface is that which has length and breadth only.*

D6. *The extremities of a surface are lines.*

D7. *A plane surface is a surface which lies evenly with the straight lines on itself.*

D8. *A plane angle is the inclination to one another of two lines in a plane which meet one another and do not lie in a straight line.*

D9. *And when the lines containing the angle are straight, the angle is called rectilineal.*

D10. *When a straight line set up on a straight line makes the adjacent angles equal to one another, each of the equal angles is right, and the straight line standing on the other is called perpendicular to that on which it stands.*

D11. *An obtuse angle is an angle greater than a right angle.*

D12. *An acute angle is an angle less than a right angle.*

D13. *A boundary is that which is an extremity of anything.*

D14. *A figure is that which is contained by any boundary or boundaries.*

D15. *A circle is a plane figure contained by one line such that all straight lines falling upon it from one point among those lying within the figure are equal to one another.*

D16. *And the point is called the centre of the circle.*

D17. *A diameter of the circle is any straight line drawn through the centre and terminated in both directions by the circumference of the circle and such a straight line also bisects the circle.*

D18. *A semicircle is the figure contained by the diameter and the circumference cut off by it. And the centre of the semicircle is the same as that of the circle.*

D19. *Rectilineal figures are those which are contained by straight lines, trilateral figures being those contained by three, quadrilateral those contained by four, and multilateral those contained by more than four straight lines.*

D20. *Of trilateral figures, an equilateral triangle is that which has three sides equal, an isosceles triangle that which has two of its sides alone equal, and a scalene triangle that which has its three sides unequal.*

D21. *Further of trilateral figures, a right-angled triangle is that which has a right angle, an obtuse-angled triangle that which has an obtuse angle, and an acute-angles triangle that which has three angles acute.*

D22. *Of quadrilateral figures, a square is that which is both equilateral and right-angled; an oblong* [a rectangle] *that which is right-angled, but not equilateral; a rhombus that which is equilateral but not right-angled; a rhomboid* [a parallelogram] *that which has its opposite sides and angles equal to one another but is neither equilateral nor right-angled. And let quadrilaterals other than these be called trapezia.*

D23. *Parallel straight lines are straight lines which being in the same plane and being produced indefinitely in both directions, do not meet one another in either direction.*

From geometry classes, I am sure you are familiar with the concepts of points, lines, planes, angles, circles, and figures such as triangles, squares, parallelograms, etc. defined generally as rectilinear figures and made more precise by Euclid elsewhere in other definitions. The final listed definition of parallel lines probably seems among the clearest to you; however, because lines were assumed to be capable of being extended indefinitely, the definition of parallel lines (D23) was destined to have an important role on the development of new geometries along with Euclid's vague definition of a straight line (D4).

Of all the definitions, Aristotle paid particular attention to points and lines. He designated them as being primary in that their existence could not be proved, but must be assumed. In Aristotle's view, this contrasted with other defined items whose existence had to be proved. For example, the very first proposition proved in the *Elements* is the existence of equilateral triangles (triangles with all sides equal). Looking more closely at the definitions for points and lines and at those for a plane, I believe that you will agree that you understand what Euclid means, but not because of the clarity of the definition, but because you already have a notion of what these words mean. Indeed, a significant part of the introduction of these concepts in modern high school textbooks is based on their visualization. while recognizing that there are terms that are ultimately undefined. For example, point and line are defined in a high school textbook in the following ways:[35]

A point has no dimension. It is usually represented by a small dot.
A [straight] *line extends in one dimension. It is usually represented by a straight line with two arrowheads to indicate that the line extends without end in two directions.*

If you try to develop your own definitions, I believe that you will see that a fully satisfactory set of definitions cannot be made. The modern view, developed in part because of the discovery of non-Euclidean geometries, is that there must be a set of undefined terms whose allowed interactions are defined by the axioms of the mathematical system and that the axioms are consistent assumptions of a deductive system, rather than self-evident truths. David Hilbert (1862-1943), one of the most significant contributors to this point of view, selected point, line, and plane along with the relations of incidence, between, and congruence for his undefined terms for plane geometry.[36] He also developed a set of axioms (postulates) to replace those of Euclid to remove deficiencies caused by unstated assumptions. [37] This will be discussed further in regard to the proof of Euclid's first proposition and in Chapter 4 in regard to the need to make the axiomatic system consistent with the use of real numbers in analytic geometry.

After the definitions, Euclid lists what he termed postulates and common notions. These are the indemonstrable principles which Aristotle says must be the start of any science. He notes that without these starting points, *"the steps of demonstration would be endless."* [38] In a similar vein Proclus says,

> ``the compiler of elements in geometry must give separately the principles of the science, and after that the conclusions from those principles, not giving any account of the principles but only of their consequences. No science proves its own principles or even discourses about them: they are treated as **self-evident** (emphasis added)....''[39]

Aristotle somewhat vaguely distinguished between the common notions and postulates. Possibly the clearest of his distinctions was that common notions were unprovable principles common to all sciences while postulates were particular to the science being developed. With our current understanding in which both types of principles are assumptions and in either case not self-evident truths, they are both referred to as axioms. Listed below are Euclid's Postulates and Common Notions (with the postulates designated with the letter "E" for Euclid to distinguish them from postulates of others that we will encounter).[40]

Postulates

E1. *To draw a straight line from any point to any point.*

E2. *To produce a finite straight line continuously in a straight line.*

E3. *To describe a circle with any centre and distance.*

E4. *That all right angles are equal to one another.*

E5. *That if a straight line falling on two straight lines make the interior angles on the same side less than two right angles, the two straight lines, if produced indefinitely, meet on that side on which are the angles less than two right angles.*

Common Notions

CN1. *Things which are equal to the same thing are also equal to each other.*

CN2. *If equals be added to equals, the wholes are equal.*

CN3. *If equals be subtracted from equals, the remainders are equal.*

CN4. *Things which coincide with one another are equal to one another.*

CN5. *The whole is greater than the part.*

Looking first at the Common Notions, we see concepts that seem familiar in the context of algebra, except for CN4. The Common Notions provided by Euclid, however, are far from the set of postulates that would be necessary to define the properties of the real number system (containing zero and the positive and negative rational and irrational numbers). Moreover, Euclid would find it necessary to use other magnitude related postulates, for example, postulates of inequalities to complete some of his proofs.[41] The postulates necessary to make a one-to-one correspondence between the real numbers and points of a line would not be resolved until the late nineteeth century, even later than the discovery of new geometries. This, along with the reinterpretation of the significance of definitions and postulates, would also play its part in making Euclidean geometry fully rigorous.

Common Notion 4 appears to fit in more appropriately with the postulates as it has specific geometric applications. Euclid based his concept of congruence on coincidence of figures (superposition) as stated in CN4. He used this in Proposition 4 of Book I to prove that two triangles are "equal" if the corresponding two sides of each triangle and the angles contained by the two sides are "equal." In this context, Euclid meant by "equal" that the figures may be made to coincide by placement of one onto the other. Euclid's approach, therefore, relied upon unstated assumptions about the uniformity of space and on the rigid motion of figures allowing comparisons of figures through their superposition. Thus, his geometry can be thought of as resting on unstated empirical notions of space. In contrast, later systems in the nineteenth century would avoid this by stating explicitly a postulate for the congruence of triangles.

Regarding the postulates, the first three seem quite basic in that they simply confirm the ability to draw the familiar figures of geometry with a straight edge and a compass. Following Aristotle's view, the postulates are not concerned with the imperfect forms that can be drawn, but their ideal forms[42] (also, see quote by Plato below). Heath notes that the fourth postulate, all right angles are equal, provides a standard for measuring other angles and again requires homogeneity of space.[43]

Through the centuries that followed Euclid's *Elements*, there were many clarifications of the first four postulates, but it was from concerns about his fifth postulate that new geometries and a new understanding of mathematics would eventually be formed. Euclid's fifth postulate differed significantly from the first four postulates in being much less acceptable to many as self–evident. If you undertake to explore its meaning with paper, pencil, and a straightedge, I believe you may come to accept it as ``true;'' However, it is certainly less basic than the other postulates. As we shall see, many mathematicians tried to prove it from the other postulates, but inevitably only discovered postulates, which while being clearer to some, could be shown to be equivalent to Euclid's Parallel Postulate (Chapter 2). Euclid, perhaps having his own concerns about this postulate did not invoke it until after he had given proofs of twenty-eight propositions.

Comparing the form of Euclid's definitions to the postulates, the tension between concepts which are idealized forms and those that are more closely grounded in experience is quite apparent. For example, we are told in the first two definitions that a point is that which has no part, and a line is breadthless length; whereas, the first postulate assumes as self–evidently true the capability to draw a straight line from any point to any point. This raises the question, how do you draw a breadthless line from that which has no part to that which has no part. A possible resolution of this contradiction for the ancient Greeks may have been found in the philosophy of Plato in which the ultimate realities rested in eternal perfect forms of which their physical counterparts were imperfect representations. Of geometers, Plato said:

> *"You know as well that they make use of visible shapes and objects and subject them to analysis. At the same time, however, they consider them only as images of the originals: the square as such or the diagonal as such. In all cases the originals are their concern and not the figures they draw... And all the while they seek a reality only the mind can discover."*[x]

Ultimately, the discovery of non–Euclidean geometries with the resulting realization that the geometric forms are abstractions logically

---

[x]   Plato, *The Republic*, Book VI (510 e), p. 200., translated by R. W. Sterling and W. C. Scott. While quoting from Plato, it is worth noting that he also said, "Further, we know that a man who has studied geometry is a better student across the board than one who has not." Ibid. (Book VII (527 c), p. 221.

following a set of assumptions, rather than self-evident truths, would make moot the question of whether geometry represented Plato's eternal forms. Still we retain a sense of Plato's eternal forms when we say that a mathematician has made a discovery. Let us make the discussion more concrete by taking a look at some of Euclid's proofs.

1.2 Consequences of the first four truths

The foundations of Euclid's geometry, as described in the previous section, are its Definitions, Postulates, and Common Notions. These function in the *Elements* in roughly the same way as they had been described by Aristotle. However, there remains the question of how they can be put to use to produce new geometric conclusions. We know that the Pythagoreans had proved many of the propositions of the first two books of the *Elements* so they had already developed rules to deduce conclusions from accepted or previously proved statements. However, it is Aristotle who is recognized as the organizer of such rules of logic through his *Analytica Posterior* written in the fourth century BC.[44]

Two of the key elements of the classic logic of Aristotle are the law of the excluded middle and law of contradiction. The former says that a statement must be either true or false while the law of contradiction says that a statement cannot be both true and false.[45] With this understanding of statements, the key engines for deducing conclusions are the syllogistic laws of logic. Syllogisms may be concisely expressed by using symbolic references to statements. As an example, let p, q, and r represent three different statements in the following syllogism: if p implies q, and q implies r, then p implies r. This is often rendered in modern notation as:

$$(p \Rightarrow q) \wedge (q \Rightarrow r) \Rightarrow (p \Rightarrow r)$$

Here, $\Rightarrow$ is the symbol for implication, and $\wedge$ is the symbol for the conjunction "and". The law of the syllogism can be shown to be a valid argument, that is, one in which the conclusions follow logically from the premises. Note that the validity of an argument has to do with its form not its truth; however, if the premises are true then, the conclusions are true.[xi] Since the law of syllogism is a valid argument, the truth of its conclusion depends, in the modern view, on premises being *accepted* as true. The Greeks saw the premises as being absolutely true rather than just assumptions - a difference which would lead to a new understanding of geometry and mathematics.

In the proofs presented in this chapter and subsequent chapters, statements given in the right hand columns are accepted statements (Definitions, Postulates, Common Notions, or previously proved statements) leading to the statement in the left hand column, all building to the conclusion. I have ended the proofs with the traditional abbreviation Q.E.D. for the Latin phrase, *quod erat demonstrandum* - that which was to be demonstrated.

Another approach to a deductive proof is the indirect proof or proof by contradiction. Recall that such a proof was used by Pythagoreans to show that $\sqrt{2}$ is not a rational number (see proof in Appendix A). In this method, the desired conclusion is assumed to be false; that is for example, $\sqrt{2}$ is assumed to be rational. If this leads to a contradiction, then the desired conclusion has been shown to be true.[xii]

The propositions of the *Elements* are of two types: those that prove the constructability of a geometric object, that is, proof of its existence, and those that prove geometric relationships. A surprising aspect of construction proofs is that the existence of squares or rectangles cannot be proved without Postulate E5. Euclid's first twenty-eight propositions, making use of only the first four postulates, remain valid even in

---

[xi] An example of an invalid argument is: if statement A implies statement B, then B implies A. A brief introduction to the laws of logic is given by Eves, pp. 5-9, 243-257.

[xii] For two thousand years, the laws of logic remained unquestioned, like assumptions about Euclidean geometry. However, as in geometry, revolutionary developments occurred in logic in the nineteenth century due to the development of symbolic logic. Furthermore, in the twentieth century, mathematicians came to realize that that they could use different laws of logic, just as geometries could be changed by using different postulates. For example, multi-value logics were developed not accepting Aristotle's law of the excluded middle. These developments, however, take us beyond our path. An introduction to this subject is given by Eves, pp. 257-262.

the geometry discovered in the nineteenth century that replaced the fifth postulate with a new one. For this reason, the first twenty-eight propositions are referred to as forming a neutral geometry.[46] In Appendix B, I have listed the first twenty-eight propositions of Euclid from Book I. I refer to the proofs as propositions in deference to traditional usage when discussing Euclid's *Elements*; however, the meaning of proposition is equivalent to the term theorem.

Let us start with the first proposition, IE1 which proves the existence of equilateral triangles. (I have numbered the propositions following the order of Euclid's proofs in Heath. For example, IE1 corresponds to Proposition 1 (Book I). of Euclid's *Elements*. In the proofs that follow, the following traditional symbolic notation will be used: $\triangle$ for triangle, $\angle$ for angle, $\cong$ for congruence of figures, and = for equality of magnitudes (angles, lengths of line segments, areas). Equality between line segments or angles can also be considered as indicating that the figures are congruent; that is, the figures can be superimposed on each other. When Euclid speaks of equality of triangles or quadrilaterals in regard to magnitudes, he means equality of areas. I have therefore used the symbol of equality for lengths, angles, and area for which a magnitude may be assigned (see e.g., Stahl see pp.41- 42).

Proposition IE1: On a given finite straight line [AB] construct an equilateral triangle.[47]

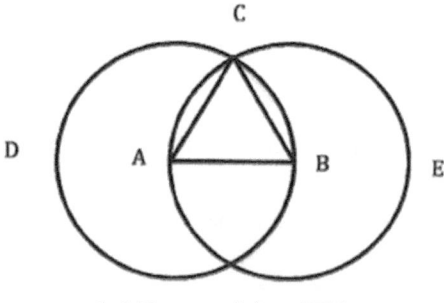

1-1 Proposition IE1

Let AB be the given straight line (see figure above)	By hypothesis
Construct the circle BCD of distance AB, centered at A	Postulate E3
Construct the circle ACE of distance AB, centered at B	Postulate E3
Construct the straight lines AC and BC with C being the point of intersection of the circles	Postulate E1
AC = AB, BC = AB	Definition 15
AC = BC	Common Notion 1
△ABC is an equilateral triangle	Definition E20

Q.E.D.

Proposition IE1 creates the equilateral triangle using only the defined concepts of point, line, and circle and the allowed operations of Postulates E1 and E3. Alternatively, given the line AB and circles of distance (radius) AB to their centers A and B, it is easily proved using the definitions and postulates that △ABC, with C being the intersection of the circles, is equilateral. The proof seems straightforward; however, you should try to understand the proof without reference to the figure above. When you do this, the following question may arise, how do we know that the circles intersect in a point C? Indeed, there is nothing in the definitions and postulates that assure the existence of point C. This is an example of a logical deficiency in Euclid's geometric system. Here Euclid needs additional postulates to complete his chain of reasoning. This would be resolved as noted by Heath by including a postulate of continuity and proofs of the existence of intersections such as that of two circles in Proposition 1.[48] The postulate of continuity is closely related to the correspondence of the real numbers with the continuous number line. Such logical deficiencies were corrected by Hilbert, among others.[49] With the addition of these assumptions Euclid's geometry forms a consistent system. Just the same, the logical necessity of including such a postulate of continuity becomes more urgent with the interpretation of geometry as coordinates in analytic geometry. For example, if quantitative descriptions are restricted to the rational numbers, then a circle presented on Cartesian axes (x, y) with a radius formed by the line from the coordinate origin (0,0) to the point (1,1) does not intersect the x-axis. The circle goes through the whole in the x-axis at the $\sqrt{2}$. To avoid this deficiency, the number system must be extended to include the irrational numbers. The basis of the extension as part of the real numbers is discussed in Chapter 4.[xiii]

The next proposition that I have selected to look at, IE16, provides both an example of a proof establishing geometric relationships as well as an example of a proposition that is limited in its conclusion due to Euclid's decision to refrain from using his Parallel Postulate. As a result of not using the Parallel Postulate, this proof, like those prior to IE29, are valid in any geometry which uses Euclid's first four postulates and assumptions, but modifies the fifth.

In the following proof, I refer to propositions that have been proved by Euclid prior to IE16. These are listed in Appendix B. Of particular note is Proposition IE4 which establishes conditions for the congruence of triangles, that is conditions for which the triangles can be made to coincide (can be superimposed on each other). In IE4 this is established when the respective two sides and the included angle of triangles are equal (often abbreviated SAS for side-angle-side). In the proof below (see accompanying figure), the triangles, $\triangle AEB$ and $\triangle CEF$ are shown to be congruent $\triangle AEB \cong \triangle CEF$). This is proved by establishing that the corresponding two sides of the triangles are equal (AE = CE, BE = EF), and included angles are equal, ($\angle AEB = \angle CEF$). Thus, by Proposition IE4, the remaining sides and angles of the two triangles are equal. Another proposition used to prove IE16 is IE15 in which it is proved that the angles formed by intersecting straight lines are equal (called

---

[xiii] A related approach to prove the existence of the intersection of the two circles in Euclid's first proposition is to introduce the Postulate of Separation (Stahl, pp. 41, 49.) which postulates conditions for intersection of lines. The Postulate of Separation states that the infinitely extended straight line, the triangle, and the circle separate the plane into two portions such that any line joining a point of one portion to a point of the other portion intersects the separating figure, In the case of the line the two portions are called the line's sides, In the case of a triangle or a circle the portions are called the interior and exterior.

vertical angles). In the proof below the proposition is invoked to prove ∠AEB = ∠CEF.

Proposition IE16: : In any triangle [△ABC], if one of the sides [BC] is produced [extended], the exterior angle [∠DCE] is greater than either of the interior and opposite angles ∠BAE, ∠ABC].

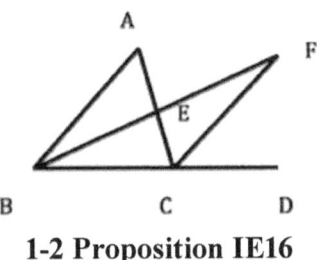

1-2 Proposition IE16

Let △ABC be the given triangle (see figure above)	By hypothesis
Produce base of △ABC to D	Postulate E2
Bisect AC at E	Proposition IE10
Produce BE	Postulate E1
Produce BE to F with BE = FE	Propositions IE2
Produce FC	Postulate E1
∠AEB = ∠CEF	Proposition IE15
∠BAC = ∠FCA	
(△AEB ≅ △CEF; SAS: AE = CE, BE = FE, ∠AEB = ∠CEF)	Proposition IE4
∠FCA < DCE	Common Notion 5
∠BAC < ∠DCE	Common Notion 1
Following the same approach by bisecting BC, it is proved that:	
∠ABC < DCE	

Q.E.D

In proposition IE16, the exterior angle ∠DCA) is proved to be greater than either of the interior and opposite angles (∠BAE, and ∠ABC).

However, with the use of the Parallel Postulate in IE32, Euclid was able to show that the exterior angle of a triangle is equal to the sum of the two interior and opposite angles∠ DCA = ∠ BAC + ∠ABC). Furthermore, in IE32, it is proved that the sum of the angles of the triangle is equal to two right angles (180°). A hallmark of non–Euclidean geometries developed through modifications of Euclid's Parallel Postulate is that the sums of the angles of the triangles of these geometries are not equal to two right angles.

Finally in regard to IE16, I note that as in IE1, a spatial assumption has been made which is not justified by Euclid's Postulates. In the proof above, line segment EF is produced with F being assumed to be exterior to △ABC. From the figure, this seems intuitively obvious, but it is not established by Euclid's Postulates. Euclid assumes here that a line may be extended to any length. This is another example of a logical deficiency in Euclid's system which would not be fully appreciated until the nineteenth century.[xiv] Unlike the case in which the existence of an intersection point was not proved, the failure to note the possibility that a line might not be unlimited in length would hide a different geometry. We will return to this in Section 2.3.2 with the introduction to spherical geometry.

The final proposition not using the Parallel Postulate which shall be given in detail is Proposition IE27. This proposition, along with Proposition IE28, gives criteria for proving that lines are parallel and are the last propositions prior to Euclid's use of the Parallel Postulate. Proposition IE27 is proved through an indirect proof, that is, assuming the conclusion is false leads to a contradiction.

Proposition IE27: If a straight line [EF] falling on two straight lines [AB and CD] make the alternate angles equal to one another [∠AEF = ∠EFD], the straight lines will be parallel to one another.

[xiv] Some limitations of Euclid's postulates as related to IE16 are given by Heath, Vol. 1, pp. 280 – 281.

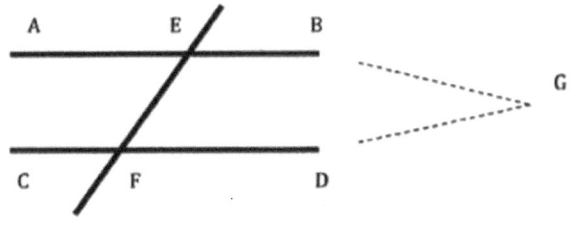

1-3 Proposition IE27

Proof by Contradiction:

Assume AB and CD (see figure above) are not parallel	By hypothesis
AB and CD when produced meet at some point G	Definition 23
EF, EG, FG form straight lines forming △GEF	By hypothesis, Postulate E1, Definition 19
∠AEF > ∠EFD	Proposition IE16
∠AEF = ∠EFD	By hypothesis

A contradiction has occurred; therefore, AB and CD must be parallel. Q.E.D.

Proposition IE28 extends the previous proposition. In addition to IE27, it makes use of Proposition IE13 which establishes that a straight line initiated on another straight line will form angles with a sum of two right angles. In the next figure, this corresponds to ∠AEF + ∠BEF = two right angles = 180°.

Proposition IE28: : [Part 1.]If a straight line [EF] falling on two straight lines [AB and CD] makes the exterior angle [∠GEB] equal to the interior and opposite angle on the same side ∠EFD], or [Part 2.] the interior angles on the same side [∠BEF and ∠EFD] equal to two right angles, the straight lines will be parallel to one another.

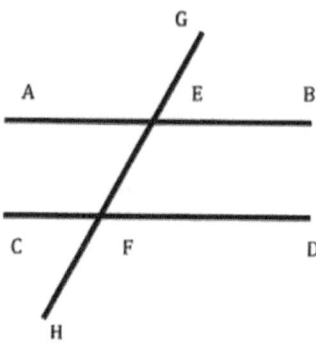

1-4 Proposition IE28

With the statement of IE28, let us now summarize the key propositions: proved by Euclid before using his fifth postulate. Some of the notable results include conditions for congruence of triangles based upon equality of corresponding sides (S) and angles(A):I E4-SAS, IE8-SSS, and IE26-ASA and AAS. Construction proofs include: the bisection of angles (IE9) and straight lines (IE10), production of a perpendicular from a point on a line (IE11), and from a point to a line (IE12). Of some importance to our understanding of the meaning of ``a straight line,' Euclid proved that in any triangle, two sides taken together in any manner are greater than the remaining one (IE20).

In addition to proving Thales' postulate of congruence of triangles (IE4), Euclid also included proofs of three of the other four propositions attributed to Thales, two of which are in the first twenty eight (IE5 and IE15). As Ball notes, Thales' proposition that a circle is bisected by any diameter ``*must have been recognized as an obvious fact from the earliest times.*''[50] Rather than being given as a proof, it is included in Definition 17. Euclid did not give a proof of Thales' proposition that an angle inscribed in a semicircle is a right angle until Book III, Proposition 31.[51]. There, Euclid used the proposition that the exterior angle of a triangle is equal to the two opposite interior angles (IE32) which ultimately depends on Euclid's fifth postulate. It is now time to look at the geometry created by including this postulate, the Parallel Postulate.

1.3 A not so self-evident truth: The Parallel Postulate and its consequences

In this section, some of the implications of the use of the Parallel Postulate are developed. For convenience, I repeat the postulate here.

Postulate E5 (Parallel Postulate*): That if a straight line falling on two straight lines make the interior angles on the same side less than two right angles, the two straight lines, if produced indefinitely, meet on that side on which are the angles less than two right angles.*

Although drawing a figure to illustrate Postulate 5 may be helpful, this postulate is clearly less obvious than the previous four. In addition, in assuming that lines may be produced "indefinitely," it assumes something that it is beyond direct experience; that is, a line may be extended to exceed any selected length. The denial of the assumption that a line could be extended to any length beyond the point of initiation would, as we shall see, be significant for non-Euclidean geometry.

The first proof taking advantage of the Parallel Postulate is IE29, which is essentially the converse of IE28. As previously discussed, the converse of a proposition is not necessarily true and in this case requires the use of the Parallel Postulate. Euclid's proof assumes two additional postulates without comment that are closely related to his Common Notions. The first is analogous to our Postulates of Order for the various number systems. Euclid assumes that in comparing two angles, only one of the following can be true: two angles must be either equal, less than, or greater than each other. Also, Euclid extends his Common Notion 2 to include inequalities. He could have stated as did latter geometers: if equals be added (or subtracted) to unequal quantities, the difference between the resulting quantities is equal to the difference between the original unequals. Some geometers evidently felt that such propositions being proved elsewhere were obvious, and thus their inclusion was not warranted. Similarly, there are situations in which an inequality is modified by multiplying both sides of the inequality with the same positive constant. Where such an explicit justification has not been supplied by Euclid, I will indicate this with the statement, Common Notion applied to inequalities. (Note the figure for IE29 is the same as that for IE28, but is reproduced below for convenience.)

Proposition IE29: [Part 1.] A straight line [EF] falling on parallel straight lines [AB and CD] makes the alternate angles equal to one another [∠AEF = ∠EFD], the exterior angle [∠GEB] equal to the interior and opposite angle on the same side [∠EFD], and [Part 2.] the interior angles on the same side [∠BEF + ∠EFD] equal to two right angles.

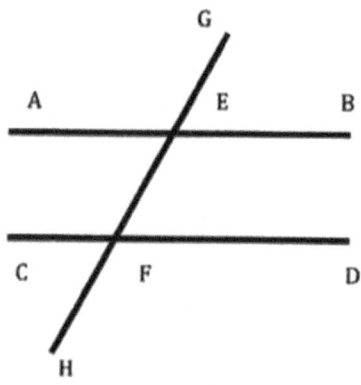

1-5 Proposition IE29

Part 1 (alternate interior angles),

Proof by Contradiction:
Assume ∠AEF ≠ ∠EFD (alternate angles not equal), then
Assume ∠AEF > ∠EFD

∠AEF + ∠BEF > ∠EFD +∠BEF	Common Notion applied to inequalities
∠AEF + ∠BEF = two right angles	Proposition IE13
∠EFD +∠BEF < two right angles	Common Notion 1
AB and CD are not parallel	**Postulate E5**
AB and CD are parallel	By hypothesis

⇒ a contradiction. Note that the same contradiction occurs in assuming ∠AEF < ∠EFD), therefore, ∠AEF = ∠EFD (alternate angles are equal)

Part 1 (exterior angle and interior opposite angle)

∠AEF = ∠GEB	Proposition IE15
∠GEB = ∠EFD	Common Notion 1

Part 2 (sum of interior angles)

∠GEB = ∠EFD	Shown above
∠GEB + ∠BEF = ∠EFD + ∠BEF	Common Notion 2
∠GEB + ∠BEF = two right angles	Proposition IE13
∠EFD + ∠BEF = two right angles	Common Notion 1

Q.E.D.

The power of the Parallel Postulate is particularly apparent in Proposition IE32 which can now provide a more precise conclusion than Proposition IE16 due to the use of the Parallel Postulate. Postulate IE16 (see previous section) can only say that the exterior angle of a triangle is greater than either of the opposite interior angles. In Proposition IE32, it is established that the exterior angle of a triangle is equal to the sum of the opposite interior angles. Moreover, it is established that the sum of the three interior angles of a triangle is equal to two right angles.

The proof of IE32 invokes the Parallel Postulate through the use of IE29 which could not be proved without it. Another proof, not encountered previously, is IE31 which proved that one can "*Through a given point... draw a straight line parallel to given straight line.*" [52] This is very similar to the more familiar statement found in high school texts which differs from IE31 by designating the drawn parallel line to be unique. This latter statement is equivalent to the Parallel Postulate. In IE31, the number of parallel lines that can be drawn is not explicitly mentioned, and its proof does not require the Parallel Postulate. This distinction will be important to the discovery of new geometries and will be discussed in Chapters 2 and 3.

Proposition IE32 In any triangle, [Part 1] if one of the sides be produced, the exterior angle [∠ACD] is equal to the two interior and opposite angles [∠BAC + ∠ABC], and [Part 2] the three interior angles are equal to two right angles.

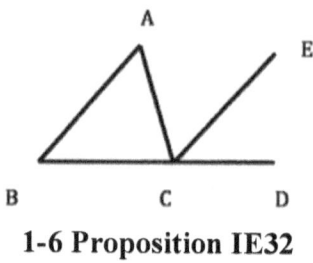

1-6 Proposition IE32

Part 1

Let △ABC be the given triangle (see figure above)	By hypothesis
Produce base of △ABC to D	Postulate E2.
Produce CE parallel to AB	Playfair's Postulate (or Proposition IE31)
∠ACE = ∠BAC (interior opposite angles)	Proposition IE29
∠ECD = ∠ABC (exterior and interior opposite angles)	Proposition IE29
∠ACD = ∠ACE + ∠ECD = ∠BAC + ∠ABC	Common Notion 2

Part 2

∠ACD + ∠ACB = ∠BAC + ∠ABC + ∠ACB	Result from Part 1, Common Notion 2
∠ACD + ∠ACB = two right angles	Proposition IE13
∠BAC + ∠ABC + ∠ACB = two right angles	Common Notion 1

Q.E.D.

Perhaps the most famous of the propositions resulting from the use of the Parallel Postulate is Proposition IE47 commonly known as the Pythagorean Theorem. In order for Euclid to prove this postulate, it was necessary for him to introduce a new notion of equality between rectilinear figures (triangles, quadrilaterals, and multilaterals – see Definition 19) based upon area. Critical to proving that the areas of non–congruent figures are equal is the comparisons of figures that are contained between common parallel lines, thus implicitly invoking the Parallel Postulate. Of particular importance to the proof of the Pythagorean Theorem is its explicit formulation depending on the existence, proved in IE46, of squares as a geometric object consistent with Euclid's Postulates and Common Notions. This is in the spirit of the proof of the existence of equilateral triangles in IE1. I have listed below other propositions that depend on the Parallel Postulate that are also needed in the chain of reasoning: [53]

IE34: In parallelogrammic areas the opposite sides and angles are equal to one another. And the diameter [diagonal] bisects the area

IE35: Parallelograms which are on the same base and in the same parallels are equal to one another.[xv]

IE37: Triangles which are on the same base and in the same parallels are equal to one another.

IE41: If a parallelogram have the same base with a triangle and be in the same parallels, the parallelogram is double of the triangle.

IE46: On a given straight line to describe a square.

To aid in the understanding these propositions, the proof of IE37 is given. Note that parallelograms are first introduced by Euclid in IE34 and are defined implicitly as a quadrilateral with parallel opposite sides. I shall use symbols such as □ABCD to indicate parallelograms and □EFGH for squares while general quadrilaterals will simply be indicated by the letters of their vertices, IJKL.

[xv]   Here for the first time Euclid uses equality in the sense of equality of areas of figures rather than congruence; Heath, Vol. 1, p. 327.

Proposition IE37: Triangles [△ABC, △DBC] which are on the same base and in the same parallels are equal [in area] to one another.

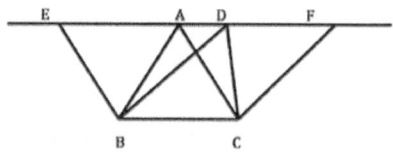

1-7 Proposition IE37

△ABC, △DBC are triangles between the parallel lines AD	
and the common base, BC	By hypothesis
Produce AD in both directions	Postulate E2
Through B produce BE parallel to C A	Proposition IE31
Through C produce CF parallel to BD	Proposition IE31
▱EBCA = ▱DBCF	Proposition IE35
△ABC = ½ ▱EBCA, △DBC = ½ ▱DBCF	Proposition IE34
△ABC = △DBC	Common Notion 1

Q.E.D.

We are now ready to follow Euclid's proof of the Pythagorean Theorem. Note that in order to interpret Euclid's proof as most commonly expressed in contemporary terms, it is necessary to take into account that the Greeks used a geometric algebra in which, as seen above, the equality of figures expressed the equality of their areas. So to express that $AB^2 = EF^2$, Euclid would express this as a square of side AB is equal to a square of side EF; for example ▱ABCD = ▱EFGH.

Proposition IE47 (The Pythagorean Theorem) : In right–angled triangles [△ABC] the square [▱BCED] on the side subtending the right angle [the hypotenuse, BC] is equal to [the sum of] the squares on the sides containing the right angle [▱ACKH and ▱ABFG]; that is

$$BC^2 = AB^2 + AC^2.$$

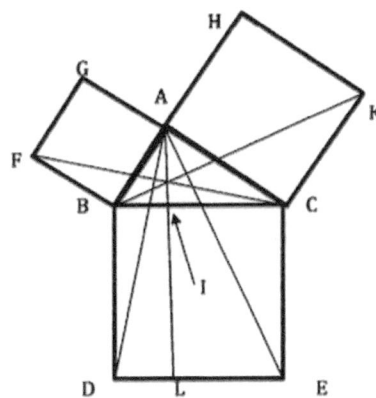

1-8 Proposition IE47 –The Pythagorean Theorem

△ABC has a right angle at ∠BAC	By hypothesis
Construct squares:	
▢BCED, ▢BAGF, and ▢ACKH on BC, BA, and AC, respectively	Proposition IE46
Draw: AD, AE, BK, and CF	Postulate E1
∠BAC, ∠BAG are right angles	By hypothesis, Proposition IE46
∠CAG =2 right angles, CG is a straight line	Proposition IE14
Similarly BH is a straight line	By hypothesis, Propositions IE46, IE14
∠DBC = ∠FBA	Proposition IE46
∠DBC + ∠ABC = ∠FBA + ∠ABC = ∠DBA = ∠FBC	Common Notion 2
DB = BC, FB = BA	Proposition IE46
△DBA ≅ △FBC, AD = FC (SAS, DB = BC, FB = BA, ∠DBA = ∠FBC)	Proposition IE4
Through point A, draw AL parallel to BD, CE	Proposition IE31
▢BILD = 2△DBA (△DBA and ▢BILD have the same base, between parallels BD and AL)	Proposition IE41

☐ABFG = 2△FBC (△FBC and ☐ABFG   Proposition IE41
have the same base, between parallels
FB and GC)

☐BILD = ☐ABFG   Common Notion 1

By a similar pattern of arguments:

∠ACE = ∠BCK; BC = EC; CK = AC; △KCB
≅△ACE; ☐CILE = ☐ACKH

☐BCED = ☐BILD +☐ CILE = ☐ABFG +   Common Notion 1
☐ACKH

(or $BC^2 = AB^2 + AC^2$)

Q.E.D.

There are two other themes in Euclid's *Elements* related to parallel lines that I wish to touch upon briefly: formulas for some areas of rectilinear figures and the idea of proportion as related to similar triangles. In regard to areas of rectilinear figures, I ask you to imagine two parallel lines separated by a distance h. On a common base of length b, of the lower parallel line, a rectangle and a parallelogram are formed with the upper parallel line. The area of the rectangle is, of course, b · h. From Proposition IE35, the area of any parallelogram that fits our description is also b · h. Imagine that the parallelogram is divided by a diagonal. Then, from Proposition IE34, the parallelogram is divided by its diagonal into two triangles of equal area, and the area of each of these triangles is 1/2 b · h. Imagining any other triangle with a base b between the parallel lines, IE37 says that all of these triangles also have an area of 1/2 b · h. Thus the propositions on these rectilinear figures between parallel lines provides the common formulas for the areas of rectangles, parallelograms, and triangles. This is a consequence of the Parallel Postulate. The areas of other rectilinear figures, such as trapezoids, can be obtained as composites of these figures.

From Euclid Book VI, similar triangles are defined as those triangles with equal angles, and corresponding proportional sides.[54] The existence of similar triangles and their properties are proved in Book VI, VIE2 and VIE4.[55] The propositions of parallel lines provide significant insight into this definition in that straight lines cutting a triangle parallel to any of the sides form similar triangles. For example, in the following figure, the straight line DE is parallel to BC, and therefore from Proposition IE29, ∠ADE = ∠ABC and ∠AED = ∠ACB. Also, as ∠BAC is common to both △ABC and △ADE, the triangles are similar. From the areas of the figure, viewed as a composite of the triangle, △ADE and the trapezoid BCDE, one can then show that b/a = h_1 /h.[xvi] This property of proportionality based upon similarity is not maintained in non-Euclidean geometries because the Parallel Postulate does not apply. In fact, similar triangles, for the non-Euclidean geometries described in Section 3.2, are congruent.

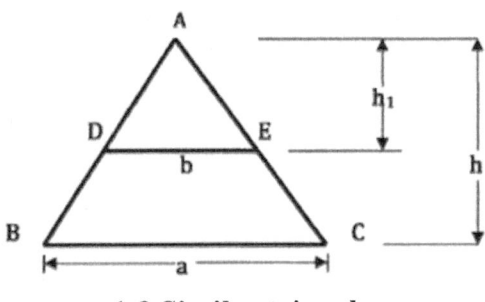

1-9 Similar triangles

The existence of similar triangles gives rise to the formulation of trigonometric functions such as the sine (sin), cosine (cos), and tangent (tan) functions. The figure below shows △ABC with the right angle at ∠B.

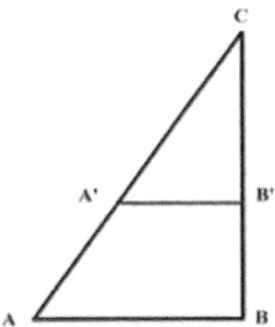

1-10 Similar right triangles: △ABC & △ A′ B′ C

---

xvi   The area of the trapezoid BCED is equal to the sum of the areas of two triangles: △BCE + △BED = 1/2(h – h_1) a + 1/2(h – h_1) b = 1/2 (h – h_1) (a + b). Then, for △ABC 1/2 ah = 1/2 bh_1 + 1/2 (h – h_1) (a + b).

The triangle is cut by a line A′ B′ parallel to the base AB. As before, from IE29, ∠A = ∠A′ ∠B = ∠B′' = right angle, and ∠C is common to both triangles. Thus, △ABC is similar to △ A′ B′ C, and corresponding sides are proportional:

$$\frac{BC}{AC} = \frac{B'C}{A'C} = \sin A$$

$$\frac{AB}{AC} = \frac{A'B'}{A'C} = \cos C$$

$$\frac{BC}{AB} = \frac{B'C}{A'B'} = \tan A$$

The trigonometric functions of an angle θ are usually recalled as:

$$\sin \theta = \frac{opposite}{hypotenuse}$$

$$\cos \theta = \frac{adjacent}{hypotenuse}$$

$$\tan \theta = \frac{opposite}{adjacent}$$

The subject of trigonometry arose from the ancient need to determine the measurements of the sides and angles of general triangles for engineering and astronomical purposes. Although trigonometry is not explicitly covered in the *Elements*, it should not be surprising that the foundations of trigonometry are found there, as it is the discipline that provides a means to determine side and angle relations of triangles. Of particular note, is Euclid's proof of the Law of Cosines given in geometric form In Book II (Proposition IIE12 and IIE13).[56] The Law of Cosines, which is a generalization of the Pythagorean Theorem, allows calculations of triangles that are not right–angled. However, the tabulation of trigonometric functions that facilitate calculations would await the trigonometric tables developed by Hipparchus of Nicaea (ca. 180 – 125 BC) over one hundred years after Euclid.[57] The trigonometric functions will be quite useful once coordinates are introduced in Chapter 4 to describe geometric curves through analytic geometry. At this point it seems useful to list some of the key characteristics of Euclid's geometry introduced by the Parallel Postulate:

- Relationships of angles formed by a straight line cutting two parallel lines.
- Exterior angles of a triangle equal to the sum of the two opposite interior angles.
- Sum of interior angles of a triangle equal to two right angles.
- The existence of rectangles and squares.
- Quantification of areas of geometric figures such as triangles, quadrilaterals, and multilaterals.
- Length of sides of right triangles interpreted geometrically through the Pythagorean Theorem.
- Existence of similar triangles with corresponding proportional sides.

These characteristics do not appear in the new geometry discovered in the nineteenth century by introducing a new fifth Postulate. However, in the intervening years, mathematicians would strive to develop a geometry which maintained these characteristics without the Parallel Postulate or with a simpler one. It is to those efforts that we now turn in the next chapter.

2   Euclid's Truths Questioned

2.1   The search for simpler truths

From a relatively early period on, there were many attempts to prove the Parallel Postulate using only the first four of Euclid's postulates or to replace it with a simpler postulate. The earliest known attempt is that of Claudius Ptolemy (d. 168), author of the *Almagest*, famous for its complex, earth-centered system for the motion of the planets.[58] Proclus, however, noted that Ptolemy had used in his proof the assumption that through a point only one parallel can be drawn to a given line in a plane.[59] As previously noted, this statement is commonly used today in high school textbooks and is referred to as Playfair's Postulate (John Playfair, 1748-1819), although it was known by Proclus and known to be an equivalent to the Parallel Postulate. Proclus made his own attempt that also used assumptions equivalent to the fifth postulate.

Heath provides an extensive commentary on some of the attempts to prove the Parallel Postulate.[60] I only note here some of the other investigators of the Parallel Postulate and their dates in order to give a feeling for the continuing mathematical interest over the centuries: Proclus (410 – 485), Nasir al-Din al-Tusi (1201–1274), John Wallis (1616 – 1703), Gerolamo Saccheri (1667 – 1733), Johann Lambert (1728 – 1777), and Adrien Marie Legendre (1752 – 1833). As finally became clear in the nineteenth century, all the proofs necessarily involved statements equivalent to the Parallel Postulate as the Parallel Postulate is independent of the first four of Euclid's postulates. However, in hindsight, such efforts made valuable contributions by clarifying the nature of the Parallel Postulate and providing equivalent statements that could be used instead.

Of particular note are the efforts in the Islamic communities which provided creative paths for exploration of the Parallel Postulate.[61] Alhazen (ibn-al-Haitham, ca. 965-1039) introduced for the study of the Parallel Postulate a quadrilateral with three right angles. Alhazen mistakenly thought he had proved the fourth angle must be a right angle and thus proved the Parallel Postulate. Nevertheless, his effort can be seen as the start of an approach in which the implications of assuming that the fourth angle was not a right angle would be explored. The quadrilateral with three right angles was the starting point for Lambert and is now known as the Lambert quadrilateral. Similarly, Umar Khayyam, (ca.1050-1123) introduced for study a quadrilateral, two sides of which are equal and are perpendicular to the base. He explored relationships of the upper angles of the quadrilateral. This figure was the starting point for the investigations of Saccheri and is now usually identified with his name. These quadrilaterals eventually helped to reveal non-Euclidean geometries.

Below is a partial list of postulates equivalent to the Parallel Postulate:[62]

- If a straight line intersects one of two parallels, it will intersect the other. (Proclus)
- If in a quadrilateral a pair of opposite sides are equal and if the angles adjacent to a third side are right angles, then the other two angles are also right. (Saccheri)
- There exist a pair of similar noncongruent triangles. (Wallis, Saccheri)
- If in a quadrilateral three angles are right angles, then the fourth angle is also a right angle. (Lambert)
- Given any figure, there exists a figure similar to it of any size we please. (Wallis)
- There exists a triangle in which the sum of the three angles is equal to two right angles. (Legendre)

And perhaps the most iconic equivalent to the Parallel Postulate:

- the Pythagorean Theorem.

The significance of the above statements that are equivalent to the Parallel Postulate are emphasized by considering that in non-Euclidean geometries, we will find that all similar triangles are congruent, there are no rectangles or squares, and the sum of the angles of a triangle is not equal to two right angles. As Playfair's postulate is probably the most well-known statement equivalent to Euclid's Parallel Postulate, I show below how it may be proved from Euclid's postulates.

Playfair's Postulate: Through a given point [P] not on a line [AB] *only one* parallel line [CD] can be drawn to the given straight line [AB] in the plane of the point and line.

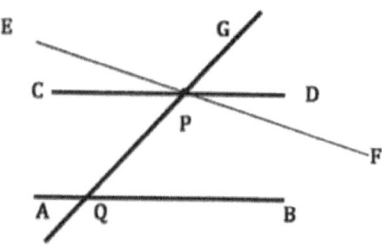

2-1 Proof of Playfair's Postulate from the Parallel Postulate

Let AB be the given straight line and P the given point	By hypothesis
Through any point Q on AB, draw the straight line QP	Postulate E1
Produce QP to G	Postulate E2
On line QG construct ∠GPD = ∠PQB(exterior and interior opposite angles)	Proposition IE23
Produce CD	Postulate E2
CD is parallel to AB	Proposition IE28
∠GPD + ∠DPQ = 2 right angles	Proposition IE13
∠PQB + ∠DPQ = 2 right angles	Common Notion 1
Draw any other line EF through P not coincident with CD	Postulate E2

Case 1: $\angle FPQ < \angle DPQ$ (shown in figure above) Common Notion 5 $\angle PQB + \angle FPQ < 2$ right angles (interior angles)	Common Notion applied to inequalities
EF is not parallel to AB	**Postulate E5**

Case 2: Similar arguments apply for $\angle EPQ < \angle CPQ$

Q.E.D.

Despite intense interest over many centuries, little progress was made in understanding the implications of the Parallel Postulate. Girolamo Saccheri would make the first real progress and discover propositions that became part of a new geometry. How did he do this? He did it by denying the truth of the Parallel Postulate and following the path where his hypothesis led him. Let us follow that path.

2.2 Saccheri vindicates Euclid – and misses a breakthrough

While none of the mathematicians mentioned in the previous section came to understand that the Parallel Postulate was independent of the first four postulates of Euclid, Gerolamo Saccheri made a significant contribution through his attempts to prove the Parallel Postulate by being the first to assume other hypotheses in place of this postulate and looking for contradictions. The vehicle for his study was the quadrilateral first investigated by Umar Khyyam. The quadrilateral consisted of a base upon which there were two equal sides perpendicular to the base. Of course if the upper side is formed by a side equal to the base and perpendicular to the sides, a rectangle is formed. Saccheri's approach was to look at the implications for other conditions of this upper side. In one case, he assumed the upper angles were acute (the hypothesis of the acute angle, HAA). As we shall see below, the geometric conclusions that he drew for this hypothesis were valid. Only his conviction that Euclidean geometry was uniquely true led him to reject the results and consequently miss the discovery of the new geometry. Later, the new geometry became known as hyperbolic. With Saccheri's failure to see in his results a new geometry, it would take almost one hundred years before this insight was revealed in a paper by Nicolai Lobachevsky in 1829 and independently by Johann Bolyai in a work submitted the same year, but not published until 1832.[63] The other hypothesis that Saccheri investigated, that of the obtuse angle (HOA), would lead, with other modifications of Euclid's postulates, to the geometry called elliptic as part of Bernhard Riemann's (1826-1866) generalization of concepts of geometry which we will encounter later in this chapter.[64]

Saccheri's hypotheses of the acute and obtuse angle led him to many conclusions in conflict with propositions proved using the Parallel Postulate however, before invoking these hypotheses, our first task will be to see with Saccheri what properties can be inferred for his quadrilateral using only Euclid's first four postulates.

2.2.1   The Saccheri quadrilateral

Saccheri's quadrilateral shown in the figure below is formed on a base AB with perpendiculars produced at points A and B of equal length. If the upper angles are right angles then a rectangle is formed. This condition Saccheri refers to as the hypothesis of the right angle (HRA). As previously noted, the existence of rectangles and squares can be shown to be equivalent to the Parallel Postulate.[xvii]

Let us prove some of Saccheri's propositions[xviii] that only require using Euclid's first four postulates and propositions prior to Proposition IE29. In the figures below, I have drawn the summit of the quadrilateral symbolically as a curve in anticipation and reminder that the straight

---

[xvii]   Recall that in the proof of the Pythagorean Theorem, squares were formed on each side of the right triangle. However, these squares, given a description in Euclid's Definition 22 of Book I, are only proved to exist using the Parallel Postulate in Proposition IE46.

[xviii]   Saccheri's propositions may be found in Girolamo Saccheri's *Euclides Vindicatus*, edited and translated in English by George Bruce Halsted. Descriptions of key propositions in modern terminology are given in McCleary, J., *Geometry from a Differentiable Viewpoint*, pp. 34 – 38. I have similarly modified the propositions for clarity; see Bibliography for all reference details.

lines drawn in the new geometries are not straight lines in a Euclidean plane.

Saccheri Proposition 1 (S1): In the Saccheri Quadrilateral [ABCD] the angles at the summit [∠ADC and ∠DCB] are equal.

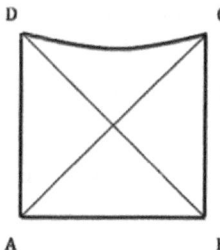

2-2 Saccheri Proposition 1

ABCD is a Saccheri quadrilateral (AD = BC, ∠DAB = ∠CBA = right angle)	By hypothesis
Produce DC, AC, BD	Postulate E1
△ADB ≅ △BCA, AC = BD (SAS, AD = BC, ∠DAB = ∠CBA, AB in common)	Proposition IE4
△DAC ≅ △CBD, ∠ADC = ∠BCD (SSS, AD = BC, AC = BD (shown above), DC in common)	Proposition IE8

Q.E.D

In the absence of the Parallel Postulate, we cannot say anything more about ∠ADC and ∠DCB), other than that the angles are equal. However if we use the conclusion of IE32 that the sum of the interior angles of a triangle is two right angles, it is easy to show that ∠ADC and ∠DCB are right angles. The use of IE32, of course means that we have accepted the Parallel Postulate.

Saccheri Proposition 2 (S2): In the Saccheri Quadrilateral the line [EF] formed by connecting the midpoint of the base and summit is perpendicular to the base and the summit.

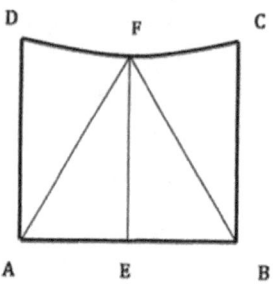

2-3 Saccheri Proposition 2

ABCD is a Saccheri quadrilateral, AD = BC	By hypothesis
Produce DC	Postulate IE1
Bisect DC at F, AB at E (DF = FC, AE = EB)	Proposition IE10
Produce AF, BF	Postulate E1
△ADF ≅ △BCF, AF = BF (SAS, DF = FC, AD = BC, ∠ADF = BCF)	Saccheri Prop. 1, Prop. IE4
△AEF ≅ △BEF, ∠AEF = ∠BEF (SSS, AF = BF AE = EB, EF in common)	Proposition IE8
∠AEF = ∠BEF = right angle	Proposition IE13

By producing lines DE and CE, similar arguments
show ∠DFE = ∠CFE = right angle.

Q.E.D.

We have created two additional quadrilaterals in the figure above, ADFE and BCFE, each containing three right angles, but as before, without the Parallel Postulate the magnitude of the ∠ADF and ∠BCF remain unknown. These quadrilaterals would be the starting point for Johann Lambert's similar investigations.

Having developed some properties of the Saccheri quadrilateral using only the first four of Euclid's postulates, it is time to see where the hypotheses of the acute and obtuse angle will take us.

2.2.2 The Saccheri Hypotheses of the Acute Angle (HAA), Obtuse Angle (HOA), and the Right Angle (HRA)

Saccheri Proposition 3 (S3): In the Saccheri quadrilateral [ABCD], depending on the assumption of HAA, HOA or HRA, (Part 1), the summit is of greater length than the lower base [DC > AB], less than the lower base [DC < AB], or equal, respectively [DC = AB];[65] and (Part 2) the sum of angles of a triangle are, respectively, less than, greater than, or equal to two right angles. [xix]

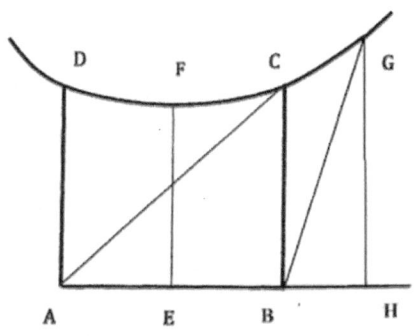

2-4 Saccheri Proposition 3

The proof below is for the HAA. The proof for the HOA follows a similar pattern

Part 1
Proof by contradiction

Assume DC < AB	Contradicts HAA
ABCD is a Saccheri quadrilateral	By hypothesis
Bisect DC with DF = FC, AE = EB, FE perpendicular to DC and AB	IE10, Saccheri Prop.2

[xix] Saccheri develops the result of Part 2 through a number of propositions, notably, his Propositions 9 and 15, Saccheri, pp. 41-42, 61-65. McCleary proves Part 2 in conjunction with Part 1; McCleary, pp. 35-36.

FC < EB	Common Notion applied to inequalities
Extend DC, produce FCG with FG = EB	Postulate E2, Proposition IE3
∠EBG > right angle	Common Notion 5
∠CGB = ∠EBG (EFGB is a Saccheri quadrilateral)	Saccheri Proposition 1
However, ∠CGB < right angle (acute angle of EFGB)	By HAA
⇒ Contradiction, therefore, DC > AB.	

Part 2

Produce AC	Postulate E1
DC > AB	shown above
∠ABC = right angle (Saccheri quadrilateral)	By hypothesis
∠DAC > ∠ACB (greater side subtends the greater angle)	Proposition IE25
∠CAB + ∠DAC = right angle	Common Notion 4
∠CAB + ∠ACB < right angle	Common Notion applied to inequalities
∠CAB + ∠ACB + ∠ABC < two right angles	Common Notion applied to inequalities

Q.E.D

Thus, we have proved that for the HAA, the sum of the angles of the right triangle, △ABC is less than two right angles. Saccheri and later Legendre went on to prove that if the sum of the angles of **one** triangle is less than two right angles, then the sum of the angles of **every** triangle is less than two right angles (Saccheri Proposition 15).[66] Notice that if the sum of the interior angles of a triangle is less than two right angles, then the sum of interior angles of any quadrilateral must be less than four right angles as the quadrilateral may be split into two triangles. This is consistent with the summit angles of the Saccheri quadrilateral being acute.

The proof for the HOA in Saccheri Proposition 3 is quite similar to that above. In the proof by contradiction, assume that DC > AB with a point G between F and C so that FG = EB. As in the previous proof, the Saccheri quadrilateral EFGB is formed, in this case interior to EFCB. In a similar manner to the previous proof, this leads to a contradiction. For the HOA, the sum of the angles of the triangle is proved to be greater than two right angles. Saccheri abandoned the HOA when he correctly proved it contradicted results using Euclid's first four postulates. We will return to the contradiction of the HOA with Euclid's postulates in the next section.

Returning to the HAA and the Saccheri quadrilateral in the figure above, the perpendicular EF forms a Lambert quadrilateral, EFCB, having three right angles.[67] Furthermore, BC and EF could be extended below AB to form, respectively, the summit and the base of another Saccheri quadrilateral. Therefore, it follows from Proposition S3, BC > EF. Note that the exterior angle at C of the Lambert quadrilateral EFCB is obtuse. Perpendiculars to the extension of AB may be formed with points along the extension of FC, e.g., HG in Figure 2-4. The exterior angles such as at G will change continuously from the obtuse at C. If the angles decrease from that at C, then for some point G, the angle would be a right angle, forming the quadrilateral EFGH. This would violate the HAA; hence the exterior angles formed at points such as G remain obtuse. It follows that the extensions of line FC and similarly FD do not intersect AB. Points further along these lines recede from AB. Lines with a common perpendicular such as AB and DC are diverging parallels and are often known simply as nonintersecting lines.[xx]

Saccheri also investigated the possibility of parallels that do not have a common perpendicular and discovered parallels that continuously approach each other in one direction (asymptotically approach) and

---

[xx] See e.g., Bonola (p. 87), McCleary (p. 39), and Wolfe (p. 68).

diverge in the other. This condition is illustrated under the HAA, with ABCD being the Saccheri quadrilateral shown in the following figure.

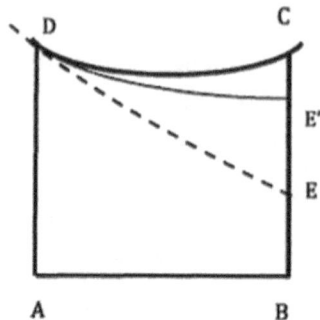

2-5 Asymptotic parallel line DE

Under the HAA, AD = BC and ∠ADC = ∠BCD, both being acute. At some point E', BE' < BC, and the line DE' will have a common perpendicular with AB. This is similar to the quadrilateral ADFE interior to ABCD in Figure 2-4. Points below E' must therefore form obtuse angles such as ∠BED, an exterior angle to △DEE' and △ DEC. Because the angle is obtuse, it is consistent with IE16 which says the exterior angle is greater than either of the interior angles.[xxi]

Lines produced from D to points below E' will also have common perpendiculars with extensions of AB; however, the perpendiculars will be located at increasing distances from B. For some limiting line extended from DE, the common perpendicular will be at infinity. This was proved by Saccheri in his Proposition 23.[xxii]

Saccheri, however, rejected the validity of his own conclusions as he considered them to be impossible because he did not have the modern understanding of asymptotic properties. Saccheri Proposition 33 states, *"The hypothesis of acute angle is absolutely false; because repugnant to the nature of the straight line."*[68] Saccheri's proof for the lines that *"...produced in infinitum...must run together at length into one and*

[xxi]   If we assume that BE > AC and ∠BED is obtuse, a contradiction with IE16 occurs (McCleary, pp. 36).

[xxii]   A modern detailed proof is given by McCleary, pp. 36-37.}

the same straight line, truly receiving, at one and the same infinitely distant point a common perpendicular...." was correct and became the limiting asymptotic parallels of hyperbolic geometry. Ironically, these asymptotic or limiting parallels would turn out to be crucial for non–Euclidean geometry. Saccheri never found a legitimate contradiction using the HAA. As we shall see in the next chapter, the HAA was found by Nicolai Lobachevsky and Johann Bolyai, almost one hundred years later, to form an entirely consistent new geometry. Before we explore their discovery, let us first look at Lambert's effort which like Saccheri's similarly failed, but added important insights into the characteristics of the new geometry.

2.3 Lambert and intimations of a new geometry on the surface of a sphere

2.3.1 Visions of the HAA on a sphere of imaginary radius

Lambert proved many of the propositions first discovered by Saccheri, but by using a quadrilateral with three right angles. Similar to Saccheri, he formed three hypotheses for the fourth angle: the HAA, HOA, and HRA. He hoped that in the cases of the acute and obtuse hypotheses a contradiction would occur, thus proving the Parallel Postulate. Travelling the path of Saccheri, Lambert proved that the HOA required that the sum of the interior angles of a triangle be greater than two right angles. He then, like Saccheri[69] and later Legendre,[70] rejected the HOA by proving that the first four postulates of Euclid were incompatible with the HOA as Euclid's first four postulates required that all triangles have the sum of their interior angles less than or equal to two right angles. A proof by Legendre is given later in this section.

One significant discovery of Lambert that followed from either the HOA or the HAA was that the area of a triangle was proportional to the difference between the sum of the interior angles of the triangle and two right angles.[71] This result would be shown by the mathematician Karl Gauss to be directly related to the curvature of surfaces. Gauss' ideas about the Parallel Postulate will be discussed in the next chapter; surface curvature will be introduced in Chapter 6.

An immediate consequence of Lambert's discovery is that the areas of all similar triangles (those having the same interior angles) are the same. But not only are the areas the same, similar triangles are congruent. In contrast, recall that in Euclidean geometry, the corresponding sides of similar triangles are proportional, rather than congruent, and therefore can be scaled. With the replacement of the HRA with that of the HOA or HAA, this no longer applies. We prove this result below by contradiction by assuming that the similar triangles are not congruent. This assumption will lead to the conclusion that the sum of the interior angles of a quadrilateral equals four right angles. As the quadrilateral may be divided into two triangles, this implies that the sum of the interior angles of a triangle is equal to two right angles, a contradiction of the HOA or the HAA

Similar Triangle Proposition for the HOA and HAA: Similar triangles are congruent.

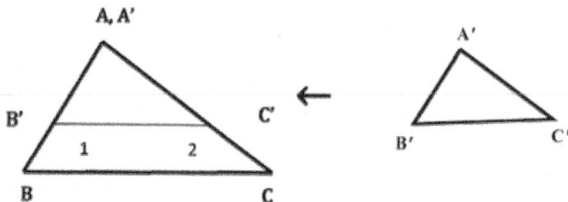

2-6 Congruence of similar triangles under the HOA or HAA

Proof by contradiction

△ABC and △A'B'C' are similar triangles   By hypothesis
(∠ABC =∠A'B'C', etc.)

Assume triangles are not congruent (AB ≠ A'B', BC ≠ B'C', CA ≠C'A'

Bring ∠C'A'B' into coincidence with ∠CAB   Proposition IE23
Note that if any one of the sides of the two triangles were equal, the triangles are congruent by Proposition IE26 (AAS). Therefore

in one of the triangles, we can select two sides to be shorter than two of the sides of the other triangle; let A'B' < AB and A'C' < AC (see figure above).[xxiii]

$\angle$C'B'B + $\angle$A'B'C' = two right angles;	
$\angle$B'C'C + $\angle$A'C'B' = two right angles	Proposition IE13
$\angle$C'B'B + $\angle$ABC = two right angles;	
$\angle$B'C'C + $\angle$ACB = two right angles	By hypothesis, Common Notion 1

$\angle$C'B'B + $\angle$ABC + $\angle$B'C'C + $\angle$ACB = four right angles;

Interior angles of BB'C'C = four right angles   Common Notion 2

Interior angles of BB'C'C ≠ four right angles   HAA or HOA

⇒contradiction; therefore

Similar triangles are congruent.

Q.E.D.

As curious as the idea that all triangle with the same interior angles are congruent is Lambert's discovery that in the HOA or the HAA, the area of a triangle is proportional to the difference between the sum of the interior angles of a triangle and two right angles, (π radians).[xxiv] The difference is a defect, δ_d, in the case of the HAA. In the case of the HOA in which the sum is greater than two right angles, it is an excess, δ_e. If the three interior angles of the triangle are α, β, and γ, and Lambert's constant of proportionality is C, then, the areas of triangles, A_Δ are given by:

$$\text{HAA: } A_\Delta = C_{HAA} \cdot \delta_d = C_{HAA} \cdot (\pi-(\alpha+\beta+\gamma)) \tag{2-1}$$

[xxiii] The line A'B' forms the exterior angle $\angle$A'B'C' equal (by hypothesis) to the interior and opposite angle on the same side $\angle$ABC; thus by Proposition IE28, B'C' does not intersect BC.

[xxiv] As a radian is the length of circular arc divided by the radius r, and a semi-circle is half the circle's circumference or πr. Therefore, 180° corresponds to π radians.

$$\text{HOA: } A_{\triangle} = C_{HOA} \cdot \delta_e = C_{HOA} \cdot (\alpha + \beta + \gamma - \pi) \qquad \text{(2-2)}$$

In Euclidean geometry, area may be thought of as adding the number of unit squares in a figure; however, in either the HAA or HOA, there are no squares to add up! It can be shown, however, that the triangle defect has the necessary property for area of being additive; that is, the sum of the defects of a triangle composed of a number of smaller triangles is equal to the sum of the defects of the smaller triangles.[xxv] An example of the additivity of triangle defect, δ_d in the HAA is shown below for a triangle which is a composite of two smaller triangles. Let the larger triangle, $\triangle ABC$ consist of the two smaller triangles, $\triangle ABD$ and $\triangle ADC$ with angles as shown below.

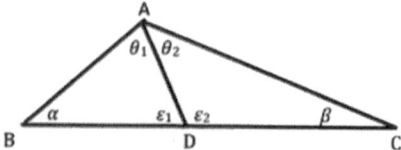

2-7 Additivity of triangle deficit, δd, or excess, δe

$$\delta_{\triangle ABD} = (\pi - \alpha - \varepsilon_1 - \theta_1);$$
$$\delta_{\triangle ADC} = (\pi - \beta - \varepsilon_2 - \theta_2)$$
$$\delta_{\triangle ABD} + \delta_{\triangle ADC} = (2\pi - \alpha - \beta - (\theta_1 + \theta_2) - (\varepsilon_1 + \varepsilon_2));$$
$$(\varepsilon_1 + \varepsilon_2) = \pi$$
$$\delta_{\triangle ABD} + \delta_{\triangle ADC} = (\pi - \alpha - \beta - (\theta_1 + \theta_2)) = \delta_{\triangle ABC}$$

Lambert had an important insight in recognizing the similarity between his relationship for the area of a triangle with the HOA and the area of a triangle on the surface of a sphere. He knew that the sum of the interior angles of a spherical triangle was greater than a right angle. For a sphere of radius R, the area of a triangle with interior angles α, β, and γ is given by:

$$A_{\triangle} = R^2 \cdot (\alpha + \beta + \gamma - \pi) \qquad \text{(2-3)}$$

[xxv]   Wolfe provides details of this proof, pp. 124-126.

Comparing the area for the sphere in equation (2-3) with the area using the HOA in equation (2-2), it is clear that they are the same if $C_{HOA} = R^2$. The formula for the area of a spherical triangle includes many concepts which have not yet been discussed, e.g., how are the lines forming the sides of the triangle defined? Details concerning spherical geometry and the HOA will be made clearer in the next section. Suffice it to say here, however, that the resemblance Lambert observed between the results for the HOA and spherical geometry made him wonder about his findings for the HAA. He had ruled out the HOA, but could find no clear inconsistencies with the HAA. Indeed, he speculated about the geometry formed by the HAA saying:

> *"I am almost inclined to draw the conclusion that the third hypothesis* [the HAA] *arises with an imaginary surface.''*[72]

The meaning of this vision becomes clearer by noting that if his constant of proportionality in equation (2-2), $C_{HOA}=R^2$ with R an imaginary radius, $R = i \cdot r$, and recalling that $i = \sqrt{-1}$, and $i^2 = -1$, then equation (2-1) is recovered with $C_{HAA}=r^2$. A full understanding of the meaning of equations (2-1) and (2-2) would await the development by Gauss of a theory of surface curvature which will be introduced in Chapter 6.

2.3.2 Relating the HOA to a real sphere

The HOA was abandoned when Saccheri, Lambert, and Legendre proved in what is now generally known as the Saccheri – Legendre Theorem that Euclid's first four postulates require the sum of the angles to be less than or equal to two right angles. However, the proof of this result contained, unknown to its creators, a clue as to the condition that would make the HOA compatible with another geometry. Legendre presented a number of proofs of what is also sometimes called Legendre's First Theorem. One of these proofs is shown next.[73]

Saccheri – Legendre Theorem: The sum of the angles of any triangle is less than or equal to two right angles.

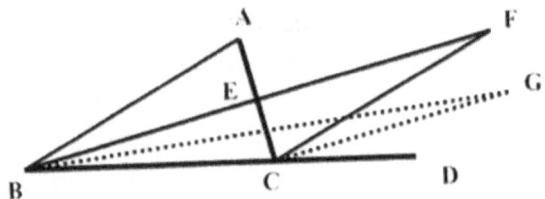

2-8 Saccheri-Legendre Theorem

Let us start off with the above figure similar to that used in proving IE16 (p. 27).[xxvi] We select ∠ABC to be the smallest angle of the triangle. The figure in the proof is constructed with AE = CE and BE = FE. Then, by IE15, ∠BEA = ∠FEC. Therefore by IE4, △ABE ≅ △CFE$ and corresponding sides and angles are equal. In particular, ∠CAB = ∠ACF and ∠ABF = ∠CFB.

A critical part of description of the construction of this figure and that used in the proof of IE16 is the assumption that the line BE may be extended so that BE = EF. The denial of this assumption, along with the denial of the Parallel Postulate, will lead us to a non-Euclidean geometry. However, ignoring this point as did Saccheri and Legendre, we now begin the proof of the Saccheri-Legendre Theorem.

Proof by Contradiction.

Assume The sum of the interior angles of △ABC is greater than two right angles, .i.e,

Sum of angles of △ABC	=	two right angles + α
	=	∠ABC + ∠BCA + ∠CAB
	=	(∠ABF + ∠FBC) + ∠BCA + ∠ACF
	=	∠CFB+ ∠FBC + ∠BCF
	=	Sum of angles of △CFB

[xxvi] Recall that IE16 states that an exterior angle of a triangle is greater than either of the interior or opposite angles ∠DCE > ∠ BAC$ and ∠DCE > ∠ABC).

Now suppose ∠ FBC < α, then the sum of the two interior angles, ∠CFB and ∠BCF is greater than two right angles. This contradicts IE17 which say that the sum of the two interior angles must be less than two right angles. If ∠FBC > α, another triangle could be produced by extending a line (shown as a dashed line) from vertex B, bisecting CE in a manner similar to the line BF that bisects AC. A new triangle is formed, △GCB with interior angles whose sum like △ABC can also be shown to be two right angles + α, This process could be continued until a triangle would be produced with an angle at vertex B less than any angle we choose - creating a contradiction with IE17. Thus, the sum of the interior angles of a triangle must be less or equal than two right angles.

Q.E.D.

It would seem that the Saccheri - Legendre Theorem shows that there is no place for the hypothesis of the obtuse angle in a geometry of a surface that is consistent with Euclid's first four postulates. However, a clue as to how the HOA could be used to create a consistent geometry is found at the very end of Legendre's proof. Legendre asks that lines be extended as necessary to meet the needs of his proof. This is the same as asking to extend the bisecting lines to any length. Euclid had done the same thing in his Parallel Postulate and Legendre had repeated this assumption in other proofs.[74] But how do we know that a line can be extended indefinitely.

In 1854, as part of a generalization of geometry to abstract surfaces, Riemann recognized that a line could be boundless yet finite.[75] The example that satisfies this condition is similar to the one that we live with every day as we travel on the earth - it is the geometry of the surface of a sphere. The circumference of the earth is a finite length, but we can go around as many times as we want which is to say that it is boundless. Thus, on a sphere, we cannot reason as Legendre did that we can extend lines indefinitely without returning to the initial position. The same reasoning limits IE16.

Consistent with the concept that such lines could be boundless but not infinite, two other modifications of Euclid's postulates can be introduced to form a new consistent geometry. These modifications associated with Riemann's generalization of geometry are that two distinct points determine **at least** one straight line as a replacement for Euclid's first postulate, and that there are no lines that are parallel as a replacement to the Parallel Postulate. Just as the geometry of a sphere gives us an example of boundlessness, the intersection of all meridians at the North and South Pole of the earth gives a concrete example for two points determining more than one line.

Let us now turn to the geometry of the surface of a sphere to confirm that it is a consistent realization (model) of what I shall term Riemann's postulates and the conclusions from the HOA.[xxvii] The surface of a sphere consists of those points which are equidistant from a given point in three-dimensional space. On this surface, the shortest distance between two points is along the curve formed by the intersection with the surface of the plane containing the two points and the center of the sphere. This curve is called a great circle. A great circle is an example of a geodesic, the term used for more general geometric circumstances. The arc of a great circle will be what we mean by a ``straight'' line on the surface of a sphere.[xxviii] The intersection of any extensive plane containing the center and the surface of the sphere forms a great circle. Thus, any line that is extended indefinitely will return to its beginning. Such a line will cut the surface of the sphere into two regions. Any other line will intersect this line if extended far enough, and there can therefore be no parallel lines in this geometry.

Our familiarity with the geometry of the earth can help us visualize effects of the HOA. In the following figure of a sphere, I have shown the great circle that is formed on the surface by the intersection of a

---

[xxvii] An excellent source on spherical geometry may be found at www.gutenberg.org/ebooks/1977Cached Nov 12, 2006, Todhunter, I., *Spherical Trigonometry for the Use of Colleges and Schools with Numerous Examples.*

[xxviii] You will no doubt realize that on a globe, a great circle provides two paths connecting the two points.

horizontal plane with the center of the sphere O. The short arc formed by the points BC is part of this great circle. In terms of our globe, the great circle corresponds to the equator. A line perpendicular to the equatorial plane from the center of the sphere meets the surface at point A, the pole of the equatorial great circle. A plane passing through points A, B, and the sphere center produces the line AB. Extending AB to the opposite pole would produce the traditional meridian of maps. The position of point B on the equatorial circle is defined by its angular position around the circle with 0° starting at an arbitrary position. Our maps of the earth traditionally select the meridian going through Greenwich, UK as that starting point. In regard to ∠ABC, it should be clear that this is a right angle as the plane forming the line AB is perpendicular to the equatorial plane. Another way to see this is to note that the tangents of the lines AB and BC at point B are perpendicular.

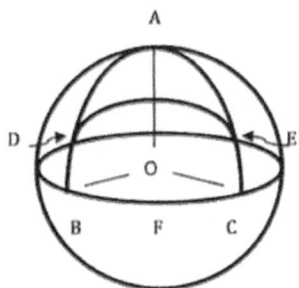

2-9 Triangles on a sphere

In the figure above, I also show another line AC forming with AB and BC a triangle, △ABC. If we locate point C a quarter of the way around the great circle from point B, then ∠BOC is a righr angle. This would correspond to the meridian just east of Calcutta, India. The line AC is formed by a plane through points A, C, and the sphere center and is perpendicular to the equatorial plane forming the line BC. Hence ∠ACB is also a right angle. To determine ∠BAC, we look at the planes forming AB and AC, (or the tangents to AB and AC at A) so ∠BAC) is also a right angle. The sum of the angles of △ABC is therefore three right angles. It should be clear that by moving point C towards B along the equatorial great circle, triangles could be formed in which the sum of the angles is any value between two and three right angles as ∠BAC can take on any value between 0°. and 90°. With the assurance of the proofs of Saccheri and Legendre, we now know that any triangle on the surface of the sphere will have angles summing to more than two right angles. By continuing to form triangles by moving C further away from B along the equatorial great circle, we can see that triangles with angle sums between two and six right angles can be formed. For example, if the line AB is extended beyond the pole to a point on the equator, then a triangle is formed by right angles where lines AB and its extension meet the equator, and the angle at A equals two right angles. The sum of the interior angles, in this case, is four right angles.

From the knowledge that the surface area of a sphere is $4\pi r^2$ with r the radius of the sphere, we can determine the area of $\triangle ABC$. As I have chosen C to be a quarter of the way around the equatorial circle, $\triangle ABC$ is a quarter of the hemisphere or an eighth of the surface area of the sphere, $A_\triangle = 4\pi r^2/8 = \pi r^2/2$. Lambert first discovered that under conditions in which the HOA is true, the area of a triangle is proportional to the difference between the sum of the angles of the triangle and two right angles. In our case this gives:

$$
\begin{aligned}
A_{\triangle ABC} &= r^2(\angle BAC + \angle CBA + ACB - \pi) \\
&= r^2(\pi/2 + \pi/2 + \pi/2 - \pi) \\
&= \pi r^2/2
\end{aligned}
$$

Notice that if $\angle BAC$ is any angle θ formed by moving the point C along the equatorial circle then $\triangle ABC = \theta \cdot r^2$ where θ is given in radians. Similar techniques[76] may be used to prove that for the general triangle with interior angles α, β, and γ, the area is given as:

$$A_\triangle = r^2(\alpha + \beta + \gamma - \pi)$$

Thus, we can see the correspondence of spherical geometry with the requirements of the HOA as discovered by Lambert.

Now we can also form quadrilaterals by passing a plane containing the center O at an angle to the equatorial plane to cut lines AB and AC symmetrically forming points D and E at the intersection. This would form the line DE in the above figure and the quadrilateral BDEC. The angles ∠BDE and ∠CED being obtuse are examples of Saccheri's requirements for a quadrilateral with the HOA. A Lambert quadrilateral can be created by connecting on the sphere's surface point A with point F taken as the midpoint of BC. Because of the symmetry of the resulting figures, the intersection of AF with DE and with BC would form right angles, and we would have a quadrilateral having three right angles and one obtuse angle.

We have now seen an interpretation for the HOA which makes a consistent geometry, although it requires changes in Euclid's postulates in addition to his fifth postulate. However, in contrast, no one openly accepted the HAA as the basis for a geometry, despite being unable to find a valid inconsistency with Euclid's first four postulates. The significance of the HAA did not become apparent until Nicolai Lobachevsky (1793-1856) and Janos Bolyai (1802-1860) found a replacement for the fifth postulate that left the other postulates in place and created a marvelous new geometry that they were willing to defend.[77]

3 The Discovery of Non-Euclidean Geometry

3.1 Gauss' insight

The description of the investigations of Saccheri and Lambert in the previous chapter should make it clear how tantalizingly close mathematicians were to understanding that the Parallel Postulate could not be proved from Euclid's first four postulates, even taking into account Euclid's unstated assumptions. The first mathematician to be generally credited with the insight that the Parallel Postulate was independent of the other postulates was the great mathematician Karl Friedrich Gauss (1777-1855.) [78] Gauss made significant contributions to the theory of prime numbers, computation of the orbits of planets and asteroids, complex numbers, the proof the Fundamental Theory of Algebra,[xxix] and convergence of infinite series, among many others. Of particular significance to us here is his development of the subject known as differential geometry in which he introduced a new approach to quantifying the local curvature of surfaces. This would become crucial to the understanding of new geometries and be the starting point for the generalized concept of geometry introduced by Riemann that was central to Einstein's formulation of General Relativity.

The development of differential geometry was not, however, in Gauss' mind at the beginning of the nineteenth century. Like his predecessors, Gauss was at first deeply involved in trying to prove the Parallel Postulate. His frustration and struggle is evident in a letter (dated

---

[xxix] Every polynomial f(x) of degree n > 0 with complex coefficients has at least one root.

December 17, 1799) that he sent to Wolfgang Bolyai (1775-1856), a friend from his student days at Gottingen.[79]

"As for me, I have already made some progress in my work. However, the path I have chosen does not lead at all to the goal which we seek, and which you assure me you have reached. It seems rather to compel me to doubt the truth of geometry itself."

"It is true that I have come upon much which by most people would be held to constitute a proof: but in my eyes it proves as good as nothing. For example, if one could show that a rectilinear triangle is possible, whose area would be greater than any given area, then I would be ready to prove the whole of geometry absolutely rigorously."

"Most people would certainly let this stand as an Axiom; but I, no!. It would, indeed, be possible that the area might always remain below a certain limit, however far apart the three angular points of the triangle were taken."[80]

Consideration of equation (2-1, p. 55), which shows that the maximum possible area of triangles under the HAA is $C_{HAA} \cdot \pi$, makes it clear that Gauss was on the verge of discovering a new geometry. His search for a proof for rectilinear triangles with unlimited area is another statement of the Parallel Postulate. Wolfgang Bolyai was also looking for a proof of the Parallel Postulate and believed he had found it only to have an error in his proof pointed out by Gauss.[81] Ironically, Wolfgang's son Johann would be one of two mathematicians who would discover and publish the truth of the independence of the Parallel Postulate.

It appears that by 1813, Gauss, no longer trying to prove the Parallel Postulate, was instead trying to develop a new geometry. This is evident from his letters to "a few trusted friends" and his unpublished papers.[82] That Gauss had indeed made the breakthrough to a new geometry is clear from a letter of November 8, 1824 to the mathematician Adolf

Taurinus (1794-1874).[83] I quote extensively from the letter as it reveals the depth of Gauss' discovery.[xxx]

> "...*The assumption that the sum of the three angles* [of a triangle] *is less than 180° leads to a curious geometry, quite different from ours (the Euclidean), but thoroughly consistent, which I have developed to my entire satisfaction, so that I can solve every problem in it with the exception of the determination of a constant, which cannot be designated* **a priori**. *The greater one takes this constant, the nearer one comes to Euclidean Geometry, and when it is chosen infinitely large the two coincide. The theorems of this geometry appear to be paradoxical and, to the uninitiated, absurd; but calm, steady reflection reveals that they contain nothing impossible. For example, the three angles of a triangle become as small as one wishes, if only the sides are taken large enough; yet the area of the triangle can never exceed a definite limit, regardless of how great the sides are taken, nor indeed can it ever reach it. All my efforts to discover a contradiction, an inconsistency, in this Non-Euclidean Geometry have been without success, and the one thing in it which is opposed to our conception is that, if it were true, there must exist in space a linear magnitude,* **determined for itself** *(but unknown to us).*[xxxi] *But it seems to me that we know, despite the say-nothing word-wisdom of the metaphysicians, too little, or nearly nothing at all, about the true nature of space, to consider as* **absolutely impossible** *that which appears to us unnatural.*"[84]

In the new geometry that was proclaimed by Johann Bolyai and by Nicolai Lobachevsky, independently, all that Gauss described of the discovery that he designated as Non-Euclidean Geometry turned out to be true.[xxxii] His agnosticism towards the nature of space could even be seen as an eerie foretelling of Einstein's radical revision of space as space-time.

[xxx] Words in bold indicate emphasis added by Gauss.

[xxxi] Here Gauss is referring to the condition under the HAA that equilateral triangles are congruent; thus an angle of a triangle fixes the length of a side. In this way, a standard length could be assigned to an angle.

[xxxii] Gauss initially named his discovery Anti-Euclidean Geometry; Bonola, p. 67.

Gauss never published his insights on Non-Euclidean Geometry. Why not? Wolfe provides a number of speculative reasons in his *Introduction to Non-Euclidean Geometry*.[85] As he suggests, it certainly would not be surprising if Gauss were intimidated by the two-thousand-year authority of Euclid's geometry as the accepted description of space. In that regard, he was probably aware of the support by many mathematicians of philosopher Immanuel Kant's view that geometry was an *a priori* truth. Finally, perhaps Gauss did not wish to risk his well-deserved reputation. His extreme reluctance to publicly announce his findings is clear from the closing of the letter quoted above: ``*...in any case consider it a private communication of which no public use or use leading in any way to publicity is to be made.*'' Gauss never published in any extensive way his revolutionary discoveries.

Before moving on to the published findings of Lobachevsky and Bolyai, it seems worthwhile to digress with some brief quotations from Kant's *Critique of Pure Reason*. From these quotes we may get some insight into the negative comments that Gauss may have anticipated he would receive from his contemporaries following an announcement of a non-Euclidean geometry. It is also interesting to read Kant's confident assertions about *a priori* intuitive truths of space and geometry in the light shed by Einstein's revolutionary, non-intuitive theories of Relativity on space and time in which length, time and simultaneity of events are dependent on the observer's motion. From the *Critique of Pure Reason:*[xxxiii]

``*Geometry is a science which determines the properties of space synthetically and yet **a priori**. What, then, must be our representation of space, in order that such knowledge of it may be possible? It must in its origin be intuition; for from a mere concept no propositions can be obtained which go beyond the concept ---as happens in geometry. Further, this intuition must be **a priori**, that is, it must be found in us prior to any perception of an object, and must therefore be pure,*

[xxxiii] Kant, Immanuel, *Critique of Pure Reason,* translated by N. K. Smith, Unabridged edition, St Martin's Press, New York, Macmillan, Toronto, 1965. Note, words in bold indicate emphasis added by Kant.

not empirical, intuition. For geometrical propositions are one and all apodeictic [beyond doubt], *that is, are bound up with the consciousness of their necessity; for instance, that space has only three dimensions. Such propositions cannot be empirical or, in other words, judgments of experience, nor can they be derived from any such judgments.'* [xxxiv]

Further on, Kant underscores this point:

*Geometry, however, proceeds with security in knowledge that is completely **a priori**, and has no need to beseech philosophy for any certificate of the pure and legitimate descent of its fundamental concepts of space. But the concept is employed in this science only in its reference to the outer sensible world---of the intuition of which space is the pure form---where all geometrical knowledge, grounded as it is in **a priori** intuition, possesses immediate evidence.'* [xxxv]

In anticipation of our discussion in Chapter 11 of Einstein's introduction of space-time and its extraordinary consequences, I also quote Kant's view of time:

"Time is not an empirical concept that has been derived from any experience. For neither coexistence nor succession would ever come within our perception, if the representation of time were not presupposed as underlying them a priori. Only on the presumption of time can we represent to ourselves a number of things existing at one and the same time (simultaneously) or at different times successively.' [xxxvi]

Whatever the meaning of Kant's *a priori* geometry, the dominance of his views were in keeping with an orthodoxy in which prestigious institutions such as the Göttingen school officially declared the necessity of admitting Euclid's Postulates.[86] The discovery of a non-intuitive, non-Euclidean geometry, not to mention Relativistic space-time would fly

---

[xxxiv] Ibid., p. 70.

[xxxv] Ibid., p. 122.

[xxxvi] Ibid., p. 74.

in the face of Kant's *a priori* truths. It is now time to reveal the new geometry that Lobachevsky and Bolyai discovered and defended.

3.2   Lobachevsky and Bolyai find a new geometry

In addition to their suspicions that the parallel Postulate could not be proved from Euclid's first four postulates, Gauss, Bolyai, and Lobachevsky all approached their search for a non-Euclidean geometry through variations of Playfair's Postulate.[87] Recall that Playfair's Postulate, equivalent to Euclid's fifth postulates, states that in a plane only one line can be drawn through a point parallel to a given line. They considered the effect on geometry of assuming, in addition to Euclid's first four postulates, that through the point either no parallel line could be drawn or more than one line could be drawn. This approach turned out to be equivalent to Saccheri's and Lambert's investigations replacing the right angles of quadrilaterals required by the Parallel Postulate with either acute angles (the HAA) or obtuse angles (HOA). As in the case of the HOA, the assumption that no parallels could be drawn was found to be inconsistent with the first four postulates of Euclid. However, as Gauss said in his letter to Taurinus in the previous section, no inconsistencies could be found assuming that more than one parallel; could be drawn.

Johann Bolyai (1802-1860), following his father's footsteps(but despite his father's misgivings) had become interested in the Parallel Postulate while attending the Royal College for Engineers in Vienna which he entered in 1817. He saw his efforts as a general investigation of parallel lines with Euclidean geometry being just a special case. By 1823, he seems to have reached his major conclusions as he wrote to his father, *"It is now my definite plan to publish a work on parallels as soon as I can complete and arrange the material and an opportunity presents itself....I can say nothing except this: that **out of nothing I have created a strange new universe**."* (Bolyai's emphasis). Johann's discoveries were published eventually in 1832 as an appendix to a work of his father's. When Johann's father had sent Gauss in 1831 parts of the appendix,

Gauss responded ungenerously saying that there was nothing new in it to him. Johann Bolyai published nothing more on the subject, but in 1848 was honored along with Nicolai Lobachevsky, of whose work he had been unaware.[88]

Nikolai Ivanovich Lobachevsky (1793-1856), a Russian mathematician, followed much the same path as Gauss and Johann Bolyai. He was an instructor and later a professor at the University of Kasan. His lecture notes of 1815 reveal attempts to prove the Parallel Postulate. In an unpublished manuscript of 1823, he noted that there had never been a proof of the Parallel Postulate indicating perhaps his doubts that it was possible. However, by 1826, he presented a paper at the University of Kasan suggesting a new geometry in which more than one parallel could be drawn in contrast to the condition of Playfair's Postulate with the result that the sum of the interior angles of a triangle is less than two right angles. He published a paper on these principles in 1829. Gauss did not learn of this work until 1841. Once again in writing to a colleague in 1846, Gauss noted there was nothing in it that was new to him; however, he praised Lobachevsky's approach. In 1848, Johann Bolyai learned of Lobachevsky's work through a book passed on to his father by Gauss. Apparently, Lobachevsky never knew of Johann Bolyai.[89]

In developing the new geometry, I will start with the postulate common to Gauss, Bolyai, and Lobachevsky that in a plane more than one parallel may be drawn from a point to a line. The mathematician Felix Klein named the resulting geometry Hyperbolic.[90] In keeping with a reasonable view that Gauss, Bolyai, and Lobachevsky could all, arguably, be credited with the discovery of the new geometry, the first postulate is here called the Hyperbolic Postulate.

Hyperbolic Parallel Postulate: Through a given point P not on a given line l more than one line may be produced parallel to the given line in the plane of P and l.

The meaning of this postulate is illustrated in the figure below.

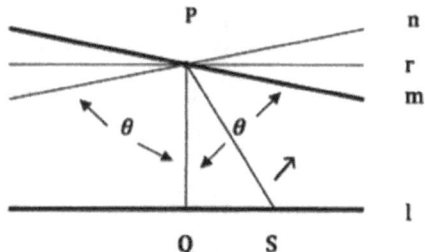

3-1 The Hyperbolic Postulate

In the figure, the given point and line are respectively, P and l. Line PQ is perpendicular to l. Line r is perpendicular to PQ, hence by Postulate IE28 parallel to l because the alternate interior angles are both right angles. As allowed by the Hyperbolic Parallel Postulate, line m is a second line parallel to l through the point P making the acute angle θ with the perpendicular from PQ. By symmetry, line n is also a parallel. Because r, m, and n are parallel to l, any line through P between m and n is parallel to l.

Now assume the line PS intersects l. Moving S to the right along l, the angle PS makes with the perpendicular PQ increases in a counterclockwise direction until the line PS no longer intersects l. Let us associate the angle θ in the figure with this limiting position. Lobachevsky called θ the angle of parallelism. which he found was a function of the distance d_{PQ} between P and Q. He denoted the function of the angle of parallelism as Π, in our example, $\Pi (d_{PQ}) = \theta$. Notice that with this understanding, line m is asymptotic with l on the right side of PQ and n is asymptotic with l on the left-hand side of PQ. These are the parallel lines that Saccheri rejected.

The Hyperbolic Postulate was given by Lobachevsky as his 16th theorem in his *Geometrical Researches on the Theory of Parallels.*[91] Prior theorems did not depend on it. From this point, he proved other theorems such as Theorem 18 in which he proved that two lines are always mutually parallel. In theorem 19, he repeated the Saccheri-Legendre Theorem.[92]

The notable consequence of the Hyperbolic Postulate that the sum of the interior angles of a triangle is less than two right angles was proved by Lobachevsky as his Theorem 22. His proof depended upon a supporting theorem, a type of theorem often called a lemma, which is given below.

Lobachevsky Angle Theorem 21:[93] From a given point (A) a straight line can be drawn that can make with a given straight line (l) an angle as small as we choose.

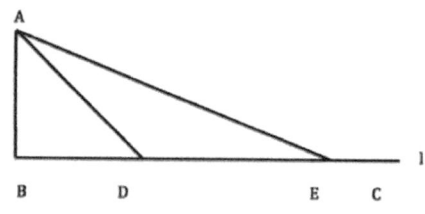

3-2 Lobachevsky Angle Theorem 21

A is a point not on line l	By hypothesis
Produce AB perpendicular to l	Proposition IE12
Choose D on l, BD = AB	Proposition IE3
∠BAD = ∠ADB	Proposition IE5
∠ADB + ∠BAD + ∠ABD ≦ 180°	Saccheri – Legendre Theorem
∠ADB + ∠BAD ≦ 90° (since ∠ABD = 90°)	Common Notion applied to inequalities
2 · ∠ADB ≦ 90°, ∠ADB ≦ 45°	Common Notion applied to inequalities
Produce on l, DE = AD; produce AE	Proposition IE3
∠DAE = ∠DEA	Proposition IE5
∠ADB + ∠ADE = 180°	Proposition IE13
∠ADE ≧ 135° (since ∠ADB ≦ 45°)	Common Notion applied to inequalities
∠ADE + ∠DAE + ∠DEA ≦180°	Saccheri – Legendre Theorem

∠DAE + ∠DEA ≦ 45°

Common Notion
applied to inequalities

∠DEA ≦ 22 ½ °

Common Notion
applied to inequalities

By similarly extending points along l, angles as small as desired may be constructed

Q.E.D.

With this preliminary proposition completed, now we can continue with an approach similar to Lobachevsky's proof that with the Hyperbolic Postulate the sum of the interior angles of a right triangle is less than two right angles.

Lobachevsky Right Triangle Theorem 22: The sum of the interior angles of a right triangle [△ABC] is less than two right angles.[xxxvii]

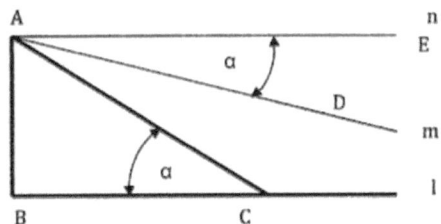

3-3 Lobachevsky Right Triangle Theorem 22

AB is perpendicular to line l at B

Proposition IE11

Line n is perpendicular to AB at A

Proposition IE11

Line m passing through A is parallel to l (does not intersect l); ∠DAE = α

Hyperbolic Parallel
Postulate

Produce ∠ACB = α with C on line l

Proposition IE23,
Lobachevsky Angle
Proposition

[xxxvii] Lobachevsky proved this as part of his Theorem 22 in which he also proved that in a Euclidean triangle the sum equals two right angles; Bonola, Appendix, Lobachevsky, N. *Geometrical Researches on the Theory of Parallels,* p. 17.

△ABC is a right triangle with base BC on l and right angle ∠ABC — Definition I21

∠BAC < 90° −α (since ∠BAE = 90°) — Common Notion 5

∠ABC +∠BAC + ∠ACB < 90° + (90° − α) + α = 180° — C. N. applied to inequalities

Q.E.D.

As Saccheri, Lambert, and Legendre proved in the past, Lobachevsky showed that if one triangle has interior angles summing to less than two right angles, than so do all other triangles (Lobachevsky Theorem 20).[94] Thus, the Hyperbolic Postulate leads us to the conclusion that all triangles have interior angles summing to less than two right angles..

The Saccheri quadrilateral used in the proof of Saccheri Proposition 2 is shown below. Recall that the Saccheri quadrilateral is defined with two perpendiculars (AD and BC) of equal length to a base (AB).

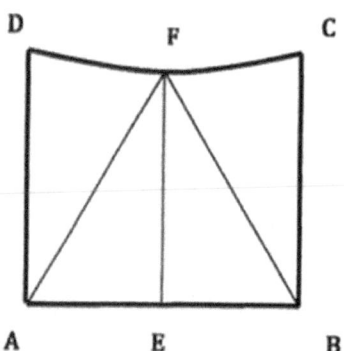

3-4 Proof of the HAA from the Hyperbolic Postulate

Saccheri propositions S1 and S2 (pp. 47, 48) were proved using only the first four of Euclid's postulates. In Proposition S1, it was proved that ∠ADF = ∠BCF. In S2, producing lines with DF = CF and AE = EB, it was proved that EF is perpendicular to AB and DC. Now in a very simple proof, we will show that the Hyperbolic Postulate leads to ∠ADF < right angle consistent with both Saccheri's and Lambert's Hypothesis of the Acute Angle.

Let us look at the quadrilateral ADFE which can be broken into △ADF and △FEA. From the Lobachevsky Right Triangle Theorem, the sum of the interior angles for △ADF and △FEA are each less than two right angles.

$$\text{Sum of the angles of } \triangle ADF = \angle FAD +$$
$$\angle DFA + \angle ADF < 2 \text{ right angles.}$$
$$\text{Sum of the angles of } \triangle FEA = \angle FAE +$$
$$\angle AEF + \angle EFA < 2 \text{ right angles.}$$

Note that using Proposition S2 which is based solely on Euclid's first four postulates:

$$\angle DFA + \angle EFA = \angle AEF = \text{right angle} = \angle FAD + \angle FAE.$$

Adding the two inequalities and simplifying based upon the relations given above, we have:

$$\angle ADF < \text{right angle,}$$
$$\text{and similarly, from quadrilateral BCFE,}$$
$$\angle BCF < \text{right angle.}$$
$$\text{Q.E.D.}$$

With the correspondence between the HAA and the Hyperbolic Postulate, all of theorems proved by Saccheri and Lambert for the HAA are valid in hyperbolic geometry, for example, the congruence of similar triangles and the relationship observed by Lambert between the area of triangles and their angle defects. This is a good point at which to stop and compare some key conclusions that are formed from the HAA, the HOA, and the Euclidean geometry inherent in what Saccheri called the hypothesis of the right angle, the HRA. The table below gives corresponding results for the three hypotheses: the sum of interior angles of a triangle and of a quadrilateral, the ratio of the length of the summit to the base in a Saccheri quadrilateral, the number of lines in a plane that may be drawn through a point parallel to a given line, and the existence of non-congruent similar triangles.

Table 3-1 Comparison of the HAA, HRA, and HOA

	HAA	HRA	HOA
Sum of ∠'s in △	< 180°	= 180°	>180°
Sum of ∠'s in ▢	< 360°	= 360°	> 360°
CD (summit)/AB (base)	> 1	= 1	< 1
# parallel lines	∞	1	0
Non–congruent, similar triangles ?	NO	YES	NO

Given the significant differences in these geometries, how are we to envision the new hyperbolic geometry associated with the HAA. This geometry was a discovery that Gauss would not publicly announce, one called "*imaginary*" by Lobachevsky, and "*a strange new universe*" by Bolyai. The meaning of the new geometry would only become clear after it was connected to the curvature of surfaces (a concept introduced in Chapter 6). From an understanding of curvature, one model of hyperbolic geometry, known as the pseudosphere, was discovered similar to the relationship between the surface of a sphere and the HOA. The discovery of this interpretation forever changed the understanding of Euclid's geometry. In developing the concept of curvature, however, we will find it necessary to leave Euclid's world of deduction, based solely on geometric forms, and begin the number-based analysis using coordinates. This approach, familiar from equations of lines and curves in high school math, is known as analytic geometry. In contrast to this approach, the method so epitomized by Euclid's *Elements* is known as synthetic geometry.

The use of coordinate systems will greatly facilitate development in the new geometries of relationships such as Lobachevsky's function for the angle of parallelism, $\Pi(d_{PQ})$ and others similar to the Pythagorean Theorem of Euclidean geometry. But before we start this new approach, I want to leave you with a preliminary view of a model of hyperbolic geometry to help you visualize its characteristics.

3.3 Modeling the hyperbolic plane – a first look

Numerous models were developed to show that Euclid's first four postulates plus the Hyperbolic Postulate formed a consistent geometry. Among the models were the pseudosphere of Eugenio Beltrami (1835-1900) partially mapping the hyperbolic plane,[95] and disk models (circular regions mapping the entire hyperbolic plane) of Felix Klein (1849-1925)[96] and Henri Poincaré (1854-1912).[97]

Poincaré first developed a model which only covers the hyperbolic half-plane. In this model, illustrated with the figures below, straight lines are segments of circles centered on the X axis. Because of the way length is defined, the "circle" lines can be shown to be the shortest distance between any two points in the half-plane. A precise mathematical definition of length, crucial to the description of non-Euclidean geometries, will be given Chapter 6 and is given for various models of the hyperbolic geometry in Section 7.2.2. However, informally, if δY is a small vertical Euclidian distance, say between points with ordinates Y_2 and Y_1, Poincaré defines the vertical model distance as $\delta S = \delta Y/Y$. Therefore, a fixed model distance δS is formed by increasingly smaller segments δY as the segment approaches the X axis (Y = 0). Although straight lines in the model are curved in the Euclidean sense, the angles at the intersection of lines are the same as Euclidean angles.

In Figure 3-5a below, a number of straight lines are shown: l, m, n, o. With the definition of lines being segments of circles with centers on the X axis, line segments, if extended to the X axis as in the figure, are perpendicular to the axis. A vertical line, perpendicular to the X axis, is also a straight line in the model and can be thought of as a line formed by a circle of infinite radius. In figure 3-5 a, the Hyperbolic Postulate is easily visualized with lines m, n, and o, all passing through point P and not intersecting line l. Lines n and o are diverging parallels, while line m is asymptotically approaching line l at the X axis. In the accompanying figure, 3-5 b, an example of a hyperbolic triangle is given by $\triangle ABC$.

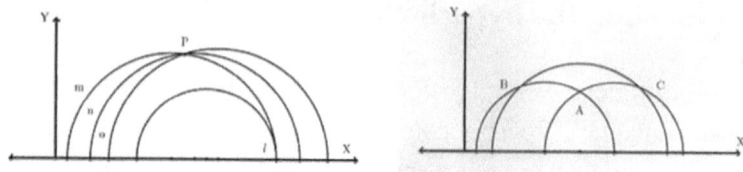

3-5 Poincaré half-plane hyperbolic geometry model

a) Lines m, n, & o parallel to *l*   **b) Hyperbolic △ABC**

Poincaré extended his model to cover the whole hyperbolic plane with a model in which the plane was represented inside a circle of Euclidean radius R (hence the term disk model). The boundary of the circle in the model represents points infinitely far away from any point within the circle. As in the half-plane model, straight lines are segments of circles which if extended to the boundary of the disk intercept it at 90° angles due to Poincaré's definition of length. The model is shown in figure 3-6 with parallel lines similar to those shown in the half-plane model. Through point P, lines m and n are parallel to line *l* with m asymptotic to *l*.

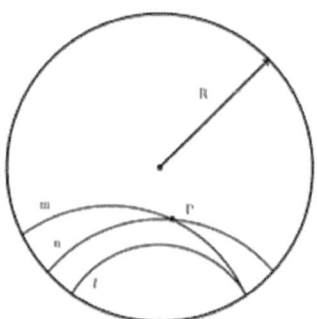

3-6 Poincaré disk model with parallel lines

Similar to the discussion of length in the half plane model, insight into the definition of length may be given informally. If δr represents the Euclidean distance between close points r_2 and r_1 on a radius of the disk (with r measured from the center of the disk), then the distance between the points, S, is given as $S = \delta r/(R-r)$. As in the discussion of the half plane, a radial segment with fixed distance S between its radial points becomes ever smaller as it approaches the boundary of the disk. In the Poincaré disk models, angles of intersecting line segments retain their Euclidean magnitudes.

Other disk models, for example, that of Felix Klein (see discussion on p. 204), preceded Poincaré's. In Klein's model straight lines remained Euclidean straight lines' however, angles have a more complicated definition. One thing that all of the models have in common is that they can be shown to represent curved surfaces with a constant negative curvature in contrast to a sphere of radius R which has a constant positive curvature equal to $1/R^2$. The concept of surface curvature, developed by Gauss, is critical to understanding new geometries. This concept is not easily developed following the deductive approach to geometric forms introduced by the Greeks. It is now time to look at the method, introduced in the seventeenth century, of developing geometric concepts from coordinates expressed in numbers. This new path would eventually lead to a more precise understanding of geometry and make possible Einstein's formulation of General Relativity.

PART II

FOLLOWING THE PATH OF NUMBER

4 Geometry Meets Algebra: Descartes' Insight and Beyond

4.1 Prelude to a revolution – symbolic algebra

The Renaissance, taking place in Europe from the late 14th to the early 17th century, was a period with an extraordinary outpouring of new modes of expression in architecture, sculpture, painting, and writing. The impact on civilization is central to any reading of history. However, virtually ignored in discussions of the period is the development of a new mode of expression with an impact as great as, or arguably greater, than these -- the development of symbolic algebra. The advances in mathematics, science, and technology following the Renaissance are almost unimaginable without this breakthrough in mathematical expression. Before describing these developments, I will highlight some of the advances in the understanding of numbers and their representation from the time of Euclid up to the beginnings of the Renaissance.

The Greeks, as we have discussed, developed a geometric algebra, what I have called the path of form, rather than focus on magnitude in numeric form, the path of number. Geometrical representations of numbers and of their algebraic relations were favored rather than the explicit acknowledgement of numerical approximations such as employed by the Babylonians. We retain some of this viewpoint in our expressions: x–square and x–cube for respectively, x^2 and x^3. They were led to this approach through their concern over the discovery of geometric magnitudes that could not be expressed as the ratio of whole numbers. As an example of geometric algebra, the distributive property of numbers, $a(b + c + d) = ab + ac + ad$ is described in Postulate 1 of Book II of the *Elements*.

If there be two straight lines, and one of them be cut into any number of segments whatever, the rectangle contained by the two straight lines is equal to the rectangles contained by the uncut straight line and each of the segments.[98]

Although, the geometric expressions and visualized proofs of algebraic relations are ingenious, the advantage of the modern algebraic notation is obvious. Despite the drawbacks of geometric algebra, the Greeks continued to make significant discoveries in mathematics. I will mention just two mathematicians and some of their notable achievements. Archimedes of Syracuse (287-212 BC) with methods anticipating integral calculus and limits estimated the value of π with polygons of 96 sides as 6336/2017 1/4 (3.1409...)$< \pi <$14688/4673 1/2 (3.1428...), the area under a parabola, and the proof that the ratio of the volumes of a cylinder and an inscribed sphere is the same as the ratio of their areas, 3 to 2. He was so proud of the last result that he requested the image of the inscribed sphere in a cylinder be carved on his tombstone;[99] Apollonius of Perga (ca. 260-200 BC) developed extensive and detailed properties for the curves known as conics (because they are curves formed on the surface of a cone by an intersecting plane). Relating the curves to their geometric algebraic forms, he introduced the names of the curves by which they are still known: ellipses, hyperbolas, and parabolas.[100]

Exceptions to geometric approaches could be found at the Schools of Alexandria in Egypt from about 300 BC to 600. Particularly noteworthy for this discussion is Diophantus (ca. 350) who developed a primitive symbolic approach to algebra in which words used to specify quantities and operations were abbreviated.[101] This approach referred to as syncopated notation, is a step forward. By condensing the description of the problem, it allows the key elements and their relationship to be more easily grasped. However, manipulation of the notation did not significantly enter into developing the problem solution.

Diophantus provided an approach to solving some forms of quadratic equations, solutions for one form of cubic equation, and algebraic solutions to a variety of word problems. In all cases the solution method was supplied for problems with specific numerical values rather than as general solutions. When the roots were irrational or negative, the solutions were rejected.[102] Others of this period would also continue the development of algebra using the practical approach begun by the Egyptians and Babylonians. However, little interest was taken in its logical foundations in sharp contrast to deductive geometry. No proofs were given for the solution techniques. The development of an axiomatic approach to numbers and their properties would not begin until the nineteenth century.

With the end of the Western Roman Empire in the fifth century and the relative stagnation of mathematics in the continuing Byzantine Empire (Eastern Roman Empire), innovation in mathematics would move eastward to the communities of the Indian sub–continent and those of the Islamic civilization developed after the seventh century. Major developments in mathematics in south Asia included a positional decimal system in the sixth century and the addition of zero, perhaps about 200 years later.[103] In addition to the positive integers and fractions, negative and irrational numbers were used. Some initial steps were taken towards symbolic algebra similar to Diophantus; however, as in the past, the rules were developed pragmatically and with analogies to the operations of the natural numbers.

Muslims came into contact with the mathematics of the Indian sub-continent in the seventh and eighth century. By the end of the eighth century, they were familiar with the numerical notation, arithmetic, and algebra that had been developed in South Asia. Advances in rules for square roots, products of terms such as $(x \pm a)(x \pm b)$ (modern notation) and quadratic and cubic equations represented geometrically continued with solutions given as verbal recipes. Despite rejecting negative solutions to equations, rules were established for signed numbers: for example, $(-a) \cdot (-a) = a^2$ (modern notation). Among the greatest of the contributors to Islamic mathematics was Muhammad ibn Musa al–Khwarizmi (ca. 780–ca. 850).[104] His name would come down in English as algorithm and his work *Hisob al–jabr wa'l muqabalah* (the title referring roughly to algebraic processes)[105] would give us the word 'algebra'. The lack of a symbolic approach greatly obscured general insights into algebraic problems. A noteworthy step in numeration outside of the context of geometric algebra was the use of decimal representations for fractions by Jamshid al–Kashi (ca. 1380 – 1429).[106]

Europeans first became aware in the twelfth century of the mathematics of the Islamic world through the schools of Granada, Cordova, and Seville in the Iberian Peninsula.[107] For the next 300 years, the primary mathematical activity was absorption and spread of the knowledge that had been developed in Greece, South Asia, and the Islamic world. By the middle of fifteenth century, the dawn of the Renaissance, this was accomplished throughout much of central and western Europe through books prepared by European scholars from Greek and Islamic sources.

From the mid fifteenth to the mid seventeenth century, advancements were made by European scholars along the previously established directions of algebra. For example, Nicholas Tartaglia (1500 – 1557) solved some forms of cubic equations with positive coefficients.[108] Girolamo Cardano (1501 – 1576) published Tartaglia's solution for cubic equations even though he had promised Tartaglia that he would not publish them. However, Cardano did provide a significant analysis of the solutions.[109] Following the tradition of geometric algebra, he interpreted cubic equations in terms of volumes, although he used some syncopated notation that took a step towards more general solutions. Cardano discussed the nature of the solutions, including negative and what became known as complex numbers (solutions involving real terms and imaginary numbers - square roots of negative numbers). He noted that the complex numbers would appear in pairs although he described them as *"sophistic"* and that the result *was* ``as subtile as it is useless.''[110] Cardano's solutions with *"sophistic"* numbers would not begin to be clarified for another two hundred years when John Bernoulli and Leonard Euler took up the study of imaginary numbers.

With this brief summary of the state of mathematics going into the Renaissance, it is time to follow the fitful path of progress leading to symbolic algebra. The 15th century gave rise to numerous notations for algebraic expressions with no standardization. Among these were the notations of the Frenchman Nicolas Chuquet (1445-1488, see below), Rafael Bombelli (1526-1572, Florence) [111] and Simon Stevin (1548-1620, Bruges).[112] To give a feeling of the type of notation in use, examples are provided from the notation of Chuquet, comparing modern and period notation[113]. Chuquet used a syncopated notation with the operations of addition, subtraction, and division simply indicated, respectively, by the words (in period French): *plus, moins, and partyr par.* Addition and subtraction were sometimes abbreviated as $\bar{p}$ and $\bar{m}$.[xxxviii] The operation of taking a square root was indicated by $\mathbf{R})^2$. Unknowns were expressed by the power of the unknown and the unknown's coefficient. For example, $6x^2$ was expressed as $.6.^2$. Exponents could be positive or negative and included zero: x^{-2} was rendered as $.1.^{2\bar{m}}$. As noted by Merzbach and Boyer, Chuquet's notation made clear the use of exponents; the result of simplifying x/x^3 was written as $.1.^{2\bar{m}}$. Other syncopated algebras would similarly clarify the role of exponentials. Examples of expressions using the notation of Chuquet are shown below. Equality is indicated by its counterpart in period French, *egaulx.*

Table 4-1 Comparisons of Algebraic Notations

Modern notation	Chuquet's notation
$\sqrt{14 - \sqrt{180}}$	$\mathbf{R})^2.14.\bar{m}.\mathbf{R})^2180$
$\dfrac{72x}{8x^3} = 9x^{-2}$	$.72.^1\ partyr\ par.8.^3\ egaulx.9.^{2.m}$
$4x = -2$	$.4.^1\ egaulx\ \bar{m}.2.^0$

Apparently, the last example is the first incidence of the use of an isolated negative number in an equation. In a recurring theme of the Renaissance mathematicians, Chuquet recognized that the solution of some equations implied solutions with square roots of negative

---

[xxxviii] The use of the symbols + and - goes back to the 15th century, and their use may have originated as symbols used in German states to mark warehouse containers as having an excess or deficit in weight. However, the symbols were not used to express the operation of addition or subtraction until the 16th century when they were used in syncopated notations by, among others, Michael Stifel (ca. 1497-1567) -- see Merzbach and Boyer, p. 254; Ball, pp. 215-217.

numbers, but did not know what to do with them. When Cardano later in the Renaissance encountered imaginary numbers, they were still expressed in a manner similar to that Chuquet, for example, $5 + \sqrt{-15}$ was expressed as 5p:R̄m̄:15. [114]

Similar in approach, but better known due to the availability of printing, were the efforts of Luca Pacioli (1445-1514), the holder of a chair of mathematics in Milan.[115] A major advance of Pacioli was the explicit designation of the unknown, called the thing, *cosa* in Italian, sometimes abbreviated *cos*, or *res* in Latin, sometimes abbreviated as R.[xxxix] Powers of the unknowns were represented by their own term, for example *cuba* for the cube. Pacioli's equations only used specific numerical coefficients of powers of the unknown, so, as with other algebraists, no progress was made towards general solutions. In any case solutions were presented as verbal instructions.[xl]

Numerous syncopated systems by others were developed; however, the symbols were still primarily abbreviations for words, and their use was still not fundamental to solving problems. In addition to the aforementioned mathematicians, I mention several other notable contributors and the notation they introduced: Christopher Rudolff (ca. 1500 - ca. 1545), $\sqrt{}$;[116] Robert Recorde (1510-1558), =;[117] Thomas Harriot (1560-1621), < and >;[118] and William Oughtred (1575-1660) × (multiplication).[119]

Clearly in this period, there were many insightful approaches, but as Merzbach and Boyer remind us, *"Mathematics is a form of reasoning, not a bag of tricks."* [120] With hindsight, one can see that among the most important missing pieces necessary to drive the algebraic reasoning process was an approach that would allow insight to general solutions rather than simply solutions to problems with specific numerical coefficients. This requires a clear approach generalizing the unknown

---

[xxxix] The use of the term ``thing'' for the unknown recalls the use of the term ``heap'' for the unknown used by the ancient Egyptians (Bunt, Jones, and Bedient, p. 30.)

[xl]   In addition to Pacioli' s mathematical accomplishments, he is also generally recognized as the father of double-entry bookkeeping (Merzbach and Boyer, p. 252.).

and unknown quantities. Although no one person can be credited with devising the notation of symbolic algebra, Ball notes ``[That] *so far as the credit of inventing symbolic algebra can be put down to any one man we may perhaps assign it to Vieta....*'' [121]

François Viète (1540-1603) was a lawyer and member of the king's council under Henry III and later Henry IV.[122] His mathematical pursuits occurred in leisure moments of which apparently he had an abundance during the six year period before Henry IV came to the throne, and Viète was out of favor. During this period he made contributions to trigonometry, geometry, and promoting decimal over sexagesimal notation, but it is his impact on symbolic algebra that interests us here. Viète's major innovation was to use different symbols for the unknown and to generalize the known coefficients. He used vowels for the unknowns and consonants for the known. He also adopted the German symbols for addition and subtraction and a fraction line for division while continuing to indicate multiplication and equality by, respectively the Latin words *in* and *aequatur*. Although he did not use numerical exponents as in the notation of Stifel or Bombelli, he used a single symbol for the unknown with the exponent given in words. This was a significant advance over the practice of expressing different powers of the unknown as a different symbol. Viète 's notation (with A the unknown) is illustrated by the following comparison with modern notation:[123]

$$3BA^2 - DA + A^3 = Z \text{ (modern)}$$
B3 in A quad. − D plano in A + A cubo aequatur Z (Viète)

Viète used his notation to develop a new approach to cubic equations and also solved biquadratic equations (x^4 the as highest power). He contrasted deductive geometry in which, starting from postulates and proven propositions, one reasons to a conclusion with algebraic reasoning. In algebra the existence of the unknown is assumed with the reasoning process leading to a value. Viète therefore referred to algebra as ``*the analytic art*'' in contrast to geometry's deductive synthetic approach.[124]

As Merzbach and Boyer wisely state, *'It is not given for one man to make the whole of a given change; it must come in steps."*[125] Harriot extended the clarity of the notation by repetitively writing the unknown to represent the unknown to different powers, e.g., AAAA for x[4] and Descartes took the next step of combining Viète's approach with Stifel's approach to exponentials, that is AA became A^2.[126] With much of algebraic notation in place, Descartes was in position to invent analytic geometry, the algebraic interpretation of geometry.

4.2 Euclid finds his place on the Cartesian plane – analytic geometry

Arguably the most significant innovation in mathematics of the first half of the seventeenth century was the creation of a bridge between geometry and algebra through the development of analytic geometry. As is often the case, major new insights may be independently discovered. However, priority appears to be given to René Descartes (1596 – 1650) as attested to, at least in terms of common attribution, through the naming of coordinates referred to perpendicular axes as Cartesian. Descartes recognized the possibility of describing geometric curves in the plane through equations of these coordinates and thus, geometric properties would be embedded in algebraic descriptions.[127] In *La Géométrie*, Descartes refers to the loci (paths) of points forming curves such as the parabola, ellipse, or hyperbola in this way:

"The solution of any one of these problems of loci is nothing more than the finding of a point for whose complete determination one condition is wanting . . . . In every such case an equation can be obtained containing two unknown quantities." [128]

In forming a relationship between the points of the Cartesian plane and numerical coordinates, Descartes assumed, certainly without appreciation for his farsighted assumption, that there was a one-to-one correspondence between points of a line and numbers. Indeed,

Descartes did not even include negative numbers. We will explore the properties of the real numbers that make this possible in the next section.

Prior to Descartes, Pierre de Fermat (1601 – 1665) had similar insights which were closer in application to modern analytic geometry; however, his results were not published in his lifetime.[129] Both men investigated geometric properties such as the tangent (slope of a curve at one point) and its normal (perpendiculars to the tangent lines) to curves. The reduction of geometry to algebraic functions would ultimately lead to common foundations in numbers for geometry and algebra.

Much of the effort through the sixteenth century in the development of algebraic methods was related to finding methods for solving equations, for example, Cardano's work in the sixteenth century on cubic equations such as $x^3 + qx = r$. Descartes and Fermat recognized that such equations could be made visual as curves on a coordinate system, with the benefit of providing insight into their solution, for example, by representing them as $y = x^3 + qx - r$.

The idea of coordinate systems envisioned by Descartes in the seventeenth century seems rather obvious to us now. We are used to it not only as a means of plotting the relationship of two variables on a flat piece of paper in school, but in our experience of expressing the location of a place on a map of the world through the coordinates of latitude and longitude. In both cases, we notice that to express the location of a point on a surface we need two variables. Following Descartes, we routinely make use of a coordinate system with one variable referred to an axis called the abscissa and with the other variable referred to an axis perpendicular to the abscissa called the ordinate. In the case of the straight line, $y = mx + b$, the abscissa is conveniently taken as the x axis and the ordinate as the y axis. A point in the plane of the x and y axes is noted as (x, y). The familiar graph of the straight line $y = mx + b$ is shown below.

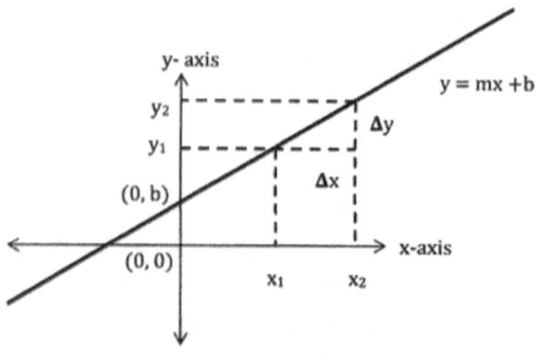

4-1 Graph of y = mx + b

In the figure above, the curve, shown in bold, is referenced to x and y axes. The origin (0, 0) is the intersection of these axes. The straight line intersects the y-axis where x is equal to 0 at the point (0, b). This point is known as the y-intercept. Two general points (x_1, y_1) and (x_2, y_2) on the curve are selected to remind you how the points are referred to the axes. The slope of the line can be calculated from these two points and is defined as the change in the y coordinate compared to the x coordinate, sometimes described as the rise over the run.

Slope $= (y_2 - y_1) / (x_2 - x_1) = \Delta y / \Delta x$ where Δy and Δx refer symbolically to the change in y and x, respectively.

Because the curve is a straight line, the slope is the same for every pair of points (x, y) and (x_1, y_1) on the line. Therefore:

$(y - y_1) / (x - x_1) = $ constant $= k$ or solving for y,
$y = k(x - x_1) + y_1 = kx + (y_1 - kx_1)$.

Now, comparing this result with the equation, y = mx + b, we see that m = k, thus m is just the slope and b = $(y_1 - kx_1)$. We have recovered the usual school result that a straight line can be defined by its slope and the y-intercept or equivalently by any two points.

While the Cartesian coordinate system is being introduced, I note that the use of axes perpendicular to one another allows the Pythagorean

Theorem to be readily used to define distances in terms of the coordinates. In the figure above the projections of the coordinates to the x and y axes forms a right triangle with the hypotenuse formed by the line connecting (x_1, y_1) and (x_2, y_2). The square of the distance between the two points is therefore, $(x_2 - x_1)^2 + (y_2 - y_1)^2$ with the distance being $\sqrt{(x_2 - x_1)^2 + (y_2 - y_1)^2}$.

The power of a coordinate representation of geometric objects such as a straight line is that we can use all of our algebraic properties of the number system to determine geometric properties. The equivalence of an algebraic description of geometry and Euclidean geometry may be made clear from a list of algebraic interpretations of Euclid's definitions and postulates. I will not be comprehensive in pointing out this correspondence, but I hope to make it plausible and to draw your attention to some of the main features.[xli] The first eight definitions in Euclid's geometric system refer to points, lines, surfaces, and angles. I will give algebraic interpretations for these, as well as the circle (the fourteenth definition), because of its prominence in Euclid's third postulate and its use in the proof of the first proposition. The definition of angles will be delayed until the Law of Cosines is proved using an algebraic approach. Euclid's proof of the Law of Cosines is given in geometric form in Book II (Proposition IIE12 and IIE13). The Law of Cosines, which is a generalization of the Pythagorean Theorem, allows calculations of triangles that are not right–angled. However, the tabulation of trigonometric functions that facilitate calculations would await the trigonometric tables developed by Hipparchus of Nicaea (ca. 180 – ca. 125 B.C) over one hundred years after Euclid.[130]

We have already identified a point as the coordinate (x, y) on the plane formed by the x-y axes. I will further take the unhistorical leap of identifying the points x and y with the real number system. Euclid's first Postulate (IE1) states that in his geometric system, it is always possible to draw a straight line from any point to any point. Given two points (x_1, y_1) and (x_2, y_2), we have shown that we can identify the straight line,

---

[xli] Eves discusses the correspondence of the analytic geometry approach to Hilbert's system in which the deficiencies of Euclid's geometry have been removed (pp. 94-98).

$y = m(x - x_1) + y_1$, with the slope $m = (y_2 - y_1)(x_2 - x_1)$. Moreover, the equation of the straight line being made up of the real numbers can be drawn continuously as required by the second postulate and indeed indefinitely as allowed in the fifth postulate. The parallel lines of the fifth postulate may be identified as line of equal slope.

In regard to surfaces, the Cartesian plane formed by the x and y axes can be considered to be within a three-dimensional space with a third z-axis perpendicular to the x and y axes and having a common origin. In this sense all figures formed by closed boundaries (in the sense of Euclid's 14th Definition) and described by an algebraic relation by the variables y and x, are surfaces in the plane defined by z =0. Such an interpretation helps to illuminate Euclid's more obscure view that *"A plane surface is that which has length and breadth only."* (Definition 5.).

We may use the Pythagorean Theorem, which we know is a consequence of Euclid's fifth postulate, to determine the algebraic expression for a circle, and also expressions for the trigonometric functions (sometimes called circular functions). In the following figure, a circle of radius r has been centered at the origin of a coordinate system with the x-axis horizontal and y-axis vertical. A right triangle is constructed with its hypotenuse, AC as a radius r of the circle with center at the point (0, 0). In terms of the coordinate system, the length CD = x, and the length AD = y. From the Pythagorean Theorem: $CD^2 + AD^2 = AC^2$ or associating the coordinate (x, y) with the general point A on the circle, we have $x^2 + y^2 = r^2$. This is the equation for the circle of radius r. with its center at the origin.

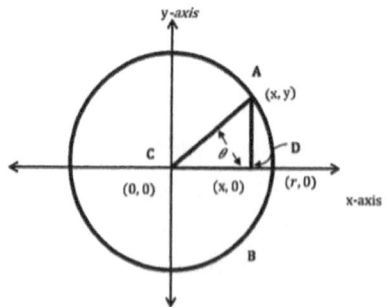

4-2 Graph of a circle

In addition to determining the equation of a circle, the above figure may be used to establish trigonometric identities. From the discussion of similar triangles (pp.38-40) and the definitions of sin θ, cos θ, and tan θ, the relations shown below follow:

$$\sin \theta = \text{opposite} / \text{hypotenuse} = AD / AC = y/r.$$
$$\cos \theta = \text{a djacent} / \text{hypotenuse} = CD / AC = x/r.$$
$$\tan \theta = \text{opposite} / \text{adjacent} = AD / CD = y/x.$$

Reciprocals of these functions are also defined: secant (sec θ) = 1/cosine, cosecant (csc θ) = 1/sine, and cotangent (cot θ) = 1/tangent. Taking into account the equation of the circle, $x^2 + y^2 = r^2$, then,

$$\sin^2\theta + \cos^2\theta = 1.$$
$$1 + \tan^2\theta = \sec^2\theta.$$
$$1 + \cot^2\theta = \csc^2\theta$$

The angle θ is traditionally taken as increasing as the angle moves counterclockwise from the x–axis. A full turn around the circle from the x–axis is taken arbitrarily as 360 degrees, a practice due to the Babylonians with their base 60 number system. Recalling that the circumference of the circle, $C = 2\pi r$, the arc of any circular sector of angle θ degrees may be expressed as $S_\theta = 2\pi r \cdot (\theta/360) = (\theta/180)\,\pi r$. Alternatively, since $C/r = 2\pi$, 2π is an angular measure corresponding to 360°. The angular measure is given the unit name radian which is dimensionless (length per unit length). As examples, the positive x-axis corresponds to 0 degrees (0 radians), and 90 degrees ($\pi/2$ radians) corresponds to the positive y axis.

With this background, we are now in a position to establish the Law of Cosines and the Law of Sines. This will allow us to solve for unknown side lengths of triangles given either two sides and an angle or two angles and a side. I prove these laws algebraically using the triangle in the figure below. For convenience, capital letters refer here to the angles at the vertices with the opposite side lengths denoted by the associated lower case letters (a notation begun by Euler).[131] Thus, the traditional notation ∠BAC is replaced by ∠A, and BC = a.

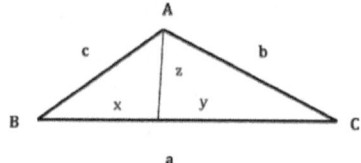

4-3 Law of Cosines

Using the above notation, we have:
Law of Cosines: $c^2 = a^2 + b^2 - 2ab \cos(C)$,
Law of Sines: $a/\sin(A) = b/\sin(B) = c/\sin(C)$.

Proof
First in the figure above, produce the perpendicular from point A to the line BC. Two right triangles are formed with sides of length c, x, z and b, y, z. Since BC = a, y = a − x. Now using the Pythagorean Theorem:

$b^2 = y^2 + z^2 = (a - x)^2 + z^2$ or $z^2 = b^2 - (a - x)^2$
$c^2 = x^2 + z^2$, substituting the above expression for z^2 gives,
$c^2 = x^2 + b^2 - (a - x)^2 = x^2 + b^2 - a^2 + 2ax - x^2 = b^2 - a^2 + 2ax$.
Using the definition of $\cos(\theta)$, y = b cos (C) and, x = a − y = a − b cos(C), therefore,
$c^2 = b^2 - a^2 + 2ax = b^2 - a^2 + 2a(a - b \cos(C))$
$c^2 = a^2 + b^2 - 2ab \cos(C)$. Q.E.D.
The Law of Sines may easily be proved by noting:
z = c sin (B) = b sin (C), therefore,
sin (B) / b = sin (C)/c.

The Law of Sines for angle A may similarly be proved by producing a perpendicular from C to the extended line BA and noting that sin (180° − A) = sin A.[xlii]

The laws of sine and cosine can be written for all the different possibilities of angles and sides, by cycling the letters a, b, c and A, B, C:

$a^2 = b^2 + c^2 - 2bc \cos(A)$, and

[xlii] Denoting the length of the perpendicular as d, then: d = a sinB = bsin(180 − A) = bsinA.

$b^2 = c^2 + a^2 - 2ca \cos(B)$.

Note that for a triangle with $\angle C = 90°$, the Pythagorean Theorem is recovered $c^2 = a^2 + b^2$.

If we associate x-y coordinates for the three vertices, A, B, and C, we can establish a general relation for the angle θ between two lines (in this example, AC and BC). First solving for $\cos(C)$

$\cos(C) = (a^2 + b^2 - c^2) / (2ab)$:

Let the coordinates at C, B, and A be, respectively, (x_1, y_1), (x_2, y_2), and (x_3, y_3), then:

$$a = \sqrt{(x_2 - x_1)^2 + (y_2 - y_1)^2},$$
$$b = \sqrt{(x_3 - x_1)^2 + (y_3 - y_1)^2},$$
$$c = \sqrt{(x_3 - x_2)^2 + (y_3 - y_2)^2}.$$

After substituting the above expressions for a, b, and c into the above equation for $\cos(C)$ with C the angle vertex with coordinates (x_1, y_1), representing the general angle θ and simplifying:

$$\cos\theta = \frac{(x_3 - x_1)\cdot(x_2 - x_1) + (y_3 - x_1)\cdot(y_2 - y_1)}{\sqrt{(x_2 - x_1)^2 + (y_2 - y_1)^2} \cdot \sqrt{(x_3 - x_1)^2 + (y_3 - y_1)^2}} \qquad (4\text{-}1)$$

The above equation becomes the coordinate-based definition of an angle. Its consistency with Euclidean geometry is assured as its follows from the Law of Cosines. In regard to figure 4-3, the above equation for $\cos\theta$ ($\cos$ C) may described as the product of the difference of the x coordinates at point A and C and that of B and C plus the similar product for the y coordinate all divided by the product of the length of AD and BC.

One consequence of this equation is the simple result that occurs for the case when the angle θ is a right angle, that is the line from (x_1, y_1) to (x_2, y_2) is perpendicular to the line from (x_1, y_1) to (x_3, y_3) In this case,

$$\cos\theta = 0 = \frac{(x_3 - x_1) \cdot (x_2 - x_1) + (y_3 - x_1) \cdot (y_2 - y_1)}{\sqrt{(x_2 - x_1)^2 + (y_2 - y_1)^2} \cdot \sqrt{(x_3 - x_1)^2 + (y_3 - y_1)^2}}, \text{ or}$$

$$(x_3 - x_1) \cdot (x_2 - x_1) + (y_3 - x_1) \cdot (y_2 - y_1) = 0.$$

This may be rearranged as:

$$\frac{(y_3 - y_1)}{(x_3 - x_1)} = -\frac{(x_2 - x_1)}{(y_2 - y_1)} = -1 \Big/ \frac{(y_2 - y_1)}{(x_2 - x_1)}$$

The ratio, $\frac{(y_2-y_1)}{(x_2-x_1)}$ is the slope m of a line from the point (x_1, y_1) to the point (x_2, y_2). Therefore, the slope of the perpendicular line from (x_3, y_3) to (x_1, y_1) has a slope $\frac{(y_3-y_1)}{(x_3-x_1)} = -1/m$. This is an algebraic version of Euclid's Proposition 11 from Book I that says, ``[One can] *draw a straight line at right angles to a given straight line from a given point on it.*''[132]

4.3 Some truths unknown to Euclid and Descartes: the real numbers

4.3.1 The Postulates

Despite the revolutionary nature of developments that would lead to analytic geometry by Descartes and Fermat, they were still far from understanding the relationship between numbers and the number lines that formed the axes of their coordinate systems. A rigorous approach to analytical geometry depends upon establishing a set of postulates that ensures a one-to-one correspondence between numbers and the points of a continuous number line. In the absence of such a set of postulates for numbers, neither the holes in the number line nor those in Euclid's logic could be filled. For example, the problem of defining the meaning of irrational numbers remained essentially in the same state as it had been since the disturbing discovery by the Pythagoreans two thousand years before. As for negative numbers, Descartes never used negative abscissas.[133] While the common notions of Euclid remind us of some of the rules of algebra, no significant attempts were made to put number systems and their associated algebras on an axiomatic basis like geometry until late in the 19th century.

The axiomatic basis of the one-to-one correspondence between points of the continuous number line and the real numbers can be simply stated by a set of underlying postulates similar to Euclid's postulates and common notions. These postulates are listed below. Another approach, however, is to build up the real number system starting with the natural numbers (1, 2, 3,...) and then adding properties to create the integers with zero and the negative whole numbers, followed by the rational and irrational numbers. Such an approach provides much insight, but can only be adequately given at great length. Such an approach is given with much detail by Landau.[xliii] Here, however, without further ado, the canonical postulates for the real numbers are listed below.

Canonical Postulates for the Real Numbers:[134]

$\mathcal{R}_c$**1 (Closure Postulate for Addition):** If a and b are real numbers, there is a unique real number c such that $a + b = c$.

$\mathcal{R}_c$**2 (Closure Postulate for Multiplication);** If a and b are real numbers, there is a unique real number c such that $a \cdot b = c$.

$\mathcal{R}_c$**3 (Associative Postulate of Addition):** $(a + b) + c = a + (b + c)$.

$\mathcal{R}_c$**4 (Associative Postulate of Multiplication):** $(a \cdot b) \cdot c = a \cdot (b \cdot c)$.

$\mathcal{R}_c$**5 (Commutative Postulate of Addition):** $a + b = b + a$.

$\mathcal{R}_c$**6 (Commutative Postulate of Multiplication):** $a \cdot b = b \cdot a$.

$\mathcal{R}_c$**7 (Distributive Postulate):** $a(b + c) = a \cdot b + a \cdot c$.

$\mathcal{R}_c$**8 (Identity Element for Addition):** There exists a real number, 0, with $a + 0 = a$.

$\mathcal{R}_c$**9 (Identity Element for Multiplication):** There exists a real number, 1, with $a \cdot 1 = a$.

$\mathcal{R}_c$**10 (Additive Inverse Postulate):** For every real number a, there is a real number (-a) such that $a + (-a) = 0$.

[xliii] A similar approach, perhaps more sympathetic (but less rigorous), for the non-specialist, is provided in my book *Images of Mathematics Viewed through Number, Algebra, and Geometry* ((pp. 44-105) with additional references.

$\mathcal{R}_c$ **11 (Multiplicative Inverse Postulate):** For every real number a, there is a real number (1/a) such that a · (1/a) = 1. (a ≠ 0)

$\mathcal{R}_c$ **12 (Order Postulate):**[xliv] For real numbers a, b, c and d:

I. Only one of the following three cases is true: 1) a = b, 2) a < b, or 3) b > a;[xlv]

II. If a < b, and b < c, then a < c;

III. If a < b, then a + c < a + c;

IV. If a > 0, and b > 0, then ab > 0.

$\mathcal{R}_c$ **13 (Postulate of Continuity):** If a set of real numbers has an upper bound, then it has a least upper bound.

The first twelve postulates are the same as the postulates for the rational numbers and form what is known as an ordered field.[xlvi] The thirteenth postulate, the Postulate of Continuity, is the one that will allow us to include in the number system the irrational numbers such as $\sqrt{2}$ which caused so much pain to the Pythagoreans and subsequent misunderstanding for two thousand years. With the addition of Postulate 13, the resulting mathematical structure is known as a complete ordered field. The seemingly innocuous thirteenth postulate required a systematic understanding of limits of infinite sequences that was beyond the grasp of even such great mathematicians as Isaac Newton and Leonard Euler. The concept of limit, introduced in the next two sections to give meaning to the Postulate of Continuity and the real numbers, will be vital to describing changes to curves and surfaces, and therefore is necessary for the analytic description of non-Euclidean geometries in subsequent chapters.

[xliv] Eves (p. 180) provides an alternative statement of the Order Postulate. There exists a subset P, not containing 0, of the set S (the real numbers) such that if a ≠ 0, then only one of a or (-a) is in P. Eves defines the elements of P as the positive real numbers with the other nonzero elements defined as the negative numbers.

[xlv] The real number a is less than b (a < b), if for some real number c greater than 0, a = b +c. If a < b, then symbolically, b > a.

[xlvi] The postulates for the real numbers have been stated in terms of the operations of addition and multiplication; however, any collection of elements with operations that are governed by the same postulates forms a field. Eves has described the postulates in this manner (p. 180.).

Of the thirteen postulates of the real numbers, the first seven and Postulates R_c 9 and R_c 12 are satisfied by the natural numbers: 1, 2, 3, 4.... The integers are created by the addition of the identity element for addition, 0, and additive inverses (-a) forming the negative integers and the mathematical structure known as a group. With these additions, if a and b are integers, any equation a + x = b may be solved for the unknown x (a tradition started by Descartes)[135] with the solution, x = b + (-a) = b - a. Solving the general equation ax + b = c, requires the further expansion of the numbers with multiplicative inverses, 1/a, allowing the general solution x =(c-b)/a. The addition of the multiplicative inverses forms a second commutative group within the rational and real numbers. Thus, the solution methods given above apply equally for the rational and real numbers.

The following rules of school algebra for solving for the unknown x can all be proved as theorems from the first 12 postulates and apply to both the rational and the real numbers.[136]

$a + b = a + c \Rightarrow b = c.$
$ab = ac \Rightarrow b = c.$
$(-a)(-b) = ab.$
$(-a)(b) = -ab.$
$ab = 0 \Rightarrow a = 0 \text{ or } b = 0.$

Although the natural numbers, integers, rational and real numbers as collections possess an unlimited number of members, that is they form infinite sets, only the rational and real numbers share the property of denseness. By denseness, it is meant that between any two numbers, there is another number. In contrast to the denseness of the real and rational numbers, there are no natural numbers between, say, 5 and 6 or integers between -49 and -50. To illustrate the property of denseness, suppose in the rational or real numbers a < b. Then a number c between a and b is:

$$a + (b - a)/2 = c.$$

A choice for a number between a and c could similarly be a + (c - a)/ 2, and so forth. Thus, between any two real or rational numbers, there is also an infinite set of numbers. The way to compare sets with an infinite number of members is crucial to the difference between the real numbers and the rational numbers. However, before we attempt such comparisons (Section 4.3.4), we need to see how the thirteenth postulate, the Postulate of Continuity, allows us to include numbers such as $\sqrt{2}$ and π to form the real number system and fill in the holes in Descartes' axes and Euclid's line segments. One starting point is to look at how rational numbers are represented as decimals and what are the implications of this approach for the irrational numbers.

4.3.2 Rational numbers as decimals – a starting place for the real numbers

The decimal number system that we are so accustomed to has ancient roots, as discussed in the first chapter. The Babylonians developed a sexagesimal positional system including fractions over four thousand years ago, while the decimal positional system was introduced in the Indian sub-continent over a thousand years ago, with the predecessors to our ten numerals. Recall that in the decimal system any integer greater than zero can be written with each digit multiplied by a positive integer power of ten, depending on its position. For example, the number 1596 (the year of Descartes' birth) is an abbreviation for: $1596 = 1 \cdot 1000 + 5 \cdot 100 + 9 \cdot 10 + 6 \cdot 1$. Or explicitly using powers of 10, $1596 = 1 \cdot 10^3 + 5 \cdot 10^2 + 9 \cdot 10^1 + 6 \cdot 10^0$. More generally any 4 digit integer $a_3 a_2 a_1 a_0 = a_3 \cdot 10^3 + a_2 \cdot 10^2 + a_1 \cdot 10^1 + a_0 \cdot 10^0$. For example in the year of Descartes birth: $a_3 = 1$. $a_2 = 5$, $a_1 = 9$, and $a_0 = 6$.

The fractions expressed in the rational numbers are similarly expressed with decreasing powers of ten as shown in the familiar example:

$$\frac{3}{8} = 0.375 = 3 \cdot 10^{-1} + 7 \cdot 10^{-2} + 5 \cdot 10^{-3}$$

One may wonder if all rational fractions may be written in this manner. The answer is yes, and I shall make that plausible with a few informal arguments.[xlvii] Let us first recall, with the fraction 3/8 as an example, how we form decimal numbers using the long division technique we learned in elementary school.

In long division we use the following process. Since 8 is greater than 3, we first look for the number of tenths in the fraction, that is $10 \cdot \frac{3}{8} = 3$ plus 6 left over. We continue, looking for the number of hundredths, that is $10 \cdot \frac{6}{8} = 7$ plus 4 left over.. And finally looking for the number of thousandths, we have $10 \cdot \frac{4}{8} = 5$ and 0 left over. Therefore, $3/8 = 0.375$

We could write this more explicitly:

$3/8 = a_{-1} \cdot 10^{-1} + a_{-2} \cdot 10^{-2} + a_{-3} \cdot 10^{-3}$.
$10 \cdot 3/8 = 30/8 = 3 + 6/8 = a_{-1} + 6/8$. Therefore,
$a_{-1} = 3$.

$10 \cdot 6/8 = 60/8 = 7 + 4/8 = a_{-2} + 4/8$. Therefore,
$a_{-2} = 7$.

$10 \cdot 4/8 = 40/8 = 5 + 0/8 = a_{-3} + 0/8$. Therefore,
$a_{-3} = 5$.

The long division process comes to an end when the remainder is 0 as shown above. But what would happen if the remainder never equals 0. Here we note that the remainder must then be one of the nine digits, 1 through 9. Eventually the remainder will be repeated and the pattern of digits formed in the long division process will simply repeat endlessly. As an example:

$$3/11 = 0.27272727\cdots$$

From these examples, it should be clear that the fractional part of any rational number can be expressed either as a decimal with a finite number of digits or one with an endless repeating pattern of digits. Any repeating pattern may easily be transformed into a rational number

---

[xlvii] A more formal discussion is given by Hamilton, pp. 43-49.

expressed as the ratio of two integers. This may be shown informally in going from the decimal representation 0.2727272... to 3/11. A proper analysis requires an understanding of the limits of infinite sequences, but at this point we only need an intuitive approach.

$S = 0.27272727\cdots.$
$100 \cdot S = 27.272727\cdots.$
$100S - S = 99S = (27.272727\cdots) - 0.272727\cdots.$
$99S = 27.$
$S = 27/99 = 3/11.$

The decimal 0.272727... is an example of a geometric series, that is, one in which each successive term in a sum is multiplied by the same number. This may be easily seen if it is written as:

$3/11 = 0.272727\cdots = 0.27 + 0.0027 + 0.00000027\cdots.$

$$3/11 = 0.27\{1 + 0.01^1 + 0.01^2 + 0.01^3 + \cdots\} \qquad (4\text{-}2)$$

Looking just at the sum within the brackets, each term is the product of the previous term and, 0.01. The sum starts with the first term equal to 1 followed by $1 \cdot 0.01$, $0.01 \cdot 0.01$, etc. The infinite, that is endless, sum can be analyzed as what is known as a sequence of partial sums. The nth partial sum is designated as S_n:

$$S_0 = 1. \ S_1 = 1 + 0.01, \ S_2 = 1 + 0.01 + 0.0001, \cdots$$

The nth sum (starting with 0) :

$$S_n = 1 + 0.01 + 0.0001 + 0.000001 + \cdots + 0.01^n. \qquad (4\text{-}3)$$

Therefore, $0.01 \cdot S_n = S_n - 1 + 0.01^{n+1}.$

Solving for S_n,

$$S_n = (1 - 0.01^{n+1})/0.99 \qquad (4\text{-}4)$$

Unlike our intuitive approach to determine a fraction for the infinitely repeating decimal 0.272727..., we are only working with a finite number of terms; however, now it is time introduce the concept of a limit to formally evaluate the sum with an infinite number of terms, that is for an n without limit.

A formal definition is possible using the concept of the limit of a sequence which was introduced by Augustin-Louis Cauchy (1789 -1857), who also among many others accomplishments, made major contributions to the theory of complex variables and differential equations.[137] The concept of a limit was vital to a precise understanding of infinite sequences and concepts of the calculus. Of the occurrence of a limit in a sequence, Cauchy stated:

"When the successive values attributed to a variable approach indefinitely a fixed value so as to end by differing from it as little as one wishes, this last is called the limit of all the others."

Taking advantage of Cauchy's definition of limit let us look at the difference between the partial sums S_n and the next sum S_{n+1} that follows from equation (4-3).

$$S_{n+1} - S_n = 0.01^{n+1} \tag{4-5}$$

The sequence of partial sums S_n clearly fits Cauchy's definition of a limit in that the *"successive values"* of the partial sums may be made to differ *"as little as one wishes."* No matter how small you want the difference to be, you can always choose an n in the sequence of partial sum so that the difference 0.01^{n+1} is smaller. In other words, the limit of 0.01^{n+1} as n increases without bound (goes to infinity) is 0. This is expressed symbolically as:

$$\lim_{n\to\infty} 0.01^{n+1} = 0.$$

Therefore, from equation (4-4):

$S = \lim_{n\to\infty} S_n = \lim_{n\to\infty} (1 - 0.01^{n+1})/0.99 = 1/0.99$.

Taking into account the coefficient of 0.27 in equation (4-2),
$S = (0.27)/0.99 = 3/11$.

More generally for the infinite geometric series shown below with the constant c and for x between 0 and 1, it may similarly be proved that:

$$S = \lim_{n\to\infty} c(1 + x + x^2 + x^3 + \cdots + x^n) = \frac{c}{1-x}.$$

If $\sqrt{2}$ were a rational number, it could be written as a decimal with either a finite number of digits, or a repeating pattern with an infinite number of digits. Since at least the time of Aristotle, we have known that $\sqrt{2}$ is not rational, which means that even allowing for an infinite number of digits as in our example above, it cannot be represented in this way. Thus, we have returned to the situation faced by the Greeks of an inexpressible quantity. In the next section, the Postulate of Continuity will allow us to understand how infinite decimals can be used to define these missing numbers to form the real numbers.

4.3.3 Filling the holes in Euclid's line with the infinity of real numbers

With the capability of expressing any rational number as a decimal, one might stubbornly try to succeed where the Pythagoreans had failed by trying find a decimal representation for $\sqrt{2}$. The Pythagoreans famously encountered this problem with a right triangle with sides equal to 1 so the hypotenuse c, must be such that $c^2 = 1^2 + 1^2 = 2$ in order to satisfy the Pythagorean Theorem. Ignoring the Pythagorean's proof that this is not possible for a rational number (see Appendix A), we can try to find a decimal such that $c_n^2 = 2$. Although this is not possible, it will give us insight to the meaning of $\sqrt{2}$.

We start with the greatest integer such that $c_0^2 < 2$ and improve the accuracy of the approximation by increasing the number of decimal places in steps, for example c_1^2 is the largest approximation with one decimal place.[xlviii] We can bracket the accuracy of the c_ns, by at the same time finding the smallest numbers d_n having the same number of decimal places as c_n with $d_n^2 > 2$. Our futile but stubborn hope is that some pattern (but not strictly repeating) will occur.

A table with some results of this approach is shown below.

Table 4-2 Approximating $2^{1/2}$ with rational numbers

n	c_n	$c_n^2 < 2$	d_n	$d_n^2 > 2$
0	1	1	2	4
1	1.4	1.96	1.5	2.25
2	1.41	1.9881	1.42	2.0164
3	1.414	1.999396	1.415	2.00225
4	1.4142	1.99996164	1.4143	2.00024449

[xlviii] Simon Stevin (1548 – 1620), who spread the use of decimal fractions, anticipated our understanding of irrational numbers by approximating them with increasing accuracy using rational numbers; Merzbach and Boyer, pp. 282 – 285.

The c_ns form a monotonically increasing sequence of rational numbers; however no matter how close c_n^2s approaches 2, the sequence can always be continued with $c_{n+1}^2 < 2$. Similarly, the d_ns form a decreasing sequence of rational numbers. These infinite sequences illustrate that the defining difference between the rational and real numbers is the Postulate of Continuity restated here:

$\mathcal{R}_c$ **13 (Postulate of Continuity):** If a set of real numbers has an upper bound, then it has a least upper bound.

In the set of rational numbers with $c_n^2 < 2$, there is no least upper bound. To be clear there are many upper bounds, in fact an infinite number of upper bounds, for example, any of the d_ns. But no matter which one you pick, there is a smaller one. Alternatively, none of the c_ns are upper bounds as there is always another that is greater, but whose square is still less than 2. The irrational number $\sqrt{2}$ can therefore be defined informally as the unique number that is the limit of these infinite sequences as defined by Cauchy and stated in the previous section.

The mathematician Karl Weierstrass (1815 – 1897) developed the now traditional mathematical expression for Cauchy's definition of limit (p. 105).[138]

A sequence a_n has a limit L, if for any number ε, there exists a number N such that for all n greater than N, the absolute value[xlix] of the difference between $|a_n - L| < \varepsilon$. Symbolically this is expressed as:

$$\lim_{n \to \infty} a_n = L, \text{ if } \forall \text{(for every) } \epsilon, \exists \text{(there exists) an N such that } |a_n - L| < \epsilon \,\forall\, n > N.$$

Here, Cauchy's qualitative description is made precise through identification of ε with the quantity which is *"as little as one wishes"* and all n > N with the *"successive values."*

[xlix] The absolute value of a number x is: x if x > 0, or -x if x < 0. Symbolically, the absolute value of x is designated as $|x|$, so $|3| = 3$, $|-3| = 3$

Regarding the sequence of c_ns, we define the $\sqrt{2}$ as the sequence's limit. Its existence, as part of the real number system, is required by The Postulate of Continuity as it is the least upper bound. Similarly, the d_ns form an infinite sequence in which $\sqrt{2}$ is defined as the limit of the sequence, in this case the greatest lower bound. In both cases, if we are restricted to the rational numbers, there is nothing for the sequence to converge to! It can be proved that the c_ns and d_ns have the same limit by noting that they form what is known are a Cauchy sequences.

A Cauchy sequence, such as the c_ns, is one in which to any level of selected precision ε one can find a number N associated with c_N such that for all m, n > N the absolute value of the difference between c_m and c_n is less than the selected precision, ε, that is, for m, n > N, $|c_m - c_n| < \varepsilon$. The two Cauchy sequences of the c_ns and d_ns may be shown to be equivalent as the sequence $|c_m - d_n| <$ has 0 as a limit.

All of the real numbers can be defined as Cauchy sequences of rational numbers. Thus any decimal of an infinite number of digits is a real number, not only the irrational numbers such as $\sqrt{2}$, but also the rational numbers with a repeating pattern of digits such as 0. 27272727..... Even numbers such as 1/2 may be represented in this way as 0.499999... with 0.5 as a limit.[1] An example of a less familiar irrational number is the number 0.1001000100001... Joseph Liouville (1809 - 1882) proved numbers like this, which exhibit a pattern, but not a repeating pattern, had a limit, and therefore were real numbers.[139]

One more representation of the real numbers, that clarifies the role of the Postulate of Continuity is that developed by Richard Dedekind (1831 - 1916) who gave the first definition of a field as a mathematical structure (see footnote xlvi and preceding discussion p. 100).[140] As with the path to the real numbers of Cauchy sequences and infinite decimals, Dedekind started with the rational numbers. He defined a cut in the rational numbers as one which divided all of the numbers into two sets. I will designate the two sets as L(eft) and R(ight). The cut is defined with

---

[1]   A formal presentation of Cauchy sequences and decimals as real numbers is given by Hamilton, pp. 26 – 49.

every member of L being less than every member of R. It may help to visualize the cut in a continuous number line. If L has a largest number, then R has no smallest number. If R has a smallest number, then L has no largest member. In either case the cut is a rational number. If L has no largest number and R has no smallest number, the cut defines an irrational number. By defining the set of all cuts as the real numbers, Dedekind established a one-to-one correspondence between the real numbers and the points on a line.[141]

As the methods of defining the real numbers start with the rational numbers, the postulates of the rational numbers, which form an ordered field, and the resulting theorems, such as the cancellation property for addition and multiplication, can be shown to apply to the real numbers.[142] After more than 2,000 years, the holes in the number line were closed ending the crisis of missing numbers first uncovered by the Pythagoreans.

One important part of symbolic algebra that has only been mentioned in passing, but can now be fully defined with the real numbers is exponentiation. Recall that Descartes had taken the step of symbolizing the products of unknowns such as $x \cdot x \cdot x$ as x^3 and that it was understood that such terms as $x/x^2 = x^{-1} = 1/x$. As noted above. such insights clearly follow from the application of exponents in the natural numbers and integers; however, the results are not always valid within the rational numbers when the restriction to natural numbers and integers is lifted. For example, $2^{5/2}/2^2 = 2^{1/2} = \sqrt{2}$, which does not exist in the rational numbers. Now, however, we can make the following definitions for real numbers as extensions to definitions for the rational numbers:

Definition $\mathcal{R}_c$1: $x^a \cdot x^b = x^{a+b}$.
Definition $\mathcal{R}_c$2: $x^a \cdot y^a = (x \cdot y)^a$.
Definition $\mathcal{R}_c$3: $1/x = x^{-1}$ $(x \neq 0)$.
Definition $\mathcal{R}_c$4: $x^0 = x^a/x^a = x^{a-a} = 1$.

Definition $\mathcal{R}_c$5: $\left(x^{a/b}\right)^b = x^{((a/b)b)} = x^a$.

It follows that if $y^b = x^a$, then $y = x^{a/b}$.[li] We can extend the exponents to all real numbers with the exception discussed below while noting that 0^0 is not defined.[lii] Furthermore, we can solve the problem that so vexed the Pythagoreans: if the sides of a right triangle are of length x and y, what is the length of the hypotenuse z. By the Pythagorean Theorem: $z^2 = x^2 + y^2$. Therefore, $(z^2)^{1/2} = (x^2 + y^2)^{1/2}$ and $z = (x^2 + y^2)^{1/2}$. We know that whatever the lengths x and y, the real numbers will provide us with a length for z. But there is still one more problem with the exponents that needs to be resolved.

For all positive real numbers, Definition $R_c 5$ is valid; however as the product of a negative number and a negative real number is positive, $x^{1/q}$ does not exist in the real numbers for $x < 0$ and q an even integer. As perhaps the most commonly cited example, there is no real number equal to $(-1)^{1/2}$ or a solution to $x^2 = -1$. To finally define a set of numbers that will allow us to solve all algebraic equations, we will need to introduce a number whose square is negative. This will lead to the complex numbers. These numbers were used by Cardano in the solution of cubic equations, but without any understanding of their significance. Recalling Lambert's vision of a geometry formed on a sphere with an imaginary radius, it should not be surprising that the complex numbers will be useful in the analysis of non-Euclidean geometries. However, before introducing these numbers, we will focus on one more property of the real numbers that differentiates them from the rational numbers - that is the difference between them as infinite collections. It is this difference that fills out the number line. We will find that even though there are an infinite number of rational numbers, there are even more irrational numbers!

4.3.4   Counting to infinity

We have encountered infinity in our various collections of numbers, or to use the terminology of mathematics, sets of numbers. Starting with

---

[li]   Let $y = x^{a/b}$, $y^b = (x^{a/b})^b = x^a$.

[lii]   If f(x) and g(x) both have a limit of 0 then, $f(x)^{g(x)}$ will have different limits as x approaches 0 depending on how f(x) and g(x) approach 0.

the natural numbers (1, 2, 3...), we know that no matter how great the number that we choose, say n, there is always a number that is greater. Since the natural numbers are embedded within the integers, and the integers within the rational numbers, each of these is also an infinite set.

While reflecting on these infinite sets, it is natural to ask if there are fewer even or odd numbers than the entire set of natural numbers. Or since every natural number has an additive inverse within the integers forming the negative numbers, do the integers form a larger set than the natural numbers? Or again, how does the set of all rational numbers compare with either the natural numbers or the integers? If you find the idea of comparing infinities unsettling, you are in good company. Of such comparisons the great scientist Galileo Galilei had this to say:

"...all numbers are infinitely many ; all their roots infinitely many; all squares infinitely many; that the multitude of squares is not less than that of all numbers, nor is the latter greater than the former. And in final conclusion, the attributes of equal, greater, and less have no place in the infinite, but only in bounded quantities."[liii]

Georg Cantor (1845 – 1918) approached this problem by applying the principle that two sets are the same size, termed equivalent, if a one-to-one correspondence can be established between the members of the infinite collection of numbers. A one-to-one correspondence for two sets of objects $\mathcal{A}$ and $\mathcal{B}$ is established if each object in $\mathcal{A}$ is paired with a single object in $\mathcal{B}$ and conversely. Cantor identified the size of the set of objects with what he termed its cardinal number. For finite collections, these are our familiar numbers. (The cardinal number associated with the set of states in the United States is 50) Cantor's genius was to extend this concept to infinite sets. In doing so, he extended the idea of infinity from merely a potential that was never attained as widely accepted since the time of Aristotle[143] to the concept of a set which contains the

---

[liii]   Galileo showed an early appreciation for the difficulties inherent in the concept of infinity in his *Two New Sciences*, but could not resolve the apparent paradoxes of there being as many squares as natural numbers (Galileo, G., *Two New Sciences*, translated with new introduction, notes, and, *History of Free Fall: Aristotle to Galileo* by S. Drake, Wall and Emerson, Inc., Toronto, 1989; p. 41.

entire infinite collection of numbers. This concept was rejected by many mathematicians; however, with time it has become one of key approaches to mathematical analysis.[liv] We will now take a first look at the implications of Cantor's concept of infinite sets. Let us start with some infinite sets within the natural numbers and integers.

With a little imagination, the natural numbers $\mathcal{N}$ can be easily matched up with the positive even integers $\mathcal{E}$, positive odd integers $\mathcal{O}$, the negative numbers $\mathcal{N}$- and integers, I, as shown below:

Table 4-3 Matching infinite sets of integers with the natural numbers

$\mathcal{N}$	1	2	3	4	5	. . . .
$\mathcal{E}$	2	4	6	8	10	. . . .
$\mathcal{O}$	1	3	5	7	9	. . . .
$\mathcal{N}$-	−1	−2	−3	−4	−5	. . . .
I	0	1	−1	2	−2	. . . .

A more explicit method of showing the one-to-one correspondence is by defining a relationship between the sets. We can do this through an expression in which substitution of the natural numbers, n = 1, 2, 3,··· will generate the other sets of numbers. Such an expression is known as a function. The following functions will generate the desired sets:

$$\text{For } n \quad = \quad 1, 2, 3,\cdots$$
$$\mathcal{E}\,(n) \quad = \quad 2 \cdot n$$
$$\mathcal{O}\,(n) \quad = \quad 2 \cdot n - 1$$
$$\mathcal{N}\text{-}\,(n) \quad = \quad -n$$

[liv] A fascinating account of the development of the concept of infinity and, in particular, Cantor's struggle to gain acceptance of his views in the face of tremendous opposition is given by A. D. Aczel, *The Mystery of the Aleph; Mathematics, the Kabbalah, and the Search for Infinity*, Washington Square Press, Pocket Books, New York, 2000.

$$\mathbf{I}(n) \quad = \quad n/2, \text{ for } n = 2, 4, 6, \cdots$$
$$(1 - n)/2, \text{ for } n = 1, 3, 5, \cdots$$

We can see by either approach that there is the "same" number of natural numbers as even numbers, odd numbers, or integers, that is, the sets are termed equivalent. A defining characteristic of infinite sets is that it is possible to select only portions (subsets) of the set to form a new set which is still infinite – the even and odd sets as subsets of the natural numbers are examples. When an infinite set such as those above can be put into a one-to-one correspondence with the natural numbers, it is said to be countably infinite or denumerable. A set that is finite or countably infinite is sometimes termed countable. Since the integers are embedded within the rational numbers, it is natural to compare the integers and natural numbers with the rational numbers. One major difference is that the rational numbers have the previously discussed property known as denseness, that is, there is an infinite number rational numbers between any two rational numbers. The property of denseness that characterizes the rational numbers would seem to imply that there are more rational numbers than the natural numbers. To test this thought, however, we need to have a way to tabulate the set of rational numbers similar to the tables above for the natural numbers and integers. Then we can see if it is possible to have a one-to-one correspondence with the natural numbers.

All of the rational numbers are generated from pairs of integers, for example 2/3. This suggests that a systematic scheme to list all possible pairs of integers would do the job. Such a scheme was developed by Cantor to meet this need for the positive rational numbers. The scheme in the next figure goes through all the combinations of pairs of positive integers; hence all positive rational numbers are represented. A systematic way to match the natural numbers $\mathcal{N}$ with the positive rational numbers $\mathcal{R}(p)$ is indicated by the directed lines.

$$\rightarrow \qquad \rightarrow$$

Start ⇨
1/1 2/1 3/1 4/1 5/1 . . .
1/2 2/2 3/2 4/2 5/2 . . .
1/3 2/3 3/3 4/3 5/3 . . .
1/4 2/4 3/4 4/4 5/4. . .
1/5 2/5 3/5 4/5 5/5. . .

.   .   .   .

4-4 Cantor counts the rational numbers

Counting the rational numbers in the order indicated and leaving out repeated numbers establishes the one-to-one correspondence of natural numbers with the positive rational number as shown in the table below:

Table 4-4 Matching the natural and positive rational numbers

$\mathcal{N}$	1	2	3	4	5	6	7	8	. . .
$\mathcal{R}(p)$	1/1	1/2	2/1	3/1	1/3	1/4	2/3	3/2	. . .

The same kind of pairing could also be generated between the natural numbers and just the negative integers. To provide a pairing of the natural numbers and all of the positive and negative rational numbers, each pairing of a positive rational number can be followed by a pairing with its additive inverse. The only missing rational number then is zero. If we think of the natural numbers as being the room numbers of a hotel with a countably infinite number of rooms then in this hotel (known as the Hilbert Hotel for David Hilbert's use of this imagery),[144] we can always make room for zero as another guest by asking each guest to move up one room, as there are plenty of rooms. Thus, Cantor proved that the rational numbers can be put into a one-to-one correspondence with the natural numbers and are therefore countably infinite.

The addition of the least upper bound property to the properties of the rational numbers generates the real numbers and fills in the holes in the number line left by the rational numbers, for example the holes such as $\sqrt{2}$ and π. The surprising result that there are no more rational numbers than integers or natural numbers, may lead you to wonder about the comparative size of the real numbers and irrational numbers as related to their ability to fill up the number line. Cantor established a way to make the comparison of the real and natural numbers through an indirect route, proof by contradiction. He assumed that the real numbers between 0 and 1 could be listed in their representation as infinite decimals to establish the one-to-one correspondence with the natural numbers. In other words, he assumed the real numbers were countably infinite. He then looked to see if this assumption led to a contradiction.

In Cantor's list of all the real numbers between 0 and 1, let the first decimal number be: $0.a_{11}a_{12}a_{13}a_{14}a_{15}\cdots$ where notation such as a_{12} indicates the second digit in the first row in the listing of all the real numbers. For example, if the first decimal number in our listing is 0.1754..., then $a_{11} = 1$, $a_{12} = 7$, $a_{13} = 5$, and $a_{14} = 4$. Cantor's hypothetical list of the real numbers is shown below.

$0.a_{11}\ a_{12}\ a_{13}\ a_{14}\ a_{15}\ \cdot\ \cdot\ \cdot$
$0.a_{21}\ a_{22}\ a_{23}\ a_{24}\ a_{15}\ \cdot\ \cdot\ \cdot$
$0.a_{31}\ a_{32}\ a_{33}\ a_{34}\ a_{35}\ \cdot\ \cdot\ \cdot$
$0.a_{41}\ a_{42}\ a_{43}\ a_{44}\ a_{45}\ \cdot\ \cdot\ \cdot$
$0.a_{51}\ a_{52}\ a_{53}\ a_{54}\ a_{55}\ \cdot\ \cdot\ \cdot$

4-5 Cantor attempts to count the real numbers

By changing one digit in each row on the diagonal (as indicated by the arrow), Cantor was able to show that such a listing encompassing all real numbers was impossible.[145] Suppose that the list of decimals forms a one-to-one correspondence between the rows of the list (each row being a decimal number) and the natural numbers. Cantor asked if a decimal number between 0 and 1 could be formed that is not on the list. Let this new decimal be $b_1b_2b_3b_4b_5$.... The first digit b_1 is selected to be any number that differs from the digit a_{11}; b_2 is selected to differ from a_{22}. We go down the diagonal of our list as shown by the arrow above continuing to select our b_ks to differ from the value of the a_{kk}s on the diagonal. Our new decimal, $b_1b_2b_3b_4b_5$.... cannot be on the list since it is not equal to the first decimal starting with a_{11}, or the second because it differs from a_{22}, and so forth. So if we assume that such a list could be formed of all the real numbers between 0 and 1, we see that this leads us to the contradiction that we can construct a decimal that is not on the list. Therefore, our assumption that the real numbers are denumerable between 0 and 1 is false. The argument can be extended to any segment of the number line or the whole number line as there are one-to-one functions that map the interval from 0 to 1 to other intervals including the entire number line. The real numbers are said to be uncountable or non–denumerable. This uncountability accounts for the real numbers ability to fill in the number line.

With Cantor's proof that there are "more" real numbers than rational numbers, one might assume that irrational numbers such $\sqrt{2}$ are the dominant source of the additional numbers that complete the number line. This, however, is not the case. Numbers such as $\sqrt{2}$, $\sqrt{5}$ are known as algebraic numbers as they are solutions to equations such as $x^2 - 2 = 0$, and $x^2 - 5 = 0$. Any rational number b/a is also an algebraic number as a solution to $ax - b = 0$. In general, algebraic numbers are solutions to:

$$a_n x^n + a_{n-1} x^{n-1} + a_{n-2} x^{n-2} + \cdots + a_1 x^1 + a_0 x^0 = 0.$$

where the coefficients such as a_n are integers and n is a positive integer.

Cantor developed a proof that the algebraic numbers are also countably infinite.[146] One can see the plausibility of Cantor's conclusion by noting that the highest power of the equation, n, sets the limit for the number of solutions for the equations. Therefore, one could in principle systematically establish a list of all algebraic solutions. Thus, even the algebraic numbers are only countably infinite.

So, if the real numbers are uncountable (nondenumerable), there must be other irrational numbers, besides the algebraic numbers that are responsible for completing the number line. These numbers are called transcendental numbers as they transcend algebraic methods. With the number line "mostly" represented by transcendentals, it behooves us to give examples. We are already very familiar with one of these numbers, π, the ratio of the circumference of a circle to its diameter. Archimedes determined approximations to the circumference of a circle using a bracketing approach with calculations of the perimeters of regular polygons of increasing numbers of sides. Another transcendental number, e, which is the base of natural logarithms[lv] and will be important in the next section, was defined by Leonard Euler (1707-1783) as:

$$e = \lim_{x \to \infty} (1 + \frac{1}{x})^x = 2.71828182845 \cdots$$

Note: $(1 + \frac{1}{10^6})^{10^6} = 2.71820 \cdots$

We previously mentioned the class of Liouville numbers such as 0.100100010001... which is also a transcendental number.[147] Although Cantor proved the transcendental numbers dominate the number line as they form an uncountable infinite set, the proof that any particular number is transcendental is quite difficult. For example, it was not until 1882 that Ferdinand Lindemann (1852 – 1939) proved that π is transcendental.[148]

Before leaving the work of Cantor, I must briefly mention his fascinating development of what he termed transfinite numbers. These are the numbers that characterize infinite sets. Cantor designated the Hebrew letter aleph ($\aleph_0$) (subscript 0 for the smallest transfinite number) as the cardinal number for the infinite countable sets. This includes the integers, rational, and algebraic irrational numbers. Recalling that, for example, the even, odd natural numbers, integers, and rational numbers are equivalent sets, each therefore characterized by $\aleph_0$, leads to the strange arithmetic of transfinite numbers:

[lv] The irrational number e is the base for natural logarithms (ln) in the same way 10 is the base for common logarithms (log). (For example, $\ln e^2 = 2$ and $\log 10^2 = \log 100 = 2$. The expression for e was first encountered by Jacques Bernoulli (1654-1705) in the problem of continuously compounded interest (Merzbach and Boyer, pp., 390, 393, 406, 408; see also p. 124 below.)

$\aleph_0 + \aleph_0 = \aleph_0$; $\aleph_0 + \aleph_0 + \aleph_0 = \aleph_0$; $\aleph_0 + \aleph_0 + \aleph_0 + \aleph_0 = \aleph_0$, ..., and

$\aleph_0 \cdot \aleph_0 = \aleph_0$.

As the real numbers are uncountable, their associated transfinite cardinal number, designated as C (for the continuum of real numbers), must be greater than $\aleph_0$. A similar strange arithmetic applies, for example $\aleph_0 + C = C$, as the natural numbers and equivalent sets are embedded in the real numbers. The relationship between $\aleph_0$ and C is suggested by considering finite decimals. The number of decimal numbers with three digits such as 0.123 is 10^3. By analogy, the number of positive decimal real numbers less than 1 is $10^{\aleph_0}$. Now we could use a binary system, just as well, in which case the only numerals are 0 and 1. Thus, the number of positive real numbers less than 1 could be calculated as $2^{\aleph_0}$.[lvi]

$$2^{\aleph_0} = 10^{\aleph_0} = C$$

There are even more wonders with questions such as whether there is a transfinite cardinal number between $\aleph_0$ and C (The Continuum Problem[149]) and the infinities of transfinite ordinal numbers taking into account order, for example the infinite set of even numbers followed by the odd numbers, $(2, 4, 6,\cdots 1, 3, 5,\cdots)$; however, this would take us too far afield.

We have now seen many of the characteristics of the uncountability of the real numbers that allow them to represent the continuous the number line. However, to find all the solutions to the algebraic equations such as $x^2 + 1 = 0$, it is necessary to leave the number line and enter the complex plane. The complex plane with its real and imaginary numbers will be significant for the description of non-Euclidean geometries

---

[lvi]   More rigorous arguments with an extensive discussion of set theory is given by Hamilton. As a final point of interest to be found in Hamilton (p. 78), there is no greatest transfinite cardinal number as each new set of numbers makes possible more numbers as the following ascending sequence makes clear:

$$\aleph_0, 2^{\aleph_0}, 2^{2^{\aleph_0}}, 2^{2^{2^{\aleph_0}}}, \cdots$$

4.4   The real joins the imaginary – the complex plane

I noted that in the sixteenth century, mathematicians such as Cardano were aware that additional solutions to algebraic equations could be found by allowing for numbers whose square was negative (see p. 86). He had found that the square root of negative numbers appeared as preliminary results in the solutions to cubic equations in which the solutions were real. However, in general the use of such solutions was unclear. Terms involving the square root of negative numbers were called imaginary numbers by Descartes because he did not believe they had the characteristics of numbers.[150] Euler created the now standard notation with $i = \sqrt{-1}$. Far from finding them "useless" in Cardano's term, Euler showed that $e^{i\theta} = \cos\theta + i\sin\theta$.

Although the simple equation, $x^2 = -1$ certainly illustrates the need for an extension of the real numbers, the solution to the quadratic equation, a staple of your early school mathematics, makes it clear as in Cardano's experience that these imaginary numbers are closely bound to solutions with real numbers. The quadratic equation, $ax^2 + bx + c = 0$, has the well-known solutions, $x = \frac{-b \pm \sqrt{b^2 - 4ac}}{2a}$. If $b^2 - 4ac < 0$, then the solution has a real and an imaginary part, that is, it is the complex number, $x = -\frac{b}{2a} \pm i\sqrt{4ac - b^2}/(2a)$.

Caspar Wessel (1745 – 1818), Robert Argand (1768 – 1822), and later Gauss recognized, complex numbers could be represented as Cartesian coordinates of real numbers in a plane with the ordinate being imaginary as in the following figure with the complex number r + *i*s represented as the point (r, s).[151]

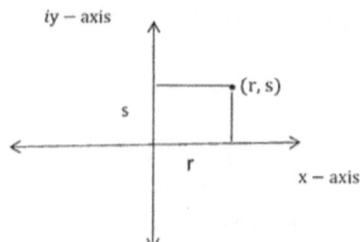

4-6 The complex plane

With this representation, the imaginary numbers should seem less mysterious than those numbers that made Cardano so suspicious. We begin with definitions for the complex numbers and the operations of addition and subtraction. We are guided in our choices by the desire for the real numbers (and therefore, irrational numbers, integers, and rational numbers) to keep all of their properties.

Definition C1: The complex numbers **C** are defined as the ordered pairs of the real numbers, (a, b).

Definition C2: Two complex numbers, (a, b) and (c, d) are equal if and only if a = c and b = d.

Definition C3: The operation of addition between pairs of complex numbers is defined by:

$$(a, b) + (c, d) = (a + c, b + d).$$

Definition C4: The operation of multiplication between pairs of complex numbers is defined by: $(a, b) \cdot (c, d) = (ac - bd, ad + bc)$.

Of particularly importance is the result that follows from Definition C4 of the complex product $(0, 1) \cdot (0, 1)$.

$$(0,1) \cdot (0,1) = (0 \cdot 0 - 1 \cdot 1, 0 \cdot 1 - 1 \cdot 0) = (-1, 0).$$

Identifying (0, 1) with i and (–1, 0) with –1, then, $i^2 = -1$.

With these definitions the complex numbers (a, b) can be shown to satisfy, like the real numbers, the closure; commutative; associative; and distributive properties of addition and multiplication. The identity elements for addition and multiplication are (0, 0) and (1, 0). The additivity inverse of (a, b) is (-a, -b). The multiplicative inverse of (a, b) is) can easily be calculated using what is known as the complex conjugate, $(a, b)^* = (a, -b) = a - ib$.:

$$\frac{1}{(a,b)} = \frac{1}{(a+ib)} = \frac{1}{(a+ib)} \cdot \frac{(a-ib)}{(a-ib)} = \frac{1}{(a,b)} \cdot \frac{(a,-b)}{(a,-b)},$$
$$= \frac{a-ib}{a^2+b^2} = \frac{(a,-b)}{(a^2+b^2,0)},$$

The extension of the number system to complex numbers allowed definite statements to be proved concerning the number of complete solutions to algebraic equations such as that given below:

$$a_n x^n + a_{n-1} x^{n-1} + a_{n-2} x^{n-2} + \cdots + a_1 x^1 + a_0 x^0 = 0.$$

First Gauss proved the following theorem on the existence of at least one complex solution:

Fundamental Theorem of Algebra: Every polynomial of degree n > 0 (integer powers) with complex coefficients has at least one complex solution.[152]

From the Fundamental Theorem of Algebra, the theorem on the total number of solutions follows:

Theorem (Roots of Polynomial Equations): Every polynomial equation of degree n > 0 (integer powers) with complex coefficients a_n has n complex solutions (not necessarily distinct).[153]

The development of the real numbers and complex described in this chapter should make it clear that the lack of a deductive system for algebra over millennia, in contrast to geometry, was not simply the result of a lack of imagination. The path to the real numbers involved the complex development of symbolic algebra in which notation would facilitate and clarify algebraic processes; the identification of algebra with geometry through the use of coordinates; the understanding of the differences between the mathematical structures of numbers systems such as the integers and rational numbers; and the concept of a limit. This last breakthrough finally made the meaning of the real number system precise and connected the line segments of Euclid with a continuous number line and made it possible to fill in the holes in many of Euclid's proofs. In the next chapter we will introduce the techniques known as the calculus allowing us to extend our understanding of geometry with its ability to calculate tangents, curve lengths, areas, and curvature. As discussed in the next chapter, the limit concept is vital to putting the calculus on a rigorous footing.

Understanding the real numbers system is also enhanced by analyzing them as being embedded within the complex numbers as in the enumeration of the solutions to polynomials. We will complete this chapter with a discussion of the relationship between trigonometric and complex functions discovered through the use of infinite series by Euler and their relation to the hyperbolic functions defined by Lambert. All will have a role is the description of non-Euclidean geometries.[154]

Infinite series have played an important role in understanding the real numbers. Despite the lack of a precise definition of the limit process and hence criteria for the convergence of infinite series, Euler made such remarkable discoveries[lvii] as:

$$\frac{\pi^2}{6} = \frac{1}{1^2} + \frac{1}{2^2} + \frac{1}{3^2} + \frac{1}{4^2} + \cdots$$

Euler was aware of the following infinite series[lviii] convergent for all x (with $n! = n(n-1)(n-2)\cdots 3 \cdot 2 \cdot 1$):

$$\sin x = x - \frac{x^3}{3!} + \frac{x^5}{5!} - \frac{x^7}{7!} + \cdots$$

(4-6)

[lvii]   In addition to remarkable discoveries, the lack of a precise understanding of convergence of infinite series sometimes led Euler astray. For example, using the geometric series relation, $1/(1-x) = x + x^2 + x^3 + \cdots$ with $x = 2$ led Euler to the conclusion that $-1 = 1 + 2 + 4 + 8 + \cdots$, and that -1 is larger than infinity (Kline, pp. 143-144).

[lviii]   Ball (p. 314) credits the Scottish mathematician James Gregory (1638-1675) with the first publication of infinite series for sin(x) and cos(x); see also Merzbach and Boyer, pp. 354-354.

$$\cos x = 1 + \frac{x^2}{2!} + \frac{x^4}{4!} + \frac{x^6}{6!} + \cdots$$

$$(4\text{-}7)$$

$$e^x = 1 + \frac{x^2}{2!} + \frac{x^3}{3!} + \frac{x^4}{4!} + \cdots$$

$$(4\text{-}8)$$

Note that the series for sin(x) and cos(x) both have terms alternating in their signs with sin(x) only having terms with odd powers, while cos(x) only has even powers. In contrast the infinite series for e^x consists of the addition of terms of all powers, even and odd. If you substitute in equation (4-8) $i\theta$ for x with $i^2 = -1$, $i^3 = -i$, $i^4 = 1, \cdots$, and similarly for $-i\theta$, you will not be surprised that Euler could show that:

$$e^{i\theta} = \cos\theta + i\sin\theta \qquad\qquad (4\text{-}9)$$

$$e^{-i\theta} = \cos\theta - \sin\theta \qquad\qquad (4\text{-}10)$$

From the above relations, the trigonometric functions of sin (θ) and cos (θ) are easily expressed as exponential functions:

$$\cos\theta = \frac{e^{i\theta} + e^{-i\theta}}{2} \qquad\qquad (4\text{-}11)$$

$$\sin\theta = \frac{e^{i\theta} - e^{-i\theta}}{2} \qquad\qquad (4\text{-}12)$$

The expression of the sine and cosine functions as infinite series liberates them from being solely interpreted geometrically as being generated from the coordinates of the unit circle, $x^2 + y^2 = 1 = \cos^2\theta + \sin^2\theta$. Lambert systematically developed functions associated with the hyperbola $X^2 - Y^2 = 1$ introducing the modern notation for hyperbolic functions:[155]

$$\cosh x = \frac{e^x + e^{-x}}{2}. \qquad (4\text{-}13)$$

$$\sinh x = \frac{e^x - e^{-x}}{2} \qquad (4\text{-}14)$$

The relationship between the hyperbolic functions and the hyperbola is clarified by the substitution of ix for θ into equations (4-11) and (4-12).

$$\cos ix = \frac{e^{i^2x} + e^{-i^2x}}{2} = \cosh x \qquad (4\text{-}15)$$

$$\sin ix = \frac{e^{i^2x} - e^{-i^2x}}{2} = i\sinh x \qquad (4\text{-}16)$$

Thus,

$$\sin^2(ix) + \cos^2(ix) = 1 = i^2\sinh^2 x + \cosh^2 x,$$

or

$$\cosh^2 x - \sinh^2 x = 1.$$

The relationship between the hyperbolic functions and the curve $X^2 - Y^2 = 1$ with coordinates (X, Y) has been made explicit in the above equation and is analogous with the relation between the sinusoidal functions and coordinates of the unit circle. Recalling Lambert's vision of a sphere with an imaginary radius, you should expect to see these relations again.

We have spent some time reviewing the sources for the real numbers and their geometric relations, but now it is time to see how to use the real numbers to express the changing curvature of curves and surfaces that we will find characterize new geometries. This will require an introduction to concepts of the calculus and vectors.

Analyzing Change, Summation, and Direction: the Calculus and Vectors

5.1 Overview

Following the paths of number and form has led us to many milestones in the development of mathematics: numeration in Egypt, Babylonia, India, and in the Islamic community; the deductive geometry of the Greeks; symbolic algebra of the Renaissance; and the analytic geometry of Descartes and Fermat. However, no milestone in mathematics is of greater importance than the discovery of the calculus by Isaac Newton (1642-1727)[156] and Gottfried Wilhelm Leibniz (1646-1716).[157] Their calculus and the extensive developments of it that followed are crucial to the description of curved lines and surfaces by enabling the analysis of lengths, slopes, and areas of enclosed geometric figures on non-Euclidean surfaces. One aspect of the development of the calculus is its extension from functions of a single variable, the calculus of one dimension, to functions of multiple dimensions. Furthermore, a natural outgrowth of the consideration of functions in three-dimensional space is the mathematical entity known as a vector, a quantity having both magnitude and direction. We will need to understand some basics of the calculus and vectors to continue our journey.

Newton and Leibniz discovered the calculus independently (although this was greatly disputed by Newton and his followers). While Leibniz's approach and notation were closer to modern calculus, Newton's use of the calculus, as the language to express his extraordinary development of the Laws of Motion and the Universal Law of Gravitation, was the impetus for enormous breakthroughs in the understanding of the physical universe.

126

One entrée to the calculus, is its capability of quantifying change. Our present concern will be to quantify the slopes of curved lines and surfaces and how they change. The curves that Newton was interested in were the trajectories of objects, such as the legendary apple falling from a tree, or more certainly, the orbits of the planets about the sun. To greatly oversimplify at this point and think in a single spatial dimension, let us take as our curve the plot of distance, indicated by the variable s, that an object, say your car, has traveled in time t. In other words distance is a function of time, s = s(t). In a plot of this function, the ordinate is distance s and the abscissa is time t, instead of the usual y and x. Suppose we want to know the velocity of the car at a particular time. We know that the average velocity in the time period between t and t + Δt is the distance the car moved Δs divided by the time period of its movement Δt. Or thinking about the plot s = s(t), the average velocity v_{av} is the slope of the line on a graph going from the points (s, t) to (s + Δs, t + Δt), or $v_{av}(t) = \Delta s/\Delta t$. This is an estimate of the velocity at time t. Similarly, average acceleration a_{av} is the slope of the velocity versus time curve; $a_{av}(t) = \Delta v/\Delta t$. If our car moves smoothly enough (no abrupt step changes), then in the limit as Δt goes to 0, we obtain the instantaneous velocity and acceleration along the car's path at time t. In mathematical notation, this is expressed as $v = \lim_{\Delta t \to 0} \Delta s/\Delta t$ and $a = \lim_{\Delta t \to 0} \Delta v/\Delta t$.

Newton's Second Law of Motion relates the acceleration of a body to the net force acting upon it. So knowing the force on a body, such as a planet under the influence of the sun by Newton's Universal Gravity, one should be able to reverse the process described above and go from acceleration to velocity, and finally to a description of the planet's path. From the work of Johann Kepler (1571-1630), it was known that the paths of the planets could be described as ellipses with the sun at one focus.[158] Newton's friend Edmund Halley (1656-1742) had guessed the form of the force law that would become Newton's Universal Law of Gravitation, but he could not prove that it would result in elliptical orbits. Newton proved it using the calculus in his world-changing *Principia*.[lix]

[lix]   An excellent translation and guide to *The Principia* is given by Cohen, I. B. and Whitman, A., *Isaac Newton, The Principia, A New Translation*, preceded by *A Guide to Newton's Principia* by I.B. Cohen (see Bibliography for full publication details).

Halley was so impressed that he published the manuscript at his own expense.[159]

The process of determining the slope of a curve is call differentiation. Velocity is the derivative of distance, in Leibniz's notation $v = \frac{ds}{dt} = \lim_{\Delta t \to 0} \frac{\Delta s}{\Delta t}$. Similarly, acceleration $a = \frac{dv}{dt} = \lim_{\Delta t \to 0} \frac{\Delta v}{\Delta t}$. To invert the process, that is to go from acceleration to velocity and from velocity to find the path of a moving object, a mathematical operator that is the inverse of the derivative is used. Reasonably enough, it is sometimes called the anti-derivative.

Let us now relate these ideas to a curve $y = F(x)$, for example, the parabola $y = ax^2$ in which case $F(x) = ax^2$. For simplicity here, I associate the process of taking the derivative with the operator $\mathcal{D}$ and the anti-derivative operator with S. Then if $\mathcal{D} F(x) = f(x)$, $S f(x) = F(x)$. The operators $\mathcal{D}$ and S are inverse operators: $\mathcal{D} S g(x) = g(x)$ and $S \mathcal{D} g(x) = g(x)$. Differentiation has been identified with calculating the slope of a curve; however, the anti-derivative also has a geometric interpretation. Quite remarkably, it is the area under the curve $F(x)$. I chose the symbol S for the anti-derivative as finding the area is a Summing process. As we shall see Leibniz chose a somewhat different looking S. The anti-derivative is usually referred to as the integral. The inverse relationship that exists between differentiation, that is, calculating slopes, and integration, calculating areas, is in essence the Fundamental Theorem of Calculus discovered by Newton and Leibniz. Details on differentiation are given in the next section, followed by a discussion of integration.

While our simple example assumed a car moving in only one direction; the need to describe motion in three directions leads to the concept of vectors, the means of specifying both magnitude of a quantity and an associated direction. A common example of a vector, as in our example, is velocity. When we want to know something about the wind, we ask for its magnitude and its direction. Similarly, Newton's Laws of Motion are vector equations. In Cartesian coordinates (x, y, z), there is an equation relating force and acceleration in each direction. We will need vectors to specify the directions of the tangents to lines and curved surfaces and vectors perpendicular to these tangents. With the use of vectors to describe spatial relations, an obvious need is to extend the calculus beyond the case of a function of a single variable such as s(t) or y(x). To be useful for evaluating the geometric characteristics of lines or curved surfaces in three dimensional space, calculus concepts must be extended to three dimensions and even four, in the case of the space-time of Special and General Relativity. We will extend the idea of the derivative of a single variable to partial derivatives. Partial derivatives are similar to derivatives in that they are taken with respect to one coordinate, while holding the other coordinates constant. This concept, part of what is known as multivariate calculus, involves considerable complexity including the relationships between the partial derivatives, but we will be able to extend our understanding sufficiently to meet our needs by making analogies to the derivatives of function of a single variable.

Putting all of this together In the next chapter, we will be able to describe surfaces in terms of their curvature, and to see as Gauss did that curvature forms the defining characteristic that distinguishes Euclidean and non-Euclidean geometry.

5.2 Changes along the path: differential calculus

With the background from the previous section, a formal definition of the derivative is given below, followed by some discussion and a number of examples.

Definition of Derivative: For a function f(x) defined over the interval $a \leq x \leq b$, the derivative at $x = c$ with $a < c < b$ is defined as:[160]

$$\frac{df}{dx}\Big|_{x=c} = \lim_{x \to c} \frac{f(x)-f(c)}{x-c} \text{ if the limit exists.}$$

Or in general identifying c with any point x, this may also be expressed as:

$$\frac{df(x)}{dx} = \lim_{\Delta x \to 0} \frac{f(x+\Delta x)-f(x)}{\Delta x}$$

The notation f'(c) is frequently used for $\frac{df}{dx}|_{x=c}$ or in general for any x, f'(x). Note that if $\frac{df}{dx} = g(x)$, then $\frac{dg}{dx} = \frac{d}{dx}\frac{df}{dx} = \frac{d^2f}{dx^2}$, the second derivative of f(x), and so forth. As an example, if the distance s an object has moved is given as a function of time s(t), then velocity $v = \frac{ds}{dt}$, and acceleration $a = \frac{dv}{dt} = \frac{d^2s}{dt^2}$.

The limit concept is crucial to the definition of the derivative for if $\Delta x = 0$, $\frac{f(x+\Delta x)-f(x)}{\Delta x} = 0/0$ which is meaningless. This was the source of much of the criticism of the calculus in the eighteenth century before the concept of a limit was defined. Bishop George Berkeley (1685-1753) argued that the analysis using the increment Δx was indeed meaningless saying:

"[The increments] *are neither finite quantities, nor quantities infinitely small, nor yet nothing.*"

Emphasizing his disbelief in Newton's technique, he asked, *"May we not call them ghosts of departed quantities."*[161]

The limit process has previously been described in regard to infinite sequences and series. A similar description applies in regard to the derivative. In order for a limit L to exist for the derivative f'(x) at a point x, it must be possible that for any selected level of precision ϵ that there exists a region δ around x, $0 < |x| < \delta$, in which $|f'(x) - L| < \epsilon$ (see footnote xlix p. 108 defining |x|, the absolute value of x). In Cauchy's language, we can find regions δ around x, so that the difference between f'(x) and L differs `as little as one wishes." Therefore, the limit process does not require Δx to be 0!!! Now for some examples.

Let us begin with the equation of a straight-line y(x) = f(x) = mx +b.

$$\frac{df}{dx} = \lim_{\Delta x \to 0} \frac{f(x + \Delta x) - f(x)}{\Delta x}$$

$$= \lim_{\Delta x \to 0} \frac{m(x + \Delta x) - mx}{\Delta x}$$

$$= \lim_{\Delta x \to 0} \frac{m\Delta x}{\Delta x}$$

$$= \lim_{\Delta x \to 0} \frac{m\Delta x}{\Delta x} = m$$

I emphasize that in the last step, division by Δx occurs before the limiting process. Since m is a constant and, therefore, independent of Δx, the limit is simply the slope m of the straight line. We have recovered, using the formal definition of the derivative, the intuitive relation of the derivative to the slope of a line. An example of an even simpler case is the function $y = f(x) = c = $ constant. Clearly this function has a derivative equal to 0 since $f(x + \Delta x) - f(x) = 0$, no matter the value of x or Δx.

Let is now look at a more complicated problem, finding the slope at any point x of the parabola $y = x^2/10$. The figure below is a plot of the parabola. To illustrate the derivative as a limit, the figure shows four estimates of the slope of the parabola at $x = 1$ at which $f(x) = 0.1$. Estimates of the slope $\frac{f(x+\Delta x)-f(x)}{\Delta x}$ are shown for $\Delta x = 5, 4, 3,$ and 2. The slope for the estimate of $\Delta x_4 = 2$ appears to be very close to a visual estimate of the slope of the parabola at $x = 1$. The process of approaching a limit is shown more precisely in the table below the figure.

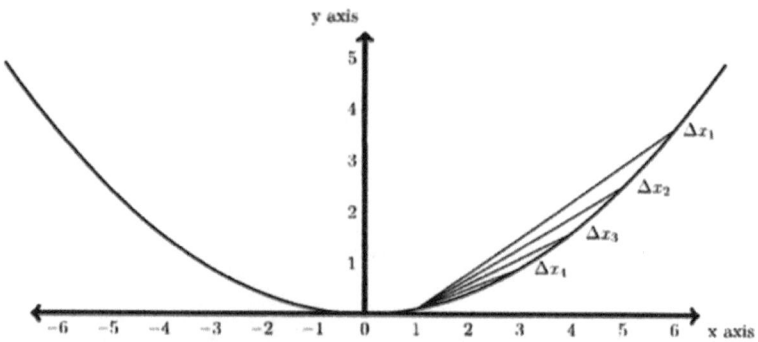

5-1 Graphical estimates of the slope of a parabola at x = 1

Table 5-1 Estimates of the Slope of a Parabola at x = 1

n	Δx_n	x+Δx	f(x+Δx)	$f'(x)_{estimate}$
1	5	6	3.6	0.7
2	4	5	2.5	0.6
3	3	4	1.6	0.5
4	2	3	0.9	0.4
5	1	2	0.4	0.3
6	0.5	1.5	0.225	0.250
↓	↓	↓	↓	↓
∞	0	1	0.1	0.2

Both the figure and table show a process in which it is plausible that the slope of the parabola at x = 1 results from estimates with decreasing Δx. The last entry in the table, however, was determined analytically through the formal limit process of taking the derivative of the function y = x²/10. The same process of differentiation previously used for the straight line y = mx +b is followed.

$$\frac{df}{dx} = \lim_{\Delta x \to 0} \frac{f(x + \Delta x) - f(x)}{\Delta x}$$

$$= \lim_{\Delta x \to 0} \frac{(x + \Delta x)^2 - x^2}{10\Delta x}$$

$$= \lim_{\Delta x \to 0} \frac{x^2 + 2x\Delta x + \Delta x^2 - x^2}{10\Delta x}$$

$$= \lim_{\Delta x \to 0} \frac{2x\Delta x + \Delta x^2}{10\Delta x}$$

$$= \lim_{\Delta x \to 0} \frac{2x + \Delta x}{10} = 2x/10$$

At x = 1, the slope is 2/10 = 0.2, as shown in the table. Because the process of differentiation and integration are inverse operations, we are also able to use the results to obtain anti-derivatives, that is the integrals of functions. Using Leibniz's notation, if $\frac{d}{dx} y = f(x)$, then $\int f(x) = \int \frac{dy}{dx} dx = y$. As discussed in the Overview, $\frac{d()}{dx}$ and $\int( )$ are inverse operators. As an example, following the result from above, $\frac{d}{dx}\left(\frac{x^2}{10}\right) = 2x/10.$ Therefore, $\int \frac{2x}{10} = x^2/10$. This will be particularly useful in the next section when a geometric interpretation of integration is given. Derivatives and anti-derivatives of common functions are given in the following table.

Table 5-2 Derivatives and integrals (anti-derivatives) of common functions

	Differentiation	Integration
1.	$\dfrac{d(cx)}{dx} = c$	$\int c\,dx = cx$
2	$\dfrac{dx^2}{dx} = 2x$	$\int x\,dx = x^2/2$
3.	$\dfrac{dx^3}{dx} = 3x^2$	$\int x^2\,dx = x^3/3$
4.	$\dfrac{dx^n}{dx} = nx^{n-1}$	$\int x^{n-1}\,dx = x^n/n$
5.	$\dfrac{d\cos x}{dx} = -\sin x$	$-\int \sin x\,dx = \cos x$
6.	$\dfrac{d\sin x}{dx} = \cos x$	$\int \cos x\,dx = \sin x$
7.	$\dfrac{d\tan x}{dx} = \sec^2 x$	$\int \sec^2 x\,dx = \tan x$
8.	$\dfrac{d\cot x}{dx} = -\csc^2 x$	$-\int \csc^2 x\,dx = \cot x$
9.	$\dfrac{d\sec x}{dx} = \sec x\tan x$	$\int \sec x\tan x\,dx = \sec x$
10.	$\dfrac{d\csc x}{dx} = -\csc x\cot x$	$-\int \csc x\cot x\,dx = \csc x$
11.	$\dfrac{de^x}{dx} = e^x$	$\int e^x\,dx = e^x$

Two important properties of the derivative that follow from its definition are the rules for taking the derivatives of sums of functions and the derivatives of products of functions. The rule for sums of functions is that the derivative of the sum is the sum of the derivatives. For example, we can prove this for the sum of two functions, f(x) and g(x).

$$\frac{d}{dx}\left(f(x) + g(x)\right) = \lim_{\Delta x \to 0} \frac{(f(x + \Delta x) + g(x + \Delta x)) - (f(x) + g(x))}{\Delta x}$$

$$= \lim_{\Delta x \to 0} \left(\frac{f(x + \Delta x) - f(x)}{\Delta x} + \frac{g(x + \Delta x) - g(x)}{\Delta x}\right)$$

$$= \frac{df}{dx} + \frac{dg}{dx}$$

Similarly, we may prove for products of functions that $\frac{d(f \cdot g)}{dx} = f \cdot \frac{dg}{dx} + \frac{df}{dx} \cdot g$.[lx] With these properties, the derivative is an example of what is known as a linear operator, that is,

$$\frac{d}{dx}\left(f(x) + g(x)\right) = \frac{df(x)}{dx} + \frac{dg(x)}{dx}, \text{ and for a constant } c, \frac{d(cf(x))}{dx} = c\frac{df(x)}{dx}.$$

Let us look at an example that illustrates the product rule and makes use of formulas 4 and 6 in the table above (Table 5-2) and a useful property called the chain rule: if g = g(v) and v = v(x), then $\frac{dg}{dx} = \frac{dg}{dv} \cdot \frac{dv}{dx}$.[lxi] We wish to find the derivative of sin²x/x. Let u = sin x, then $\frac{d(\sin x)}{dx} = \cos x = \frac{du}{dx}$.

$$\frac{d(\sin^2 x / x)}{dx} = \frac{(u^2 x^{-1})}{dx}. \text{ Then using the product rule,}$$

$$= x^{-1}\frac{du^2}{dx} + u^2\frac{dx^{-1}}{dx}.$$

$$= x^{-1}2u\frac{du}{dx} + u^2(-1)x^{-2}.$$

[lx]   The results for the derivative of a product may be proved by noting that f(x + Δx) = f + Δf and g(x + Δx) = g + Δg, Such terms in forming the derivative give, e.g., $\lim_{\Delta x \to 0} \Delta g \cdot \frac{\Delta f}{\Delta x} = 0$, as shown by Leibniz (Merzbach and Boyer, p. 386).

[lxi]   We can get insight into the chain rule $\frac{dg}{dx} = \frac{dg}{dv} \cdot \frac{dv}{dx}$ by noting that $\frac{\Delta g}{\Delta x} = \frac{\Delta g}{\Delta v} \cdot \frac{\Delta v}{\Delta x}$. Letting Δx, $\Delta v \to 0$ suggests the chain rule is a result of the limiting process. As discussed by Courant and Robbins (pp. 430 – 431), this result is valid although the argument has some subtle elements, which we will not elaborate on here.

So far the derivatives have only been obtained for equations of the form $y = f(x)$; however, they may also be easily obtained for implicit functions such as $F(x, y) = 0$. An example of such a function is the equation of a circle of unit radius, $x^2 + y^2 - 1 = 0$. The linearity of the derivative can be used to take the derivative of the equation of the unit circle.

$$\frac{d}{dx}(x^2 + y^2 - 1) = 0$$

$$\frac{d(x^2)}{dx} + \frac{d(y^2)}{dx} - \frac{d(1)}{dx} = 0.$$

$$2x + 2y\frac{dy}{dx} - 0 = 0, \text{ and } \frac{dy}{dx} = -\frac{x}{y}.$$

The derivative $dy/dx = -x/y$ is the slope of the unit circle at any point of the circle (x, y). Let (x_c, y_c) be specific points on the circle, then the equation of the tangent line to the circle at (x_c, y_c) is $y = -(x_c/y_c) x + 1/y_c$. The unit circle, tangent line, and radius at (x_c, y_c) are shown in the figure below. The intercepts of the y-axis and x-axis are, receptively, $1/y_c$ and $1/x_c$.

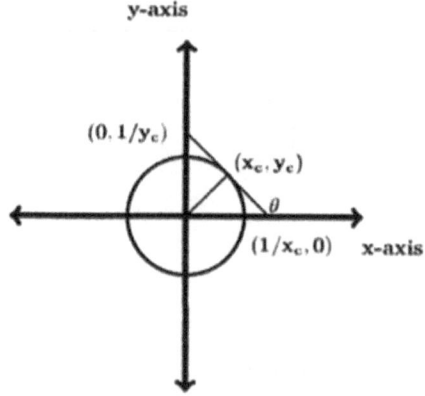

5-2 Tangent line and radius of the unit circle at the point (x_c, y_c)

The figure suggests that the tangent line through the point (x_c, y_c) is perpendicular to the circle's radius at that point. This can be confirmed

136

analytically by noting that the slope of the line through the radius at the point (y_c, x_c), that is the radial line, is $y = \frac{y_c}{x_c}x$. As proved previously (see p.98), the perpendicular to a line of slope m is $-1/m$, and indeed the slope of the radial line $\frac{y_c}{x_c} = -1/(-x_c/y_c)$.

Another way to characterize the derivative as the slope of the circle is by the angle that the tangent line makes with the x-axis measured counter clockwise. The figure indicates this angle as θ. The relation ship between the slope and the angle θ is given by the trigonometric function the tangent:

$$\tan \theta = -x_c/y_c.$$

This may be made more apparent by looking at the supplement of θ, that is, $\varphi = \pi - \theta$ radians, or $\varphi = 180° - \theta$ degrees. Using the definition of the tangent function:

$$
\begin{aligned}
\tan \varphi \quad &= \quad \text{opposite/adjacent} \\
&= \quad y_{intercept}/x_{intercept} \\
&= \quad 1/y_c/1/x_c \\
&= \quad x_c/y_c/; \text{ however,} \\
\tan \varphi \quad &= \quad \tan(\pi - \theta) \\
&= \quad -\tan \theta, \text{ therefore,} \\
\tan \theta \quad &= \quad -x_c/y_c
\end{aligned}
$$

Our knowledge of the derivatives of functions of a single variable is useful for describing motion along a curve. In the Euclidean plane suppose we wish to analyze motion along the curve $y = f(x)$, and we know motion in the coordinate x as a function of time t. Then the velocity in the y direction as a function of time is dy/dt which can be evaluated from our knowledge of the curve $y = f(x)$ and motion in the x direction x(t). From the chain rule, this simple results follows: $\frac{dy}{dt} = \frac{dy}{dx} \cdot \frac{dx}{dt}$. However, in three-dimensions, changes occur in the other coordinates, requiring the concept of a partial derivative. The partial derivative simply provides a derivative in which only one variable changes and the other two (in three dimensions) are held constant. The partial derivative of the function F(x, y, z) with respect to x is written as $\partial F/\partial x_{y,z}$. The subscripts y and indicates that these coordinates are held constant. The definition for the partial derivative of the function of three variables F(x, y, z) follows closely the definition of the derivative for a single variable. The extension to functions dependent upon more variables should be clear.

Definition of Partial Derivative: For a function F(x, y, z) defined over a neighborhood of the point x_0, y_0, z_0 the partial derivative at the point is defined as:

$$\frac{\partial F}{\partial x_{y,z}} = \lim_{x \to x_0} \frac{F(x, y_0, z_0) - F(x_0, y_0, z_0)}{x - x_0}, \text{ if the limit exists.}$$

For example, if, $F(x, y, z) = x^2 y^2 z - xy + z$, then $\frac{\partial F}{\partial x_{y,z}} = 2xy^2 z - y$.

We will have a need, later in the chapter, to describe changes along curved paths in three-dimensional space. The function F(x, y, z) could be determined along a curve described using the path length s along the curve as x = x(s), y = y(s), and z = z(s). Similar to the derivative $\frac{df}{dx}$ of a function f(x), the total derivative $\frac{dF}{ds}$ can be calculated from changes along the path with coordinates x, y, z using the partial derivatives as defined above.

$$\frac{dF}{ds} = \frac{\partial F}{\partial x_{y,z}} \frac{dx}{ds} + \frac{\partial F}{\partial y_{x,z}} \frac{dy}{ds} + \frac{\partial F}{\partial z_{x,y}} \frac{dz}{ds}. \qquad (5\text{-}1)$$

Another way to express this relation is through the differential of the function.

$$dF = \frac{\partial F}{\partial x_{y,z}} dx + \frac{\partial F}{\partial y_{x,z}} dy + \frac{\partial F}{\partial z_{x,y}} dz. \qquad (5\text{-}2)$$

Just as there is a chain rule for derivatives of functions of one variable, there is a chain rule for partial derivatives. This will be very important in Chapter 8 in the development of tensor properties. I will illustrate the chain rule for a function of two variables, but its extension to more variables should be clear.

Let $F = F(x, y)$ with x and y being the usual Cartesian coordinates which can be related to polar coordinates by $x = r \cos \theta$ and $y = r \sin \theta$, that is the Cartesian coordinates may be expressed as functions of polar coordinates, $x = x(r, \theta)$, and $y = y(r, \theta)$. Then the partial derivatives with respect to the polar coordinates may be obtained from the partial derivatives in Cartesian coordinates using the chain rule (with the variables held constant implied).

$$\frac{\partial F}{\partial r} = \frac{\partial F}{\partial x}\frac{\partial x}{\partial r} + \frac{\partial F}{\partial y}\frac{\partial y}{\partial r}. \qquad (5\text{-}3)$$

$$\frac{\partial F}{\partial \theta} = \frac{\partial F}{\partial x}\frac{\partial x}{\partial \theta} + \frac{\partial F}{\partial y}\frac{\partial y}{\partial \theta}. \qquad (5\text{-}4)$$

As an example, let the function F be the radial coordinate r. We know that $\partial r / \partial \theta = 0$ as r and θ are independent coordinates, but we can also see this with the chain rule recalling that $r = \sqrt{x^2 + y^2}$, $x = r \cos \theta$, and $y = r \sin \theta$.

$$\frac{\partial r}{\partial \theta} = \frac{\partial r}{\partial x}\frac{\partial x}{\partial \theta} + \frac{\partial r}{\partial y}\frac{\partial y}{\partial \theta}$$

$$\frac{\partial r}{\partial x} = \frac{1}{2}\frac{2x}{\sqrt{x^2 + y^2}} = \cos \theta; \quad \frac{\partial r}{\partial y} = \frac{1}{2}\frac{2y}{\sqrt{x^2 + y^2}} = \sin \theta.$$

$$\frac{\partial x}{\partial \theta} = -r \sin \theta; \quad \frac{\partial y}{\partial \theta} = r \cos \theta. \text{ Therefore,}$$

$$\frac{\partial r}{\partial \theta} = \cos \theta(-r \sin \theta) + \sin \theta(r \cos \theta) = 0.$$

Considerations of the changes that are quantified by partial derivatives in three dimensions, as well as the spatial relation between the tangent and radial lines of the circle, suggest that it would be useful to define mathematical entities that explicitly express directions, particularly once we move off the Euclidean plane into three-dimensional space and onto non-Euclidean surfaces. This language will be provided by vectors. However, before moving beyond the Euclidean plane, the other half of the calculus developed by Newton and Leibniz, integration, needs to be given a meaning beyond its rather obscure role as the anti-derivative or inverse of differentiation. This is accomplished through a geometrical interpretation.

5.3   Summing up along the path: integral calculus

Many of the details of the techniques related to the problem of tangents were initiated before Newton and Leibniz. For example, Fermat developed a technique for finding the tangents to polynomials, conceptually the same as that devised by Newton and Leibniz.[162] A similar evaluation could be made of Isaac Barrow's (1630-1677), method of tangents.[163] Barrow was the holder of the Lucasian chair at Cambridge. When Barrow became the chaplain of King Charles II in London, he was succeeded by Newton with Barrow's support.

Integration in its role as a technique for determining areas under curves (or quadrature, as it is also known) and volumes had an even longer history. Archimedes had used an approach of approximating the areas and volumes of geometric objects such as the ellipse, parabola, and the ellipsoid with increasing numbers of simple geometric figures.[164] A similar approach was employed in the works of Bonaventura Cavalieri (1598-1647),[165] Fermat,[166] and Barrow which involved a vaguely defined summation of infinitesimal figures whose totality covered the area contained by portions of a curve.[lxii] Summation would remain central

---

[lxii]   Infinitesimals along with their use were poorly defined and, in the absence of rigorous methods for their use, were employed informally as quantities smaller than any specified quantity.

to the method that Leibniz named integration; however, as noted by Courant and Robbins, "Newton and Leibniz's great merit is to have clearly recognized the intimate *connection between these two problems*, [the inverse nature of the problem of tangents and the problem of quadrature]."[167] Merzbach and Boyer suggest that Fermat and Barrow were likely to have understood the inverse nature of the operations of differentiation and integrations that would be formally proved as the Fundamental Theorem of Calculus.[168] In any case, their understanding did not have the extraordinary impact on mathematics and physics that came from the work of Newton and Leibniz. One thing that Newton and Leibniz had in common with their predecessors was their ignorance of the limit concept, which would end controversies over the validity of the calculus and make the vague use of infinitesimals unnecessary. Using the limit process, as we did with differentiation, integration will now be presented as a method to calculate area and as the anti-derivative.

Let us again look at a simple problem for which we already know the answer: find the area A under the curve y = x from x = 0 to the general point x = X. A plot of the function is given below.

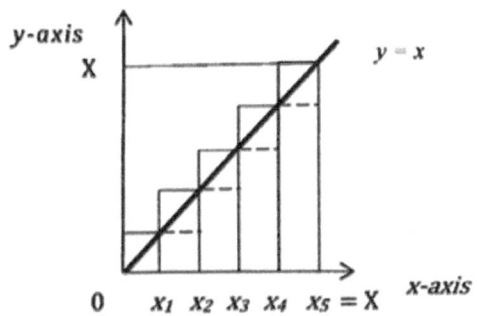

5-3 Finding the area under y = x

The curve $y = x$ from 0 to X forms a triangle whose base *b* and height *h* are X. Thus, since $A = 1/2\ bh$, $A = 1/2\ X^2$. Our goal here, however, is to employ the method of calculating the area by approximating it with an increasing number of rectangles in the spirit of Cavalieri and others, to give us a perspective on how it would work before tackling other curves such as $y = x^2$.

In the figure above, I have shown how to estimate the area under the curve $y = x$ from 0 to X by two sets of five rectangles. In both cases the bases of the rectangles on the *x-axis* are equal to X/5. The heights of rectangles are determined by their intersection with the straight line $y = x$. The upper rectangles, shown completely with solid lines, overestimate the area as the height is set by the right edge of the rectangle. Lower rectangles, with their height set by the left edge indicated by a dashed line, underestimate the area. Thus, the two sets of rectangles bracket the area under the curve.

If we increase the number of rectangles and the width of their bases decrease, our estimates should get better, and perhaps we can show that in the limit as the base width approaches 0, we get the right answer. Let N be the number of rectangles, then the base width is $\Delta x = X/N$. The x coordinate associated with any particular rectangle is equal to $k \cdot \Delta x$ in which k goes from 1 to N for the upper rectangles and 0 to $N - 1$ for the lower rectangles. At x location $k \cdot \Delta x$, the height of the rectangle is also $k \cdot \Delta x$. Let $A_{UN}^{0 \to X}$ be the area of the upper N rectangles from 0 to X and $A_{LN}^{0 \to X}$ be the corresponding area for the lower rectangles. I introduce here a convenient notation, introduced by Euler,[169] that will find frequent use: the sum of a number of terms such as $x_1 + x_2 + x_3 + x_4$ is abbreviated using the Greek letter Σ as $\Sigma_{k=1}^4 x_k$. With this notation then:

$$A_{UN}^{0 \to X} = (1 \cdot \Delta x)\Delta x + (2 \cdot \Delta x)\Delta x + (3 \cdot \Delta x)\Delta + \cdots + N\Delta x$$

$$= \sum_{k=1}^{N} (k \cdot \Delta x)\Delta x.$$

$$= \frac{X^2}{N^2} \sum_{k=1}^{N} k$$

A method called mathematical induction may be used to prove that $\Sigma_{k=1}^N k = N(N+1)/2$. In this method in which one generally wishes to prove a relationship dependent on the natural numbers as in that above, we first show that it is true for $k = 1$. Then, we assume that it is true for all k, and under this assumption, prove that it is true for $k +1$. Once this is proved, then the truth of the statement falls out like a line of dominoes in which the first domino is knocked over, knocking over the second, and continuing on to successive dominoes. The proof for $\sum_{k=1}^{N} k = N(N+1)/2$ is given below.

Let $S_N = \sum_{k=1}^{N} k = 1 + 2 + 3 + 4 + \cdots + N$.

First let us look at the simple case of N=1. The sum $S_1 = 1$.

$N(N + 1)/2 = 1(1 + 1)/2 = 1$, so our formula checks for N = 1.

Now our hypothesis is that, up to N, the sum $S_N = N(N + 1)2$ We must show using this hypothesis that for the sum up to $N + 1$, $S_{N+1} = (N + 1)$ $((N + 1)+1)/2$. Using the assumption for S_N.

$$
\begin{aligned}
S_{N+1} &= S_N + N + 1 = [N(N + 1)/2] + N + 1 \\
&= (N^2 + N + 2N+2)/2 \\
&= (N^2 + 3N + 2)/2 \\
S_{N+1} &= (N + 1) ((N + 1) + 1)/2
\end{aligned}
$$

Thus, $\sum_{1}^{N} k = S_N = N(N +1)/2$ is true for all N since it is true for N = 1, and true for N +1 if it is true for N. Continuing with the calculation of area,

$$
\begin{aligned}
A_{UN}^{0 \rightarrow X} &= \frac{X^2}{N^2} \sum_{k=1}^{N} k \\
&= \frac{X^2}{N^2} \cdot \frac{N(N + 1)}{2} \\
&= X^2 \left(\frac{1}{2} + \frac{1}{2N}\right)
\end{aligned}
$$

Similarly,

$$
\begin{aligned}
A_{LN}^{0 \rightarrow X} &= \frac{X^2}{N^2} \sum_{k=0}^{N-1} k \\
&= \frac{X^2}{N^2} \cdot \frac{(N - 1)N}{2} \\
&= X^2 \left(\frac{1}{2} - \frac{1}{2N}\right).
\end{aligned}
$$

Now we want to find the limit of the areas under the curve as $\Delta x \to 0$, or equivalently $N \to \infty$. $\lim\limits_{N \to \infty} \frac{1}{2N} = 0$ as $\frac{1}{2N} < \varepsilon$ for all $N > \frac{1}{2\varepsilon}$. We will identify the limits of $A_{UN}^{0 \to X}$ and $A_{LN}^{0 \to X}$ as the integral:

$$A_{0(y=x)}^{X} = \int_{0}^{X} x dx = \lim_{N \to \infty} A_{UN}^{0 \to X} = \lim_{N \to \infty} X^2 \left(\frac{1}{2} + \frac{1}{2N} \right) = \frac{X^2}{2.}$$

Clearly, the same result would be obtained for the lower rectangles, that is both the upper and lower estimates converge to the same limit of $\frac{X^2}{2}$. Thus, our procedure works for this simple case giving the same results as Euclid's geometry.

Now let us follow exactly the same procedure to find the area under the curve $y = x^2$ from $x = 0$ to $x = X$. As before let $\Delta x = X/N$. The x coordinate $k \cdot \Delta x$ is associated with the rectangle height $(k \cdot \Delta x)^2$. For the area of the upper rectangeles we have:

$$A_{UN}^{0 \to X} = (1 \cdot \Delta x)^2 \Delta x + (2 \cdot \Delta x)^2 \Delta x + (3 \cdot \Delta x)^2 \Delta x \Delta + \cdots + (N \cdot \Delta x)^2 \Delta x$$

$$= \sum_{k=1}^{N} (k \cdot \Delta x)^2 \Delta x.$$

$$= \frac{X^3}{N^3} \sum_{k=1}^{N} k^2$$

Using mathematical induction as described above, it can be proved that:

$$= \sum_{k=1}^{N} k^2 = \frac{(2N^3 + 3N^2 + N)}{6}. \text{ Therefore,}$$

$$A_{0(y=x^2)}^{X} = \int_{0}^{X} x^2 dx = \lim_{N \to \infty} \frac{X^3}{N^3} \sum_{k=1}^{N} k^2 = \lim_{N \to \infty} \frac{X^3}{6} \left(2 + \frac{3}{N} + \frac{1}{N^2} \right) = \frac{X^3}{3.}$$

Following the same approach, we would find that the estimate of the lower rectangles gives the same answer. This is equivalent to the result discovered by Archimedes.[170]

With the examples given above, the following definition of an integral of a function should seem reasonable: [171]

The Definite Riemann Integral: If f(x) is a bounded function on the closed interval $a \leq x \leq b$, and the interval from a to b is divided into intervals $\Delta x_i = x_i - x_{i-1}$ with $a = x_0 < x_1 < \ldots < x_{n-1} < x_n = b$, with c_i such that $x_{i-1} \leq c_i \leq x_i$, then the integral, $\int_a^b f(x)dx = \lim_{max.\Delta x_i \to 0} \sum_a^b f(c_i) \Delta x_i$ if the limit exists.

The above definition is consistent with the examples given above and is known as Riemann integration. If $\int_a^b f(x)dx = \lim_{max.\Delta x_i \to 0} \sum_a^b f(c_i) \Delta x_i$ exists then the function f(x) is said to be Riemann integrable. Unlike the examples, the definition does not require that the intervals Δx_i be equal. Any division is fine as long as the limit is taken with the largest of the Δx_i approaching 0.

The definition of the integral may be used to prove that the integral, like the derivative, is a linear operator. Here, however, I will present properties without formal proof, leaning on our understanding of the integral as the area under the curve. If $y = f(x)$ is the curve, and the area is to be calculated from $x = a$ to $x = c$, then if b is a point between a and c

$$\int_a^c f(x)dx = \int_a^b f(x)dx + \int_b^c f(x)dx$$

$$\int_a^a f(x)dx = 0 = \int_a^b f(x)dx + \int_b^a f(x)dx. \text{ Therefore,}$$

$$\int_a^b f(x)dx = -\int_b^a f(x)dx.$$

Using these properties, the definite integral is calculated over the general range of a to X using as an example the specific case of $f(x) = x^2$.

$\int_0^X x^2 dx = \int_0^b x^2 dx + \int_b^X x^2 dx$. This implies,

$$\int_b^X x^2 dx = \int_0^X x^2 dx - \int_0^b x^2 dx = \frac{X^3}{3} - \frac{b^3}{3} = \text{(often written as)} \; \frac{x^3}{3} \Big|_b^X \qquad (5\text{-}5)$$

As another example, let us use integration to calculate the area of the trapezoid A_{trapzd} produced by the line $y = x$ between the points $x = a$ and x = b. First, using simple geometric arguments, the area consists of the rectangle of sides a and (b-a) and the triangle on top of base and height (b-a). Thus $A_{trapzd} = a(b-a) + \frac{1}{2}(b-a)^2 = \frac{(b^2-a^2)}{2}$. By integration,

$$A_{trapzd} = \int_a^b x\,dx = \frac{x^2}{2}\Big|_a^b = \frac{b^2}{2} - \frac{a^2}{2} = \frac{(b^2-a^2)}{2}.$$

With this background, the Fundamental Theorem of Calculus will now be stated without proof, expressing the inverse relation of differentiation and integration.[lxiii]

The Fundamental Theorem of Calculus: Part 1, If f(x) is a continuous function over the interval a $\leq$ x $\leq$ b and $\int_a^x f(t)dt = F(x)$, then $\frac{dF}{dx} = f(x)$; Part 2, if $\frac{dF}{dx} = f(x)$ is continuous over the interval a $\leq$ x $\leq$ b, then $\int_a^x \frac{dF}{dt}dt = \int_a^x f(t)dt = F(x) - F(a)$,

By identifying in the Fundamental Theorem, F(x) with $\frac{x^3}{3}$ in the example of equation (5-5) and noting that $\frac{d(x^3/3)}{dx} = x^2 = f(x)$, the inverse nature of differentiation and integration should be clear: $\frac{d}{dx}\int_a^x t^2 dt = \frac{d}{dx}\left[\frac{x^3}{3} - \frac{a^3}{3}\right] = \frac{d}{dx}\left(\frac{x^3}{3}\right) = x^2$, and $\int_a^x \frac{d}{dt}\left(\frac{t^3}{3} - \frac{a^3}{3}\right) dt = \int_a^x t^2 dt = \frac{x^3}{3} - \frac{a^3}{3}.$

A particular definite integral which will be important in expressing some of the relationships in models of non-Euclidean geometry is the integral which defines the natural logarithm ln x. Natural logarithms were introduced in footnote lv (p. 118) to introduce the transcendental number e. Recall that e is the base of the natural logarithm, that is, ln $e^2 = 2$ in the same way that log $10^2 = 2$ for the base 10 logarithm. The natural logarithm is defined as:

$$\ln x = \int_1^x \frac{dt}{t}. \qquad\qquad (5\text{-}6)$$

[lxiii] A proof of the Fundamental Theorem of Calculus is given by Courant and Robbins, pp. 436-439.

Hence, by the Fundamental Theorem,

$$\frac{d}{dx}(\ln x) = \frac{d}{dx}\left(\int_1^x \frac{dt}{t}\right) = \frac{1}{x}.$$

Natural logarithms have the same properties as common logarithms: ln (ab) = ln (a) + ln (b), ln(a/b) = ln (a) - ln (b), and b · ln a = ln(a)^b. Natural logarithms will be important in Chapter 7.

A vast number of techniques have been developed to determine integrals. An introduction to such techniques is available in standard beginning calculus textbooks. As an example, we wish to determine the area A of the unit circle $x^2 + y^2 = 1$ above the x-axis from 0 to 1. Solving for y, then,

$$A = \int_0^1 \sqrt{1 - x^2}dx.$$

As A is the area of a quarter of a circle, we know that $A = \frac{\pi 1^2}{4} = \pi/4$. However, we can show this by making the substitution, $x = \sin\theta$, and therefore, $\frac{dx}{d\theta} = \cos\theta$, and $dx = \cos\theta\, d\theta$ (see Table 5-2). With the change of variable from x to θ, the limits of the integral are now from 0 to π/2 since sin $0 = 0$ and $\sin\frac{\pi}{2} = 1$:

$$A = \int_0^1 \sqrt{1 - x^2}\, dx$$
$$= \int_0^{\pi/2} \sqrt{(1 - \sin^2\theta)}\cos\theta d\vartheta$$
$$= \int_0^{\pi/2} \sqrt{\cos^2\theta}\cos\theta d\vartheta$$
$$= \int_0^{\pi/2} \cos^2\theta\, d\theta$$

Now $\cos^2\theta = \frac{1}{2}(1 + \cos 2\theta)$,[lxiv] Therefore,

---

[lxiv]   This trigonometric identity may be easily shown by recalling that $e^{i\theta} = \cos\theta +$ isin θ, and $(e^{i\theta})^2 = (\cos\theta + i\sin\theta)^2 = e^{i2\theta} = \cos 2\theta + i\sin 2\theta$, then expanding in terms of the sine and cosine functions and collecting the real and imaginary parts.

$$A = \int_0^{\pi/2} \cos^2\theta \, d\theta$$

$$= \int_0^{\pi/2} \frac{1}{2}(1 + \cos 2\theta) \, d\theta$$

$$=$$

$$1/2 \int_0^{\pi/2} d\theta + 1/2 \int_0^{\pi/2} \cos 2\theta \, d\theta^{\text{lxv}}$$

$$= \frac{\theta}{2} \Big|_0^{\pi/2} + \frac{1}{4} \sin 2\theta \Big|_0^{\pi/2}$$

$$= (\pi/4 - 0) + (0 - 0)$$

$$= \pi/4.$$

Integration in the example given above is not difficult, but in many cases, it requires techniques which are well beyond the scope of our current story. However, even in cases where a solution is not apparent, the fundamental definition of the Riemann integral may be used to estimate the result numerically. In our particular example, it may be used to estimate a numerical value for π since we know the integral equals $\pi/4$.

Let us estimate the integral $\int_0^1 \sqrt{1 - x^2} dx$ by approximating the area with N rectangles with bases Δx equal to $1/N$. The height at y_k of the kth rectangle at $k\Delta x$ is $\sqrt{1 - (k\Delta x)^2}$.

$$\int_0^1 \sqrt{1 - x^2} dx \approx \sum_{k=1}^{N} y_k \Delta x = \sum_{k=1}^{N} \sqrt{1 - (k\Delta x)^2} \, \Delta x.$$

There are more accurate and efficient ways of numerically calculating the area, but this method in its simplicity should help you to understand the meaning of integration. Calculations may easily be performed using a spreadsheet such as Excel with a programming capability such as Visual Basic for Applications (VBA). In the table below, the results are given for various increasing numbers of rectangles N with the sums

---

[lxv] Let u = 2θ, then du = $2d\theta$. $\int \cos 2\theta \, d\theta = \int (\cos u \, du)/2 = \sin u/2 = \sin 2\theta/2$.

given for the rectangles that underestimate the area and those that overestimate it, thus bracketing the value of π. Recall that the irrational number π is 3.141592.... It is apparent that accuracy increases with increasing N and that the sum of the N rectangles is an approximation to the integral multiplied by 4.

**Table 5-3 Numerical approximation of π
from area of a quarter circle**

N	Δx	$4 \cdot \Sigma_{under}$	$4 \cdot \Sigma_{over}$
20	0.05	3.03	3.23
200	0.005	3.13	3.15
2000	0.0005	3.140	3.143
20000	0.00005	3.1415	3.1417

Considering that the integral is determined by a limiting process in which intervals Δx_i approach 0, one might wonder if the division of the real numbers into rational and irrational numbers is of any significance to this limiting process. The answer is decidedly yes, but takes us again a bit too far afield. Nevertheless, insight into the impact of the nature of the real numbers can be gained by considering the integral from 0 to 1 of three functions:

$$f_1(x) = 1, f_2(x) = \begin{cases} 1 \text{ if } x \text{ is rational} \\ 0 \text{ if } x \text{ is irrational} \end{cases}, f_3(x) = \begin{cases} 0 \text{ if } x \text{ is rational} \\ 1 \text{ if } x \text{ is irrational} \end{cases}$$

By taking into account that the rational numbers are countable, it can be shown that the contributions to the integral from each of the Δx_i about the ith rational number can be made as small as one wishes. As the number of intervals N increases without limit and the $\Delta x_i \to 0$, the limit of the sum over the rational numbers is 0. These considerations lead to the following results:

$$\int_0^1 f_1(x)dx = 1, \int_0^1 f_2(x)dx = 0.$$

Therefore,

$$\int_0^1 f_3(x)dx = 1.$$

It is apparent that the entire contribution to the integral comes from the irrational numbers!!! In fact, since the algebraic irrational numbers are countable, the entire contribution is just from the transcendental numbers - not surprising as we have already seen that they dominate the real number line. The mathematical term for this result is that the rational numbers have measure zero.[172]

It should not be surprising that the integral, defined through a limit of the summation process, can also be used to calculate the length of curves. In this case, instead of the area estimates formed by increasing numbers of rectangles, the curve length is estimated through increasing numbers of chords. The figure below shows the ith chord $\Delta s_i = \sqrt{\Delta x_i^2 + \Delta y_i^2}$ in an estimate of the curve length from $x = a$ to $x = b$.

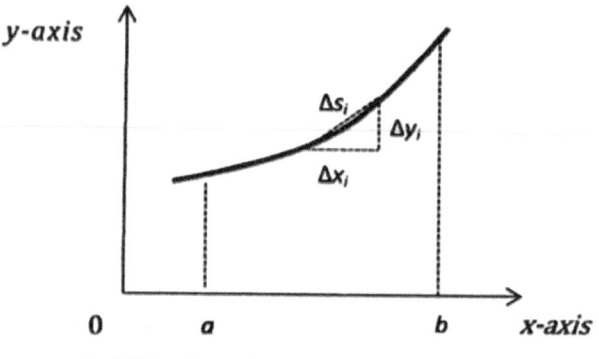

5-4 Finding the length of a curve

The total length s_a^b is given by:

$$s_a^b = \lim_{N \to \infty} \sum_{i=1}^{N} \Delta s_i,$$

$$= \lim_{N \to \infty} \sum_{i=1}^{N} \sqrt{\Delta x_i^2 + \Delta y_i^2},$$

$$= \lim_{N \to \infty} \sum_{i=1}^{N} \sqrt{1 + \left(\frac{\Delta y_i}{\Delta x_i}\right)^2} \, \Delta x_i.$$

$$s_a^b = \int_a^b \sqrt{1 + \left(\frac{dy}{dx}\right)^2} \, dx$$

(5-7)

Similar to calculating the area under the unit circle, we can calculate the length of the circle above the x-axis from 0 to 1 which is $\frac{2\pi(1)}{4} = \pi/2$. As shown in the previous section (see figure 5-2 (p. 136) and the following discussion), the slope of the circle is $\frac{dy}{dx} = -\frac{x}{y} = -x/\sqrt{1 + x^2}$. Therefore,

$$s_0^1 = \int_0^1 \sqrt{1 + \left(\frac{dy}{dx}\right)^2} \, dx$$

$$= \int_0^1 \sqrt{1 + \left(\frac{-x}{\sqrt{1 - x^2}}\right)^2} \, dx$$

$$= \int_0^1 \frac{dx}{\sqrt{1 - x^2}}.$$

As in the case of integrating the area under the unit circle, we make the substitution, $x = \sin \theta$, with the result,

$$s_0^1 = \int_0^{\pi/2} \frac{\cos \theta}{\sqrt{1 - \sin^2 \theta}} \, d\theta,$$

$$= \int_0^{\pi/2} \frac{\cos \theta}{\sqrt{\cos^2 \theta}} \, d\theta,$$

$$= \int_0^{\pi/2} d\theta,$$

$$= \theta|_0^{\pi/2}$$

$$= \pi/2.$$

We may also numerically estimate the length by approximating the curve with an increasing number N of chords Δs_i (see figure 5-4 above). This is an approach similar to the numerical estimate of area under a curve. In this case the ith chord $\Delta s_i = \sqrt{\Delta x_i^2 + \Delta y_i^2}$ with $\Delta x_i = 1/N$, $x_i = i \cdot \Delta x_i = i/N$, $y_i = \sqrt{1 - (i/N)^2}$, and $\Delta y_i = y_i - y_{i-1}$.

$$s_0^1 = \int_0^1 \sqrt{1 + (dy/dx)^2} dx \approx \sum_{i=1}^N \sqrt{\Delta x_i^2 + \Delta y_i^2} \qquad (5\text{-}8)$$

Using a spreadsheet. for N = 2000, $2 \cdot \Sigma_{i=1}^N \sqrt{\Delta x_i^2 + \Delta y_i^2} = 3.14159$.

The curve path length s in the example above may also be described by what is known as a differential:

$$ds = \sqrt{1 + (dy/dx)^2} dx,$$

or

$$ds^2 = 1 + \left(\frac{dy}{dx}\right)^2 dx^2$$

Gauss made the discovery, which he termed his *Theorem Egregium* (most excellent theorem), that knowing the coefficients and derivatives of a specified differential form of length ds was sufficient to characterize the intrinsic nature of the geometry of surfaces.[173] This will be a major theme of Chapter 6.

5.4 Numbers with direction: vectors

Before discussing Gauss's description of surface curvature, we must expand our mathematical language to include vectors as expressions that include both magnitude and direction. Vectors will be vital to describing the tangents to surfaces and to the changes in the direction of the tangents along a surface. Gauss defined surface curvature in terms of these changes. Other examples of vectors in physics are force, velocity, acceleration, and the electric and magnetic field. Thus, the vector concept, along with its generalization, tensors, would turn out to be central to the description of the physics of the universe.

Some early premonitions of vectors related to the resultants of combinations of forces were anticipated by Aristotle in the fourth century BC and Simon Stevin in the sixteenth century.[174] However, it was only in the seventeenth century that Galileo would give some precision to the concept of a vector. The vector concept, allowing motions to be analyzed in terms of their horizontal and vertical components was crucial to Galileo's description of projectile motion as a parabolic path when air resistance was neglected.[175] The concept of representing a vector sum of forces as the diagonal of a parallelogram is explicit in Corollary 1 of Newton's *Principia*. Newton states that:

"A body acted on by [two] forces acting jointly describes the diagonal of a parallelogram in the same time in which it would describe the sides if the forces were acting separately."[176]

Let us look at this geometric interpretation of vector addition. Vectors will be designated as variables in bold, for example, **u**. Furthermore, they will, for the moment, be restricted to three dimensions in Euclidean space, being designated as an ordered triple, **u** = (a, b, c). Here a, b, and c are the components of the vector along the x, y, and z axes, respectively. The figure below shows the geometrical interpretation of the addition of **v** = (a, b, 0), and w = (c, d, 0) in the x-y plane to form the resultant vector, **r** = (a + c, b + d, 0). As in Newton's description given above in which the vectors are forces, the resultant vector **r** is the diagonal of the parallelogram formed with sides equal to **v** and **w**. The length of the resultant, $|r| = \sqrt{(a + c)^2 + (b + d)^2}.$ The angle that **r** makes with the x-axis, θ_r is such that $\cos \theta_r = (a + c)/r$. Scalar multiplication can be considered as stretching the length of the vector while maintaining its direction. Scalar multiplication by a negative number indicates a stretching in which the vector is now directed in the opposite sense as illustrated by **−r**.

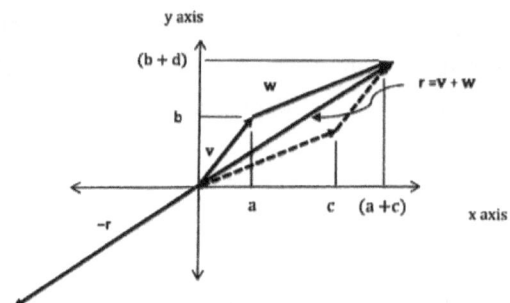

5-5 Addition of vectors

With the operation of the addition of vectors illustrated above, a more formal definition of what is known as a vector or linear space is now introduced. The definition given below is introduced for three dimensional spaces defined in terms of rectangular Cartesian coordinates; however, it can be generalized and extended to n-dimensional spaces.[177]

Vector Space
Definition V1: A Cartesian vector is defined with respect to Cartesian axes as the coordinate triple of real numbers, v = (a, b, c).

Definition V2: Two Cartesian vectors, v = (a, b, c) and w = (d, e, f) are equal if and only if a = d, b = e, and c = f.

Definition V3: The operation of addition between two Cartesian vectors, v = (a, b, c) and w = (d, e, f) is defined by: v + w = (a, b, c) + (d, e, f) = (a + d, b + e, c +f).

Definition V4: If d is a real number and v = (a, b, c), then scalar multiplication is defined as:

$$dv = (da, db, dc).$$

Definition V5: The identity element for vector addition is $0 = (0, 0, 0)$.

Note that multiplication by two vectors, **vw** is not defined. From the properties of the real numbers, vectors satisfy the closure, associative, and commutative properties with respect to addition. Each vector $v = (a, b, c)$ has an additive inverse, $-v = (-a, -b, -c)$.

In addition to these properties familiar from the real numbers, the following additional properties are satisfied by scalar multiplication:

Associativity of scalar multiplication: $a(bv) = (ab)v$.
Distributive property of scalar sums: $(a + b)v = av + bv$.
Distributive property of vector sums: $a(v + w) = av + aw$.
Identity element for scalar multiplication: $1v = v$

Further geometric interpretations can be established for vectors, Let **v** $= (v_1, v_2, v_3)$ and **w** $= (w_1, w_2, w_3)$. The figure below shows the two vectors **v** and **w** and the angle, θ between them.

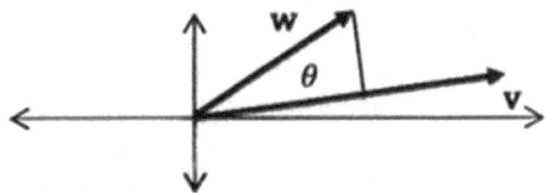

5-6 Angle between vectors

Recalling the discussion on the algebraic definition of the cosine function in equation (4-1, p. 97) the angle θ between the vectors **v** and **w** is: [lxvi]

$$\cos\theta = (v_1\,w_1 + v_2\,w_2 + v_3\,w_3)/\left(\sqrt{v_1{}^2 + v_2^2 + v_3^2} \cdot \sqrt{w_1{}^2 + w_2^2 + w_3^2}\right)$$

$$= (v_1\,w_1 + v_2\,w_2 + v_3\,w_3)/(|v| \cdot |w|)$$ with $|v|$ and $|w|$ being the magnitudes of their respective vectors.

[lxvi]   In regard to equation (4-1), note that the angle vertex in Figure 5-6 is the origin corresponding to $(x_1, y_1) = (0, 0)$.

We now introduce a new operation with vectors.

Definition of the inner product: $\mathbf{v} \bullet \mathbf{w} = |\mathbf{v}| \cdot |\mathbf{w}| \cos \theta$ (5-9)

A comparison of the definition of the inner (scalar) product with the expression for $\cos \theta$, results in the following definition for Cartesian vectors. (The inner product is also frequently called the dot product for obvious reasons.)

Definition of inner product for Cartesian vectors: For Cartesian vectors with $\mathbf{v} = (v_1, v_2, v_3)$ and $\mathbf{w} = (w_1, w_2, w_3)$, the inner product $\mathbf{v} \bullet \mathbf{w} = v_1 w_1 + v_2 w_2 + v_3 w_3 = \sum_{i=1}^{3} v_i \cdot w_i$.

Note that $|\mathbf{v}| = \sqrt{\mathbf{v} \bullet \mathbf{v}}$. Also, this is a good time to note that while we have arrived at our notion of a vector in Cartesian coordinates, the vector and vector concepts such as the inner product have an independent existence that does not depend on the coordinate system. There are many different coordinate systems, for example spherical and polar coordinates in which the relationship of the components are different than that of Cartesian vectors; however fundamental definitions such as, $\mathbf{v} \bullet \mathbf{w} = |\mathbf{v}| \cdot |\mathbf{w}| \cos \theta$, remain true in other coordinate systems.

The geometric interpretation of the inner product is that $|\mathbf{w}| \cos \theta$ is the length of the projection of the vector $\mathbf{w}$ onto $\mathbf{v}$ and $\mathbf{v} \bullet \mathbf{w}$ is the length $|\mathbf{v}|$ times this projection. In Cartesian vectors, $\mathbf{v} \bullet \mathbf{w} = (v_1 w_1 + v_2 w_2 + v_3 w_3)$ and clearly from the definition, $\mathbf{v} \bullet \mathbf{w} = \mathbf{w} \bullet \mathbf{v}$. From the definition of Cartesian inner products in component form, one may prove the following properties.

Inner Product Properties:
Commutative Property: $\mathbf{u} \bullet \mathbf{v} = \mathbf{v} \bullet \mathbf{u}$
Associative Property: $\mathbf{u} \bullet (\mathbf{v} \bullet \mathbf{w}) = (\mathbf{u} \bullet \mathbf{v}) \bullet \mathbf{w}$
Distributive Property: $\mathbf{u} \bullet (\mathbf{v} + \mathbf{w}) = (\mathbf{u} \bullet \mathbf{v}) + (\mathbf{u} \bullet \mathbf{w})$

The inner product provides a simple guide to whether vectors are perpendicular to each other. If $\mathbf{v} \bullet \mathbf{w} = 0$ ($\mathbf{v}$ and $\mathbf{w} \neq 0$), then $\cos \theta = 0$ and $\theta = 90°$. As an example, if $\mathbf{u}$ represents the velocity of a fluid and **A** is an area through which the fluid can flow with the vector **A** representing the magnitude of the area and its orientation (with **A** defined as perpendicular to the area's surface), then $\mathbf{u} \cdot \mathbf{A}$ is the volumetric flow rate, say in meter^3/second.

Let us now look at vectors as a means of describing lines and planes in three dimensional space. Suppose we want to describe a line in the direction of the vector $\mathbf{v} = (a, b, c)$ with the point (x_0, y_0, z_0) on the line. Any other point, (x, y, z) will form a line with (x_0, y_0, z_0) and the direction of that line is the vector $\mathbf{w} = (x - x_0, y - y_0, z - z_0)$. The vector **w** is in the direction of **v** if $\mathbf{w} = s\mathbf{v}$ for some constant s. Equating components of the vectors **v** and **w**, we have:

$$x - x_0 = s \cdot a, \; y - y_0 = s \cdot b, \text{ and } z - z_0 = s \cdot c$$

or solving for s,

$(x - x_0)/ a = (y - y_0)/ b = (z - z_0)/ c = s$, and we have the equation for the desired line.

Now suppose instead, we wish to describe a plane perpendicular to $\mathbf{v} = (a, b, c)$ and containing the point x_0, y_0, z_0. The vector, $\mathbf{w} = (x - x_0, y - y_0, z - z_0)$ forms a family of vectors describing all possible directions. We now select those points perpendicular to v by requiring $\mathbf{v} \cdot \mathbf{w} = 0 = a$ $(x - x_0) + b (y - y_0) + c (z - z_0)$. Collecting all constant terms this may also be written as:

$$ax + by + cz + d = 0, \tag{5-10}$$

which is the general equation of a plane perpendicular to the direction $\mathbf{v} = (a, b, c)$.

Another notation for Cartesian vectors uses unit vectors aligned with each of the Cartesian axes. Traditionally, **i**, **j**, and **k** are the unit vectors associated with the x, y, and z axes. As unit vectors, $|\mathbf{i}| = |\mathbf{j}| = |\mathbf{k}| = 1$. In this notation, $\mathbf{v} = (a, b, c) = a\,\mathbf{i} + b\,\mathbf{j} + c\,\mathbf{k}$.

We will now introduce the vector cross product in terms of the unit vectors. The cross product operation between two vectors produces a vector that is perpendicular to the plane defined by the two vectors. In the case of the unit vectors:

$\mathbf{i} \times \mathbf{j} = \mathbf{k}$, $\mathbf{j} \times \mathbf{k} = \mathbf{i}$, and $\mathbf{k} \times \mathbf{i} = \mathbf{j}$. Similarly,
$\mathbf{j} \times \mathbf{i} = -\mathbf{k}$, $\mathbf{k} \times \mathbf{j} = -\mathbf{i}$, and $\mathbf{i} \times \mathbf{k} = -\mathbf{j}$.
Also, for parallel vectors,
$\mathbf{i} \times \mathbf{i} = \mathbf{j} \times \mathbf{j} = \mathbf{k} \times \mathbf{k} = 0$.

Using these relations and the distributive property of the cross product, the following theorem follows for any two Cartesian vectors. Let $\mathbf{w} = w_1\mathbf{i} + w_2\mathbf{j} + w_3\mathbf{k} = (w_1, w_2, w_3)$ and $\mathbf{v} = v_1\mathbf{i} + v_2\mathbf{j} + v_3\mathbf{k} = (v_1, v_2, v_3)$. then:

Theorem of the vector (cross) product for Cartesian vectors: For Cartesian vectors, $\mathbf{w}$ and $\mathbf{v}$,

$$\mathbf{w} \times \mathbf{v} = (w_2v_3 - w_3v_2)\mathbf{i} + (w_3v_1 - w_1v_3)\mathbf{j} + (w_1v_2 - w_2v_1)\mathbf{k}. \quad (5\text{-}11)$$
$$= ((w_2v_3 - w_3v_2), (w_3v_1 - w_1v_3), (w_1v_2 - w_2v_1)).$$

A geometric interpretation of the cross product may be illustrated with the aid of figure 5-5(p. 154). The area of the parallelogram formed from the vectors $\mathbf{w}$ and $\mathbf{v}$ may be calculated from the components of these vectors with $\mathbf{w} = (c, d) = (w_1, w_2)$ and $\mathbf{v} = (a, b) = v_1, v_2$. (Here we take the vector $\mathbf{w}$ as initiated from the origin as shown with a dashed line.) The area of the parallelogram A can be calculated as the rectangle $(a + c)(b + d)$ minus the areas of the triangles and trapezoids outside of the parallelogram formed by the axes. The result is $A = cb - da = w_1v_2 - w_2v_1$. [lxvii]

Comparing this result with the definition of the vector product of Cartesian vectors equation (5-11), $\mathbf{A} = \mathbf{w} \times \mathbf{v}$, and $\mathbf{A}$ is a vector in the direction of $\mathbf{k}$ (that is perpendicular to the plane of the parallelogram). This result in two dimensions is an example of the general result in three dimensions, leading to the definition below that does not depend on a specific coordinate system.

[lxvii] $A = (a + c)(b + d) - \frac{1}{2}(cd) - \frac{1}{2}(ab) - \frac{1}{2}(b + 2d)a - \frac{1}{2}(2a + c)d = cb - da = w_1v_2 - w_2v_1$.

Definition vector product: $\mathbf{v} \times \mathbf{w} = \mathbf{n} \, |\mathbf{v}| \, |\mathbf{w}| \sin\theta$ with $\mathbf{n}$ a vector (called the normal) perpendicular to the plane formed by the vectors $\mathbf{v}$ and $\mathbf{w}$, and θ the angle between the vectors.

There are two possibilities for the direction of the normal $\mathbf{n}$ to the plane formed by $\mathbf{v}$ and $\mathbf{w}$. By convention, the normal is oriented as in the following figure. If your palm and fingers of your right hand are aligned with the direction of $\mathbf{v}$ and then rotated counterclockwise to the position of $\mathbf{w}$, your thumb will point in the direction of $\mathbf{v} \times \mathbf{w}$. This is called the right hand rule.[lxviii] In the figure below the axes are shown consistent with the right-hand rule, and the relations given previously for unit vectors.

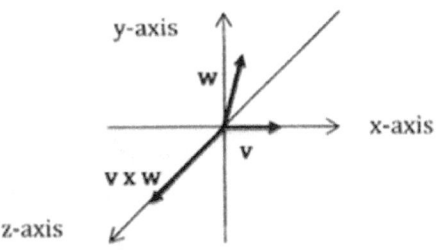

5-7 Vector cross product

The cross product will be very important in the determination of tangent planes to curved surfaces in non-Euclidean geometries. The following properties of the cross product may be proved using the vector representation with unit vectors.

Vector Cross Product Properties:
Anti-Commutative Property: $\mathbf{u} \times \mathbf{v} = -\mathbf{v} \times \mathbf{u}$
Distributive Properties: $\mathbf{u} \times (\mathbf{v} + \mathbf{w}) = \mathbf{u} \times \mathbf{v} + \mathbf{v} \times \mathbf{w}$
$$(\mathbf{u} + \mathbf{v}) \times \mathbf{w} = \mathbf{u} \times \mathbf{w} + \mathbf{v} \times \mathbf{w}$$

The associative property for the vector product is not generally valid. For example:

[lxviii] Another popular way to determine the direction of the normal $\mathbf{n}$ is by noting that it is the direction a normally threaded screw would move if turned in the same direction as the rotation of $\mathbf{v}$ towards $\mathbf{w}$.

$(i \times i) \times k = 0 \times k = 0 \neq i \times (i \times k) = i \times (-j) = -k$. However, there is an associative property with scalars:

Scalar Associative Property: $a(u \times v) = (au) \times v = u \times (av)$

Using the definitions of scalar and vector products along with the unit vectors, you can show that the volume, V of a parallelepiped formed by vectors **a**, **b**, and **c**, seen in the figure below, can be calculated as: $V = a \cdot (b \times c)$. This is known as the triple scalar product. That the triple product is equal to the volume V may be seen by noting that **b x c** is a vector with magnitude equal to the area A of the parallelogram and direction **n** normal to the area that is the base of the parallelepiped (**b x c** = An). Thus, the height h of the parallelepiped normal to the base is **a • n**. Therefore, $a \cdot (b \times c) = a \cdot An = Ah = V$. If $a \cdot (b \times c) = 0$, Then the vectors **a**, **b**, and **c** are all in the same plane.

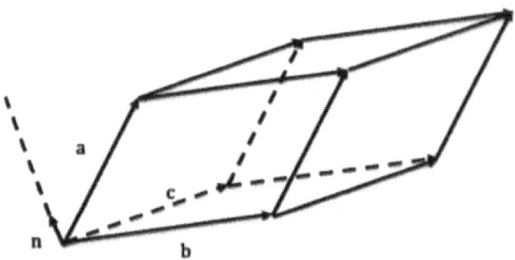

5-8 Volume of a parallelepiped

Vectors provide a concise means of describing curves, surfaces, and their characteristics in space. For example if **r** is a position vector in three-dimensional space, then a curve may be specified by the parameter t as $r = (x(t), y(t), z(t))$. In addition to the position vector, the tangent and normal vectors to surfaces and their variation, as described by the calculus, will be vital to describing surface curvature.

With our review in the previous chapter of analytic geometry and the real and complex numbers, and in this chapter, the calculus and vectors, we are now ready to develop a mathematical description of surface curvature. This description will be the key to understanding non-Euclidean geometries - a discovery that changed the meaning of mathematics.

6 Gauss Reveals Curvature as the Heart of Geometry

A fundamental geometric concept that provides a unified understanding for the differences between the non-Euclidean geometry discovered by Lobachevsky and Bolyai and that of Euclid is revealed in Gauss's groundbreaking *General Investigations of Curved Surfaces*, a dense work published in 1827. [lxix] Perhaps for the reasons discussed in Section 3.1 *Gauss' insight*, Gauss did not mention the direct relationship of his new geometric discovery to non-Euclidean geometry.[178] In this work, Gauss introduced new definitions of the curvature of surfaces, one local, that is defined at a point on the surface, now known as Gaussian curvature and the other, the integral of the Gaussian curvature over an area of the surface. These new methods of quantifying surface curvature explicitly connect surface curvature to the differences between the interior angle sums of Euclidean triangles and those discovered in the triangles of Lobachevsky and Bolyai (or for that matter Saccheri and Lambert). The concept of surface curvature, which will be developed in the following sections, was not entirely new with Gauss, Euler had introduced a definition that was completely compatible with that of Gauss, but not as broadly revelatory as to its significance.

[lxix] Gauss's *General Investigations* is available with a related work and notes on the historical background in a Dover Edition (2005), reprinted (with a new introduction and notes by Peter Pesic) from the translation published in 1902 by the Princeton University Library. Unlike the other references given, I have cited this work primarily for its historic significance rather than as a text recommended for additional mathematical details.

161

6.1 Curves on the Euclidean Plane

In order to develop an understanding of Gauss's curvature concept, we will begin with simple cases of curves in a plane and characterize them in terms of vectors that are tangent to the curve and the changes to these tangent vectors along the path of the curve. We will find that it is helpful to designate the point on the curve in question by the path length s measured from an assigned point. The tangent vector at a point on a curve is a vector with the same direction as the tangent line, as illustrated by the example of the tangent line shown in figure 5-2 (p.136). Let us start with a very simple ``curve'', the straight line in the Euclidean plane, $y = mx$ with m being the constant slope. Many of the subsequent techniques that will be used can be illustrated in this simple case.

At any point on the line $y = mx$, we can associate the Cartesian position vector $\mathbf{r} = (x, y) = (x, mx)$. As the slope at any point is $\frac{dy}{dx} = \frac{m}{1}$, the tangent vector is $\mathbf{t} = (1, m)$. The unit tangent vector is designated as:

$$\mathfrak{t} = \frac{\mathbf{t}}{|\mathbf{t}|}$$

$$\mathfrak{t} = \frac{\mathbf{t}}{\sqrt{\mathbf{t} \cdot \mathbf{t}}}.$$

$$\mathfrak{t} \cdot \mathfrak{t} = \frac{\mathbf{t}}{\sqrt{\mathbf{t} \cdot \mathbf{t}}} \bullet \frac{\mathbf{t}}{\sqrt{\mathbf{t} \cdot \mathbf{t}}} = 1$$

For the straight line with $\mathbf{t} = (1, m)$, $|\mathbf{t}| = \sqrt{\mathbf{t} \cdot \mathbf{t}} = \sqrt{1 + m^2}$, and therefore,
$$\mathfrak{t} = \frac{(1,m)}{\sqrt{1+m^2}}.$$

The position vector $\mathbf{r}$ for the straight line can be considered to be a curve parameterized by the coordinate x; however, a path length s from the origin could also be used as the parameter. By considering the right triangle formed by $y = mx$ and the x-axis with sides of length x and mx, the hypotenuse $s = \sqrt{1 + m^2}x$; however, the path length may be

determined in a more general approach by integration as in Section 5.3
Summing up along the path: integral calculus.

$$s = \int_0^x \sqrt{1 + \left(\frac{dy}{dx}\right)^2}\, dx,$$

$$= \int_0^x \sqrt{1 + m^2}\, dx,$$

$$= \sqrt{1 + m^2}\, x.$$

Therefore, $x = s/\sqrt{1 + m^2}$, and $\mathbf{r} = (\frac{s}{\sqrt{1+m^2}}, \frac{ms}{\sqrt{1+m^2}}) = \frac{1}{\sqrt{1+m^2}}(s, ms)$.

Comparing $\mathbf{r}(s) = \frac{(s, sm)}{\sqrt{1+m^2}}$ and $\mathfrak{t} = \frac{(1, m)}{\sqrt{1+m^2}}$ may suggest to you the following important relation ship:

$$\frac{d\mathbf{r}}{ds} = \frac{1}{\sqrt{1 + m^2}}\left(\frac{d}{ds}s, \frac{d}{ds}ms\right) = \frac{(1, m)}{\sqrt{1 + m^2}} = \mathfrak{t}.$$

This example illustrates two important points: the derivative of a Cartesian vector is simply the derivative of its Cartesian components,[lxx] and $\frac{d\mathbf{r}}{ds} = \mathfrak{t}$. We have not proved this, but it is true in all circumstances.[179] We can also confirm from a different point of view that $\mathfrak{t} = \frac{d\mathbf{r}}{ds} = \frac{d}{ds}(x, y, z)$ is a unit vector.

$$\mathfrak{t} \cdot \mathfrak{t} = \frac{d\mathbf{r}}{ds} \cdot \frac{d\mathbf{r}}{ds} = \frac{(dx, dy, dz)}{ds} \cdot \frac{(dx, dy, dz)}{ds} = \frac{dx^2 + dy^2 + dz^2}{ds^2} = 1,$$

or more simply, $ds^2 = d\mathbf{r} \cdot d\mathbf{r}$. (6-1)

$$\text{Therefore, } \frac{d\mathbf{r}}{ds} \cdot \frac{d\mathbf{r}}{ds} = 1.$$

Now one might wonder what happens if I take the derivative of $\mathbf{r}$ with respect to another parameter such as x. Then,

---

[lxx]   In other coordinate systems in which the orientation of the axes changes with position, this must also be taken into account when taking the derivative. This will be important in Chapter 8.

$$\frac{dr}{dx} = \frac{d}{dx}(x, mx) = (1, m).$$

The result is still a tangent $\mathbf{t}(x)$[lxxi] to the curve, but it is not a unit vector. This is clear as it may be written as $t(x) = \sqrt{1 + m^2}\,\mathfrak{t}.$.

The unit tangent vector may also be obtained in general from a tangent vector based on a parameter different from the path length s using the chain rule of differentiation (see footnote lxi, p. 135).

So far we have introduced a number of concepts and techniques that will be very useful in describing curvature; however, as the tangent vector in our straight line example is constant, it is not of much further use. Our next step, therefore, is to apply these concepts to a curve that clearly exhibits the effects of a continuously changing tangent vector, but is familiar enough that we can anticipate what the results of our investigations will be, that is a circle.

Let us use a circle of radius R with its center at the origin as our example. The equation $x^2 + y^2 = R^2$ describes any point on the circle taking into account that the square root gives positive and negative results. In this description, the curve can be parameterized by the Cartesian coordinate x. We can, however, easily switch to the path length s as the parameter by noting that $s = R \cdot \theta$ with θ being the angle between the position vector $\mathbf{r}$ and the x-axis (see figure 4-2, and following discussion, p. 94). The angle θ may be substituted with $x = R\cos\theta$ $y = R\sin\theta$. With this change of variable $\mathbf{r} = (R\cos\theta, R\sin\theta)$. Taking the derivative with respect to θ and then using the chain rule to calculate the derivative with respect to path length s.

$$\mathbf{r} = R(\cos\theta, \sin\theta).$$

$$\frac{d\mathbf{r}}{ds} = \frac{d\mathbf{r}}{d\theta} \cdot \frac{d\theta}{ds}.$$

[lxxi] The tangent vector in this case is a constant independent of x. I have written it as $\mathbf{t}(x)$ simply to call attention to its derivation from the parameter x.

$$\frac{dr}{d\theta} = R\frac{d}{d\theta}(\cos\theta, \sin\theta).$$

$$= R(-\sin\theta, \cos\theta).$$

Since $\theta = \frac{s}{R}, \frac{d\theta}{ds} = \frac{1}{R}$, therefore,

$$\frac{dr}{ds} = R(-\sin\theta, \cos\theta)/R, \text{ and}$$

$$\mathfrak{t} = (-\sin\theta, \cos\theta).$$

As expected the position vector is perpendicular to the tangent vector:

$$\mathbf{r} \cdot \mathfrak{t} = R(\cos\theta, \sin\theta) \cdot (-\sin\theta, \cos\theta).$$
$$= R(-\cos\theta \sin\theta + \sin\theta \cos\theta) = 0.$$

So far we have done nothing different with the circle than we did with the straight line; however, in the case of the circle, the tangent vector is a function of position. The curvature vector **k** is defined from the change in the tangent's direction with path length, that is,

$$\mathbf{k} = \frac{d\mathfrak{t}}{ds}. \tag{6-2}$$

In the case of the circle:

$$\mathbf{k} = \frac{d\mathfrak{t}}{ds}.$$

$$= \frac{d\mathfrak{t}}{d\theta} \cdot \frac{d\theta}{ds}$$

$$= \frac{d(-\sin\theta, \cos\theta)}{d\theta} \cdot \frac{1}{R}$$

$$= (-\cos\theta, -\sin\theta)/R$$

$$= -\frac{\mathbf{r}}{R^2}, \text{ as } \mathbf{r} = R(\cos\theta, \sin\theta).$$

The curvature vector $k = -\frac{r}{R^2}$ is in the opposite direction to the position vector $\mathbf{r}$, and is therefore perpendicular to the tangent vector $\mathfrak{t}$. This is true for all curves as shown below.[lxxii]

$$\mathfrak{t} \bullet \mathfrak{t} = 1$$

$$\frac{d}{ds}(\mathfrak{t} \bullet \mathfrak{t}) = \frac{d\mathfrak{t}}{ds} \bullet \mathfrak{t} + \mathfrak{t} \bullet \frac{d\mathfrak{t}}{ds} = 0.$$

$$2\mathfrak{t} \bullet \frac{d\mathfrak{t}}{ds} = 0,$$

$$\mathfrak{t} \bullet \frac{d\mathfrak{t}}{ds} = \mathfrak{t} \bullet \mathbf{k} = 0.$$

The curvature vector is not necessarily a unit vector. The magnitude of $\mathbf{k}$ is called the curvature, $\kappa = |\mathbf{k}|$ and in the case of the circle equals 1/R. A unit curvature vector $\mathbf{N}$ may be defined as:

$$\mathbf{k} = \kappa\mathbf{N}.$$

The radius of curvature ρ is defined as:

$$\rho = 1/\kappa$$

Consistent with intuitive expectations for the circle, $\rho = R$.

Similar analyses maybe made with other curves in the Euclidean plane; however, it is time to use and extend these concepts to curves on more general surfaces.

6.2 Surfaces in space

Surfaces may be represented in a number of ways, but in contrast to curves described with a single parameter, surfaces require two parameters. The familiar map of the earth with its two coordinates

---

[lxxii] The derivative of $\mathfrak{t} \bullet \mathfrak{t}$ with respect to s may be formed from the definition of the derivative by noting that $\mathfrak{t}(s + \Delta s) = \mathfrak{t} + \Delta\mathfrak{t}$.

of latitude and longitude is a good example. The position vector for a surface, with two parameters u and v, is expressed in three-dimensional Euclidean space with coordinates (x_1, x_2, x_3) as $\mathbf{r}(u, v) = (x_1(u, v), x_2(u, v), x_3(u, v))$, again with each component given as a function. Our intuitive notion of a smooth continuous surface is only described in this way if certain conditions are met by their derivatives.[lxxiii] We will assume that surfaces discussed here meet these conditions. Other approaches describing a surface include the use of functions, for example, $f(x_1, x_2, x_3) = 0$, or $x_3 = g(x_1, x_2)$. For example, the surface of a sphere of radius R may be described in Euclidean three-dimensional space as:

$$x_1^2 + x_1^2 + x_1^2 - R^2 = 0,$$

or

$$x_3 = \sqrt{R^2 - (x_1^2 + x_2^2)}.$$

In what follows, the vector approach is used because of its conciseness in expressing surface concepts. We begin with the familiar surface of a sphere. Spherical coordinates ϕ and θ are introduced in the figure below to facilitate the description.[lxxiv]

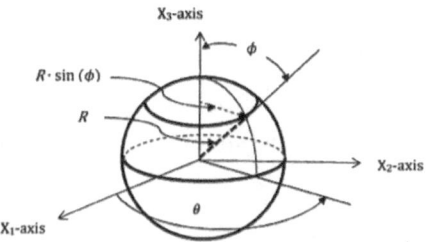

6-1 Spherical coordinates

The angle ϕ called the colatitude, is equal to 90°− the angle of latitude. For example, Boston is approximately at latitude 42°, therefore its colatitude $\phi = 48°$. A line of constant ϕ is shown in the figure which would be represented as a straight horizontal line on the familiar

---

[lxxiii] A discussion of these conditions is given by Struik (pp. 55-58).

[lxxiv] Other authors, particularly in physics, reverse the meaning of the notation for θ and ϕ.

Mercator projection map of the world. The line forms a circle of radius Rsin ϕ (assuming a perfectly spherical earth) on the surface called a small circle. As will be discussed in Chapter 8, this is not the path of shortest distance between two points on this line. For example, Rome is approximately at the same latitude as Boston, but the shortest path between the two cities, a great circle route, is shorter by about 200 miles than the path along the line of constant latitude. The familiar coordinate of longitude is represented by the angle θ, shown in the equatorial plane. Any point on the surface, such as the point at the end of the radius vector shown with a heavy dashed line, may be located by the coordinates θ and ϕ. The projected length of radius vector onto the equatorial plane is Rsin ϕ. With this understanding, the Cartesian coordinates may be expressed as functions of ϕ and θ. The position at the surface $\mathbf{r}(\theta, \phi)$ in Euclidean space has three components which are designated as x_1, x_2, x_3 that is:

$$\mathbf{r} = (x_1, x_2, x_3) = R(\sin \phi \cos \theta, \sin \phi \sin \theta, \cos \phi).$$

Note that $|r| = \sqrt{\mathbf{r} \cdot \mathbf{r}} = R$.

Curves on the surface of the sphere may described with a single parameter just as in the examples on the Euclidean plane in the previous section. For example, if $\theta = \theta_0$ a constant, then $\mathbf{r} = \mathbf{r}(\phi) = R(\sin \phi \cos \theta_0, \sin \phi \sin \theta_0 \cos \phi)$ which describes a great circle. Similarly with $\phi = \phi_0 =$ a constant, $\mathbf{r} = \mathbf{r}(\theta) = R(\sin \phi_0 \cos \theta, \sin \phi_0 \sin\theta, \cos\phi_0)$ which describes a small circle.

More generally, if the surface $\mathbf{r} = \mathbf{r}(u, v)$, then a curve on the surface may be given with the single parameter t as $u = u(t)$, $v = v(t)$. This reduces the number of parameters on the surface to one by making one of the parameters a function of the other, e.g., $v = v(u)$. The direction of the tangent to the curve at a point on the surface is determined by dv/du just as the derivative dy/dx determines the tangent of a curve on the Euclidean plane. A curve on the surface of the sphere would be determined by the relationship, $\phi = \phi(\theta)$.

The path length of a curve is the same as that given in equation (6-1, p.163):

$$ds^2 = \mathbf{dr} \cdot \mathbf{dr}$$

The differential of $\mathbf{r} = \mathbf{r}(u, v)$, analogous to that of a function (equation (5-2, p. 138), is expressed as,

$$\mathbf{dr} = (\partial \mathbf{r}/\partial u)\, du + (\partial \mathbf{r}/\partial v)\, dv$$

$$\partial \mathbf{r}/\partial u\, du = \partial(x_1, x_2, x_3)\,/\partial u\, du$$

$$= \left(\frac{\partial x_1}{\partial u}, \frac{\partial x_2}{\partial u}, \frac{\partial x_3}{\partial u}\right) du$$

$$\partial \mathbf{r}/\partial v\, dv = \partial(x_1, x_2, x_3)\,/\partial v\, dv$$

$$= \left(\frac{\partial x_1}{\partial v}, \frac{\partial x_2}{\partial v}, \frac{\partial x_3}{\partial v}\right) dv$$

Let us now apply this to the surface of a sphere of radius R.

Equations for $\mathbf{r}_\theta$, $\mathbf{r}_\varphi$ for a sphere (6-3)

$$\mathbf{r} = R(\sin \phi \cos \theta, \sin \phi \sin \theta, \cos \phi).$$

$$\mathbf{dr} = (\partial \mathbf{r}/\partial \theta)\, d\theta + (\partial \mathbf{r}/\partial \phi)\, d\phi$$

$$\frac{\partial \mathbf{r}}{\partial \theta} = \frac{R\,\partial(\sin \phi \cos \theta, \sin \phi \sin \theta, \cos \phi)}{\partial \theta}.$$

$$= R(-\sin \phi \sin \theta, \sin \phi \cos \vartheta, 0).$$

$$\frac{\partial \mathbf{r}}{\partial \phi} = \frac{R\,\partial(\sin \phi \cos \theta, \sin \phi \sin \theta, \cos \phi)}{\partial \phi}.$$

$$= R(\cos \phi \cos \theta, \cos \phi \sin \theta, -\sin \phi).$$

Therefore, recalling that $\sin^2\theta + \cos^2\theta = \sin^2\phi + \cos^2\phi = 1$, collecting terms, and simplifying:

$$ds^2 = \mathbf{dr} \cdot \mathbf{dr} = R^2(\sin^2\phi\, d\theta^2 + d\phi^2) \qquad (6\text{-}4)$$

This result should not be surprising as with a little reflection an element of path length Δs at the position (θ, ϕ), similar to figure 5-4 (p. 150), can be approximated as the hypotenuse of a triangle with one side along a line of latitude equal to $R\sin\phi\,\Delta\theta$ and the other side the meridian of length $R\Delta\phi$, or $\Delta s^2 = R^2\sin^2\phi\Delta\theta^2 + R^2\Delta\phi^2$.

If a curve on the surface of the sphere is specified by a function of $\theta = \theta(\phi)$ with derivative $\frac{d\theta}{d\phi}$, then analogous to equation (5-8), p. 152), the length of the curve from ϕ_0 to ϕ_1 may be determined from the integral:

$$s_{\phi_0}^{\phi_1} = \int_{\phi_0}^{\phi_1} R\sqrt{1 + \left(\sin\phi\frac{d\theta}{d\phi}\right)^2}\,d\phi.$$

With the specific example given above, it is now time to look at more general implications of the surface description $\mathbf{r} = \mathbf{r}(u, v)$. If the general variable v is replaced with a series of constants $v_0, v_1, v_2, v_3\cdots$, a family of curves such as $r = r(u, v_0)$ are established. Similarly, replacing the variable u with $u_0, u_1, u_2, u_3\cdots$, forms another family of curves. These curves on the surface $\mathbf{r} = \mathbf{r}(u, v)$ replace the grid that we associate with a Cartesian coordinate system or the system of meridians and latitudes on a global map. These are shown schematically in the following figure.

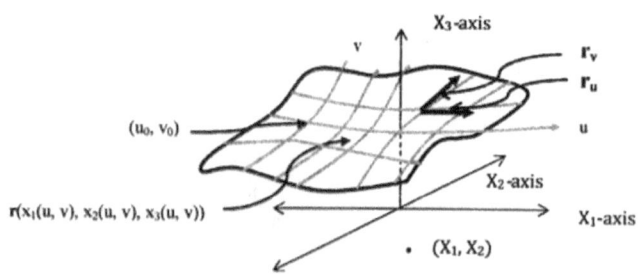

6-2 Surface $r(x_1(u, v), x_2(u, v), x_3(u, v))$

The surface is shown within a Euclidean space and a Cartesian coordinate system (x_1, x_2, x_3). The partial derivative with respect to u, $\partial r/\partial u$, at any point (u, v) on the surface is the tangent to a curve with v constant. Similarly, the partial derivative with respect to v, $\partial r/\partial v$ at a point (u, v) on the surface is the tangent to a curve with u constant.

For simplicity, we introduce the vector notation $\mathbf{r}_u = \partial \mathbf{r}/\partial u$ and $\mathbf{r}_v = \partial \mathbf{r}/\partial v$. The vectors $\mathbf{r}_u$ and $\mathbf{r}_v$ are independent vectors of the surface at the point (u, v). In this notation $d\mathbf{r} = \mathbf{r}_u du + \mathbf{r}_v dv$. They also provide an expression for the equation of the tangent plane $\mathbf{T}_p$ at that point and the normal vector to the plane $\mathbf{N}_p$:

$$\mathbf{T}_p(u, v) = \mathbf{r}(u, v) + a\mathbf{r}_u + b\mathbf{r}_v, \tag{6-5}$$

$$\mathbf{N}_p = \mathbf{r}_u \times \mathbf{r}_v {}^{\text{lxxv}} \tag{6-6}$$

We now calculate ds^2 using the new vector notation for the partial derivatives.

$$ds^2 = d\mathbf{r} \cdot d\mathbf{r} = (\mathbf{r}_u du + \mathbf{r}_v dv) \cdot (\mathbf{r}_u du + \mathbf{r}_v dv). \tag{6-7}$$

$$ds^2 = Edu^2 + 2Fdudv + Gdv^2 \tag{6-8}$$

In equation (6-8) above, $E = \mathbf{r}_u \cdot \mathbf{r}_u$, $F = \mathbf{r}_u \cdot \mathbf{r}_v$, $G = \mathbf{r}_v \cdot \mathbf{r}_v$. This follows the notation first used by Gauss (see footnote lxix (p.161)[180]. Equation (6-8) is known as the First Form of the surface. If the coordinates u and v are orthogonal, that is $\mathbf{r}_u \cdot \mathbf{r}_v = 0$ then F = 0, and $ds^2 = Edu^2 + Gdv^2$. In our example, a spherical surface of radius R with $ds^2 = R^2(\sin^2\phi d\theta^2 + d\phi^2)$, $E = R^2\sin^2\phi$, F=0, and G=R².

If a curve on the surface is specified by the single parameter t such that u = u(t), v = v(t), then the path length from t_0 to t_1 is:

$$s_{t_0}^{t_1} = \int_{t_0}^{t_1} \sqrt{E\left(\frac{du}{dt}\right)^2 + 2F\frac{du}{dt}\frac{dv}{dt} + G\left(\frac{dv}{dt}\right)^2} \, dt. \tag{6-9}$$

The coordinate vectors $\mathbf{r}_u$ and $\mathbf{r}_v$ may also be used to determine the normal $\mathbf{N}_p$ (perpendicular vector) to the surface using the vector cross product as in equation (5-11, p. 158). In Cartesian coordinates, let $\mathbf{r}_u = (a_1, a_2, a_3)$ and $\mathbf{r}_v = (b_1, b_2, b_3)$.

---

lxxv Note that $\mathbf{r}_u \cdot (\mathbf{r}_u \times \mathbf{r}_v) = \mathbf{r}_v \cdot (\mathbf{r}_u \times \mathbf{r}_v) = 0$; see discussion associated with Figure 5-8 (p. 160).

$$N_p = r_u \times r_v = (x_1, x_2, x_3) = (a_2 b_3 - a_3 b_2, a_3 b_1 - a_1 b_3, a_1 b_2 - a_2 b_1).$$

Applying this to the case of a spherical surface of radius R, from equations (6-3), p. 169),

$$
\begin{aligned}
N_p &= r_\theta \times r_\phi. \\
&= (R(-\sin\phi \sin\theta, \sin\phi \cos\theta, 0) \times R(\cos\phi \cos\theta, \\
&\quad \cos\phi \sin\theta, -\sin\phi)). \\
&= (x_1, x_2, x_3). \\
x_1 &= -R^2 \sin^2\phi \cos\theta - 0. \\
x_2 &= 0 - R^2 \sin^2\phi \sin\theta. \\
x_3 &= -R^2 \sin\phi \cos\phi. \\
&= R^2(-\sin\phi \cos\phi \sin^2\theta - \sin\phi \cos\phi \cos^2\theta) \\
N_p &= -R\sin\phi \, (R\sin\phi \cos\theta, R\sin\phi \sin\theta, R\cos\phi). \\
&= -R\sin\phi \, r.
\end{aligned}
$$

A normal unit vector **N** may be calculated in the usual manner as:

$$N = \frac{N_p}{|N_p|},$$

$$= \frac{-R \sin\phi \, r}{R^2 \sin\phi},$$

$$= \frac{-r}{R}$$

The calculation just completed is consistent with our intuition that the normal to the sphere surface is in the radial direction **r** with the negative sign indicating the direction of concavity. In general, a unit normal can be calculated as,

$$N = \frac{N_p}{|N_p|} = \frac{N_p}{|r_u \times r_v|}.$$

The mathematical machinery that we have developed is directed towards the goal of describing geometric figures on curved surfaces. In the next section, the vector analysis that we will continue to develop will provide insight into the differences between the geometries of Euclid, Lobachevsky, Bolyai, and Lambert. Of particular importance in what follows is the relationship of $\mathbf{r}_u \times \mathbf{r}_v$ to the area of geometric figures on general curved surfaces. Towards that end, I offer with your patience, a vector identity that is useful here and in more general cases to come.[181] For vectors **a** and **b**:

$$(\mathbf{a} \times \mathbf{b}) \cdot (\mathbf{a} \times \mathbf{b}) = |\mathbf{a} \times \mathbf{b}|^2 = (\mathbf{a} \cdot \mathbf{a})(\mathbf{b} \cdot \mathbf{b}) - (\mathbf{a} \cdot \mathbf{b})^2.$$

If $\mathbf{a} = \mathbf{r}_u$, and $\mathbf{b} = \mathbf{r}_v$, then,

$$|\mathbf{r}_u \times \mathbf{r}_v|^2 = (\mathbf{r}_u \cdot \mathbf{r}_u)(\mathbf{r}_v \cdot \mathbf{r}_v) - (\mathbf{r}_u \cdot \mathbf{r}_v)^2.$$

Recalling the notation for Gauss' First Form in equation (6-8, p. 171),

$$|\mathbf{r}_u \times \mathbf{r}_v| = \sqrt{EG - F^2}. \tag{6-10}$$

Note, also that equation (6-10) is consistent with $|\mathbf{r}_u \times \mathbf{r}_v| = (\mathbf{r}_u \times \mathbf{r}_v) \cdot \mathbf{N}$.

For a spherical surface with $u = \theta$, $v = \phi$, $\mathbf{r}_\theta \cdot \mathbf{r}_\theta = E$, $\mathbf{r}_\theta \cdot \mathbf{r}_\phi = F = 0$, and $\mathbf{r}_\phi \cdot \mathbf{r}_\phi = G$ (equation (6-4, p. 169). Therefore, $|\mathbf{r}_\theta \times \mathbf{r}_\phi| = \sqrt{EG} = R^2 \sin\phi$.

Equation (6-10) has particularly important application in regard to the determination of areas enclosed by curves on surfaces. A differential area on the surface of $\mathbf{r}(u, v)$ may be expressed as,

$$dA = |\mathbf{r}_u \times \mathbf{r}_v| du\, dv = \sqrt{EG - F^2}\, du\, dv. \tag{6-11}$$

This may be explained informally noting that the surface area may be considered to be covered by parallelograms of sides $\mathbf{r}_u \delta u$ and $\mathbf{r}_v \delta v$ whose vector cross product is the area δA (see definition of vector cross product and following geometric interpretation – equation (5-11, p.158).[182] Summing over the parallelograms in the region of interest $\mathcal{A}$

and taking the limit as they approach 0 leads to a double integral over u and v.

$$A = \iint \sqrt{EG - F^2} \, du \, dv. \qquad (6\text{-}12)$$

The following example of double integration is given as an aid to understanding equation (6-12) as the limit of sums over Δy and Δx (see the following figure). In Cartesian coordinates x, y, double integration is used to determine the area A from x = 0 to 1 between the line y = x and the parabola y = x^2. $A = \int_0^1 \int_{y=x^2}^{y=x} dy \, dx = \int_0^1 y\vert_{y=x^2}^{y=x} \, dx = \int_0^1 (x - x^2) \, dx = \left(\frac{x^2}{2} - \frac{x^3}{3} \right)\vert_0^1 = (\frac{1}{6} - 0) = \frac{1}{6}$. The same result would be obtained from $\int_0^1 \int_{x=\sqrt{y}}^{x=y} dx \, dy$.

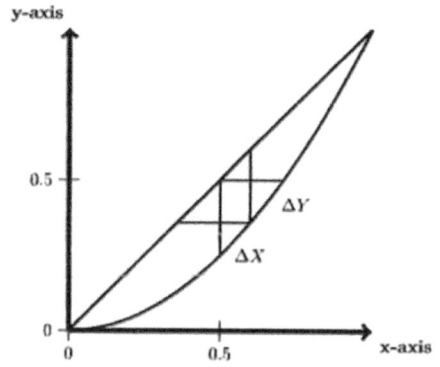

6-3 Double Integration

Recall that the areas of triangles on the surface of a sphere and in the geometry of Lobachevsky and Bolyai only depended upon the sum of the interior angles of the triangle. This should make you anticipate that a relationship for areas on general surfaces will lead to important insights into non-Euclidean geometry. Gauss developed such a relationship through his introduction of a new method to characterize surface curvature. This method, now known as Gaussian curvature, and its implications are the subject of the next section.

6.3 Gaussian curvature and the demystification of Non-Euclidean geometry

Previously, we developed the expression for the curvature of a curve in a plane, **k** [equation (6-2, p. 165). Now we must extend that concept to curves on general surfaces. The extension of curvature to surfaces discovered by Gauss will tie together the relationship between the areas of triangles and the sum of their interior angles that differentiates the geometries of Euclid, Lobachevsky (Bolyai), and Lambert (Table 3-1, p. 76).

On the general surface $\mathbf{r} = \mathbf{r}(u, v)$, curves are formed when one coordinate is a function of the other, $v = v(u)$ or equivalently when both coordinates are a function of a single parameter, $u = u(t)$ and $v = v(t)$ in which case $\mathbf{r} = \mathbf{r}(u(t))$. Just as in the case of curves on the Euclidean plane, the unit tangent at a point on the curve $\mathfrak{t} = \frac{d\mathbf{r}}{ds} = \frac{d\mathbf{r}}{dt} \cdot \frac{dt}{ds}$. We will now see how to apply this in three dimensional space to curves on a surface. The simple examples that will be used as illustrations are a great circle on a sphere formed by the constant meridian $\theta = \theta_0$ and a small circle with the constant colatitude, $\phi = \phi_0$.

The great circle $\mathbf{r} = R(\sin\phi\,\cos\theta_0, \sin\phi\,\sin\theta_0, \cos\phi)$ is just a circle of radius R centered at the origin in a plane rotated to the angle θ_0. The curvature vector **k** is therefore a radial vector of magnitude 1/R directed towards the origin (see discussion following equation (6-2, p. 165). The only difference is that **k** is a vector in three-dimensional space. However, the detailed calculation that is applicable to more complicated situations is shown below. First, a tangent vector **t** to the curve is determined, and then the curvature vector is determined from the derivative of the unit tangent vector.

$$
\begin{aligned}
\mathbf{r}(\theta_0,\phi) &= R(\sin\phi\,\cos\theta_0, \sin\phi\,\sin\theta_0, \cos\phi). \\
\mathbf{t}(\theta_0,\phi) &= d\mathbf{r}/d\phi. \\
&= R(\cos\phi\,\cos\theta_0, \cos\phi\,\sin\theta_0, -\sin\phi).
\end{aligned}
$$

Now, $\mathfrak{t} = \frac{dr}{ds} = \frac{dr}{d\phi} \cdot \frac{d\phi}{ds}$. On the surface of the sphere, $ds^2 = R^2(\sin^2\phi d\theta^2 + d\phi^2)$, equation (6-4, p. 169), so on the meridian, θ_0, ds=Rdϕ, or $\frac{d\phi}{ds} = 1/R$. Therefore, $\mathfrak{t} = (dr/d\phi) \cdot \frac{1}{R} = (\cos\phi\cos\theta_0, \cos\phi\sin\theta_0, -\sin\phi)$.[lxxvi]

The calculation for **k** continues in three-dimensional space.

$$
\begin{aligned}
\mathbf{k} &= d\mathfrak{t}/ds. \\
&= \frac{d\mathfrak{t}}{d\phi} \cdot \frac{d\phi}{ds}. \\
&= \frac{d}{Rd\phi}(\cos\phi\cos\theta_0, \cos\phi\sin\theta_0, -\sin\phi). \\
&= (-\sin\phi\cos\theta_0, -\sin\phi\sin\theta_0, -\cos\phi)/R. \\
\mathbf{k}(\phi, \theta_0) &= -\mathbf{r}(\phi, \theta_0)/R.
\end{aligned}
$$

At any point on the meridian (θ_0, ϕ), the curvature vector **k** is in the radial direction pointing towards the center of the sphere and perpendicular to the tangent plane $\mathbf{T}_p$ at that point - see equation (6-5, p.171).[lxxvii] The curvature vector can be expressed in terms of a unit vector, $\mathbf{k} = \kappa\,\mathbf{N}$. Therefore, $\mathbf{N} = (-\sin\phi\cos\theta_0, -\sin\phi\sin\theta_0, -, \cos\phi)$ and $\kappa = 1/R$. As expected, the radius of curvature $\rho = R$. Thus, the calculations in three dimensions follow those previously made for the circle in two dimensions.

A new concept, geodesic curvature, is introduced in analyzing the curvature of the small circle of constant colatitude ϕ_0. The same procedure to calculate **k** is followed as shown above. In the case of constant $\phi = \phi_0$, ds = (R$\sin\phi_0$)dθ.

$$
\mathbf{r}(\theta, \phi_0) = R(\sin\phi_0\cos\theta, \sin\phi_0\sin\theta, \cos\phi_0).
$$

$$
\mathfrak{t}(\theta, \phi_0) = \left(\frac{dr}{d\theta}\right) \cdot \left(\frac{d\theta}{ds}\right) = (-\sin\theta, \cos\theta, 0).
$$

$$
\mathbf{k}(\theta, \phi_0) = \left(\frac{d\mathfrak{t}}{d\theta}\right) \cdot \left(\frac{d\theta}{ds}\right) = (-\cos\theta, -\sin\theta, 0)/(R\sin\phi_0).
$$

[lxxvi] In the analyses that follow, radial vectors associated with spherical surfaces such as r given above and the normal vector T are frequently encountered. Recognizing their form in spherical coordinates will be helpful to understanding the implications of what follows.

[lxxvii] Calculations of the inner products will show that $\mathbf{k} \cdot \mathbf{r}_\theta = \mathbf{k} \cdot \mathbf{r}_\phi = 0$.

With the third component of the curvature vector equal to zero, the curvature vector is parallel to the plane of the x_1 and x_2 axes and perpendicular to the third axis - see figure 6-1 (p. 167). The curvature vector may be expressed as the sum of a normal curvature vector $\mathbf{k}_n$, in the direction of the unit normal $\mathbf{N}$ to the tangent plane $\mathbf{T}_p$ and a vector in the tangent plane.[lxxviii] The normal curvature vector $\mathbf{k}_n$ is therefore,

$$\mathbf{k}_n = (\mathbf{k} \cdot \mathbf{N}) \, \mathbf{N} = \kappa_n \, \mathbf{N}, \qquad\qquad (6\text{-}13)$$

with its magnitude κ_n known as the normal curvature.

The normal vector to the spherical surface $\mathbf{N} = -\mathbf{r}/R = -(\sin\phi_0 \cos\theta, \sin\phi_0 \sin\theta, \cos\phi_0)$.[lxxix] Therefore, in the case of the great circle at the meridian θ_0, $\mathbf{k} = \mathbf{k}_n$. However for the small circle at the colatitude ϕ_0,

$$\kappa_n = \mathbf{k} \cdot \mathbf{N} = \frac{1}{R\sin\phi_0}(-\cos\theta, -\sin\theta, 0) \cdot (-\sin\phi_0 \cos\theta, -\sin\phi_0 \sin\theta, -\cos\phi_0).$$

$$\kappa_n = (\sin\phi_0 \cos^2\theta + \sin\phi_0 \sin^2\theta)/(R\sin\phi_0) = \frac{1}{R}.$$

$$\mathbf{k}_n = \kappa_n \mathbf{N} = -(\sin\phi_0 \cos\theta, \sin\phi_0 \sin\theta, \cos\phi_0)/R \neq \mathbf{k}.$$

The component of the curvature vector in the tangent plane is known as the geodesic (or tangential) curvature vector $\mathbf{k}_g$. It is the difference between the curvature vector and the normal curvature vector.

$$\mathbf{k}_g = \mathbf{k} - \mathbf{k}_n \qquad\qquad (6\text{-}14)$$

For the small circle at colatitude ϕ_0,

[lxxviii] As expressed in equations (6-5) and (6-6). (p. 171): $\mathbf{T}_p(u, v) = \mathbf{r}(u, v) + a\mathbf{r}_u + b\mathbf{r}_v$, and $\mathbf{N}_p = \mathbf{r}_u \times \mathbf{r}_v$.

[lxxix] See calculation on p. 172 of $\mathbf{r}_\theta \times \mathbf{r}_\varphi$.

$$\mathbf{k}_g = \frac{-\cos\phi_0}{R\sin\phi_0}(\cos\theta\cos\phi_0, \sin\theta\cos\phi_0, -\sin\phi_0).^{lxxx}$$

The relationship of $\mathbf{k}_g$ to the tangent plane may be made explicit by noting that it can be expressed in terms of the tangent to the meridian at the point (θ, ϕ_0). From the equation for r_ϕ (p. 169), $\mathbf{k}_g = \frac{-\cos\phi_0}{R^2\sin\phi_0}r_\phi(\theta, \phi_0)$. The curvature vectors $\mathbf{k}$, $\mathbf{k}_n$, and $\mathbf{k}_g$ are shown in the following figure.

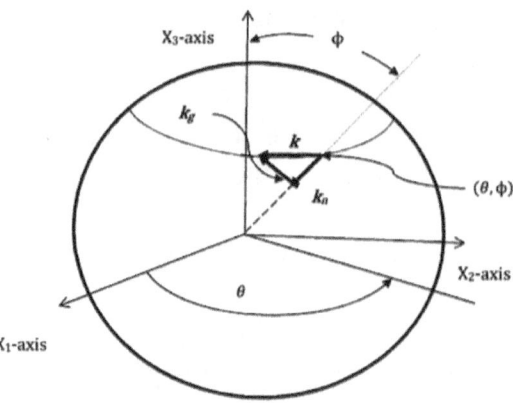

6-4 Curvature vectors at a circle of constant colatitude

The magnitude and orientation of the vectors established in the previous calculations are consistent with the conclusion that the vectors form a right triangle with $\mathbf{k} = \mathbf{k}_n + \mathbf{k}_g$ Therefore,

---

lxxx The geodesic curvature vector for the small circle at colatitude $\phi0$ is calculated below.

$$\mathbf{k}_g = \mathbf{k} - \mathbf{k}_n,$$
$$= \frac{-(\cos\theta, \sin\theta, 0)}{R\sin\phi_0} - \frac{-(\sin\phi_0\cos\theta, \sin\phi_0\sin\theta, \cos\phi_0),}{R}$$
$$= \frac{(\sin^2\phi_0\cos\theta, \sin^2\phi_0\sin\theta, \sin\phi_0_0\cos\phi_0) - (\cos\theta, \sin\theta, 0)}{R\sin\phi_0},$$
$$= \frac{(\cos\theta(\sin^2\phi_0 - 1), \sin\theta(\sin^2\phi_0 - 1), \sin\phi_0\cos\phi_0}{R\sin\phi_0},$$
$$= \frac{-\cos\theta\cos^2\phi_0, -\sin\theta\cos^2\phi_0, \cos\phi_0\sin\phi_0}{R\sin\phi_0},$$
$$= \frac{-\cos\phi_0}{R\sin\phi_0}(\cos\theta\cos\phi_0\sin\theta\cos\phi_0, -\sin\phi_0).$$

$$|k_g|^2 = |k|^2 - |k_n|^2$$

$$= \frac{1}{(R \sin \phi_0)^2} - \frac{1}{R^2}$$

$$= \frac{\cos^2 \phi_0}{(R \sin \phi_0)^2}$$

$$|k_g| = \frac{\cos \phi_0}{R \sin \phi_0}$$

Notice that when the colatitude $\phi_0 = 90°$, $k_g = 0$. This colatitude corresponds to the equatorial plane where the curve is a great circle. One definition for a geodesic is that it is a line for which the curvature vector is equal to the normal curvature vector, k_n, or $k_g = 0$.. All great circles on the surface of a sphere are geodesics. As discussed in more detail in Chapter 8, the shortest distance between two points on a surface is found on a geodesic. In this sense, geodesics are considered to be "straight" lines. This is a generalization of the straight line on the Euclidean plane for which the tangent vector is a constant, and its derivative with path length is zero. On general surfaces, the "straight lines" are those for which the derivative of the tangent with path length $dt/ds = k$ is directed entirely normal to the tangent plane. In other words, the derivative of the portion of the tangent vector in the tangent plane is zero. The curve is the "straightest" possible one.

With the motivation provided by examples of curvature on the sphere, a more general approach will now be considered. We start with the relationship between the tangential vector of a curve and the unit surface normal vector.

$$\mathbf{t} \cdot \mathbf{N} = 0.$$

$$\frac{d(\mathbf{t} \cdot \mathbf{N})}{ds} = \frac{d\mathbf{t}}{ds} \cdot \mathbf{N} + \mathbf{t} \cdot \frac{d\mathbf{N}}{ds}$$
$$= 0. \text{ Therefore,}$$

$$\frac{d\mathbf{t}}{ds} \cdot \mathbf{N} = -\mathbf{t} \cdot \frac{d\mathbf{N}}{ds}, \text{ or}$$

$$\mathbf{k} \cdot \mathbf{N} = \kappa_n = -\mathbf{t} \cdot (d\mathbf{N}/ds).$$

Since $\mathbf{t} = \frac{d\mathbf{r}}{ds}$, and $ds^2 = d\mathbf{r} \cdot d\mathbf{r}$,

$$\kappa_n = -\frac{d\mathbf{r} \cdot d\mathbf{N}}{d\mathbf{r} \cdot d\mathbf{r}} \qquad (6\text{-}15)$$

For the general surface r(u, v) with partial derivatives r_u, and r_v

$$d\mathbf{r} = \mathbf{r}_u du + \mathbf{r}_v dv \text{ and } d\mathbf{N} = \mathbf{N}_u du + \mathbf{N}_v dv.$$

Substituting for d**r** and d**N** in equation (6-15),

$$\kappa_n = -\frac{(\mathbf{r}_u \cdot \mathbf{N}_u)du^2 + (\mathbf{r}_u \cdot \mathbf{N}_v)dudv + (\mathbf{r}_v \cdot \mathbf{N}_u)dudv + (\mathbf{r}_v \cdot \mathbf{N}_v)dv^2}{(\mathbf{r}_u \cdot \mathbf{r}_u)du^2 + (\mathbf{r}_u \cdot \mathbf{r}_v)dudv + (\mathbf{r}_v \cdot \mathbf{r}_u)dudv + (\mathbf{r}_v \cdot \mathbf{r}_v)dv^2} = \frac{II}{I}. \qquad (6\text{-}16)$$

The denominator is the First Form (I), and as in equation (6-8, p. 171), $E = \mathbf{r}_u \cdot \mathbf{r}_u$, $F = \mathbf{r}_u \cdot \mathbf{r}_v$, $G = \mathbf{r}_v \cdot \mathbf{r}_v$. The numerator is known as the Second Form (II), and similar definitions are made for convenience: $e = -\mathbf{r}_u \cdot \mathbf{N}_u$, $2f = -(\mathbf{r}_u \cdot \mathbf{N}_v + \mathbf{r}_v \cdot \mathbf{N}_u)$, and $g = -\mathbf{r}_v \cdot \mathbf{N}_v$.[lxxxi] Therefore

$$\kappa_n = \frac{edu^2 + 2fdudv + gdv^2}{Edu^2 + 2Fdudv + Gdv^2} = \frac{II}{I} \qquad (6\text{-}17)$$

[lxxxi] By taking into account that $\mathbf{r}_u \cdot \mathbf{N} = \mathbf{r}_v \cdot \mathbf{N} = 0$, other convenient expressions may be obtained, for example, $f = -\mathbf{r}_u \cdot \mathbf{N}_v$. Expressions for e, f, and g may be formed involving only the partial derivatives $\mathbf{r}_u$, $\mathbf{r}_v$, and the second partial derivatives, $\mathbf{r}_{uu}$, and $\mathbf{r}_{vv}$ (Struik, p. 75).

The normal curvature κ_n is dependent on the the curve $v = v(u)$ that passes through the the point (u, v). The direction of the curve through (u, v) may be specified as $\frac{dv}{du} = \lambda$. In this direction, then equation (6-17) becomes,

$$\kappa_n = \frac{e + 2f\lambda + g\lambda^2}{E + 2F\lambda + G\lambda^2} \tag{6-18}$$

From equation (6-18) a number of properties of the normal curvature of surfaces can be determined, among them, the maximum and minimum values of normal curvature. These are known as principal curvatures which can be calculated with their associated directions. Euler in 1760 found that the directions of the lines of principal curvature are orthogonal.[183] The intersection of a plane containing the surface normal N with the surface generates a family of curves as the plane is rotated about the surface normal. Associated with each curve is a normal curvature κ_n. The relationship between these normal curvatures and the principal curvatures designated as κ_1 and κ_2 also follows from analysis of equation (6-18). In the relationship given below, known as Euler's theorem, α is designated as the angle between a curve and the curvature directions of κ_1.

$$\kappa_n = \kappa_1 \cos^2\alpha + \kappa_2 \sin^2\alpha. \tag{6-19}$$

The details of the analyses from equation (6-18) leading to Euler's theorem are given in Struik.[184] Of this equation and the concepts that formed its basis, Gauss said that *"These conclusions contain almost all that the illustrious Euler was the first to prove on the curvature of curved surfaces."* He went on to define what he called the measure of curvature, now known as Gaussian curvature, designated here as K. *"The measure of curvature at any point whatever of the surface is equal to a fraction whose numerator is unity, and whose denominator is the product of the two extreme radii of curvature of the sections by normal planes."*[185] In our notation,

$$K = \frac{1}{\rho_1 \rho_2} = \kappa_1 \kappa_2. \tag{6-20}$$

Of crucial importance to our understanding of non-Euclidean geometry, was Gauss's use of this new measure of curvature in what he termed the integral or total curvature. Gauss introduced this in analyzing the area of triangles on a curved surface composed of the shortest lines connecting the vertices of the triangle. The new curvature is defined as an integral over the surface area of the triangle, $\int_\Delta K dA$[186] He then made the remarkable discovery that,

``The excess of the sum of the angles of a triangle formed by shortest lines over two right angles is equal to the total curvature of the triangle.''[187]

Designating the interior angles of the triangle as α, β, and γ,

$$(\alpha + \beta + \gamma - \pi) = \int_\Delta K dA. \tag{6-21}$$

Our next task is to gain further insight to the concepts of Gaussian and integral curvature and its implications. Not surprisingly equation (6-18) allows the Gaussian curvature $K = \kappa_1 \kappa_2$ to be calculated since the principal curvatures may be obtained from this equation. The result is,[188]

$$K = \frac{eg - f^2}{EG - F^2}. \tag{6-22}$$

A key discovery by Gauss concerning this relationship is that the factors e, f, and g depend only on the factors making up the First Form, E, F, and G and their derivatives. Gauss called this his *"Theorema egregium,"* a most excellent theorem. Since the factors of the First Form are defined by the element of length ds^2, curvature can also be determined just from the formulation of length on the surface.[189] If the coordinate parameters u and v are orthogonal, then the Gaussian curvature may be expressed entirely in terms of the parameters of the First Form, E and G, and their partial derivatives such as $G_u = \frac{\partial G}{\partial u}$.[190]

$$K = -\frac{1}{\sqrt{EG}}\left[\frac{\partial}{\partial u}\left(\frac{1}{\sqrt{E}}\frac{\partial\sqrt{G}}{\partial u}\right) + \frac{\partial}{\partial v}\left(\frac{1}{\sqrt{G}}\frac{\partial\sqrt{E}}{\partial v}\right)\right].^{lxxxii}$$

(6-23)

The significance of Gaussian curvature is greatly clarified, however, through Gauss's more geometrical definition of K. Gauss introduced a concept which he called the auxiliary sphere, a sphere of unit radius.[191] On this sphere, the image of, for example, a triangle of shortest lines on the curved surface in question can be represented. This is accomplished by noting that at each point of the boundary of the triangle, a unit normal to the curved surface may be determined. The unit normal corresponds to a point on the auxiliary sphere, that is the point on the sphere is established as a unit vector from the origin of the sphere parallel to the unit normal at a point on the triangle. In this way, the triangle on the curved surface may be mapped onto the sphere. Of this procedure, Gauss made the following general statements,[192]

``If we represent the direction of the normal at each point of the curved surface by the corresponding point of the sphere ... in this way, to every point on the surface, let a point on the sphere correspond; then, generally speaking, to every line on the curved surface will correspond a line on the sphere, and to every part of the former surface will correspond a part of the latter. The less this part differs from a plane, the smaller will be the corresponding part on the sphere. It is, therefore, a very natural idea to use the measure of the total curvature which is to be assigned to a part of the curved surface, the area of the corresponding part of the sphere. For this reason, the author calls this area **the integral curvature** [Gauss's emphasis] of the corresponding part of the curved surface.''*

The measure of curvature is a local property of a surface and, in general, may vary over a surface. As such the ratio of the image area on the

---

[lxxxii] This result may also be expressed as: $K = -\frac{1}{2\sqrt{EG}}\left[\frac{\partial}{\partial v}\left(\frac{E_v}{\sqrt{EG}}\right) + \frac{\partial}{\partial u}\left(\frac{G_v}{\sqrt{EG}}\right)\right]$. K may also be expressed in terms of the parameters of the First Form, including their derivatives, when the coordinates are non-orthogonal (F≠0) by a more complex expression (McCleary, p. 149).

auxiliary sphere ΔA_s to the area of interest on the curved surface ΔA, a triangle for our purposes, may also be defined as the Gaussian curvature in the limit of smaller and smaller triangles.[193]

Geometric definition of Gaussian curvature (6-24)

$$K = \lim_{\Delta A \to 0} \Delta A_s / \Delta A,$$

$$= \frac{dA_s}{dA}, \text{or}$$

$$dA_s = KdA.$$

With this introduction to a geometric interpretation of Gaussian curvature, we interpret its meaning in terms of the vectorial expressions developed for surfaces. Let us start with a surface of interest defined by the position vector $\mathbf{r}(u, v)$. A coordinate system of the surface can be established with the partial derivatives of the position vector $\mathbf{r}_u$ and $\mathbf{r}_v$. Also, at each point of the surface a unit normal vector exists, $N = \mathbf{r}_u \times \mathbf{r}_v / |\mathbf{r}_u \times \mathbf{r}_v|$. Let us assume that the area of interest is a closed curve, say our triangle of shortest lines. The area of interest is designated as A. As noted in equation (6-11, p. 173), $dA = |\mathbf{r}_u \times \mathbf{r}_v| dudv = \sqrt{EG - F^2} dudv$. Now the normals $N(u, v)$ also may be considered to define the position vectors of the surface of the auxiliary sphere as it is a sphere of unit radius. For each point of the area of interest, there corresponds a point N on the sphere. Coordinates (u, v) can be introduced on the auxiliary sphere so that a point from the area of interest has the same coordinates on the sphere. In this way an image is formed on the sphere. A coordinate system for the image may be taken as the vectors N_u and N_v. The differential of the area of interest projected on the unit sphere is therefore,

$$dA_s = |N_u \times N_v| dudv$$

Once again considering the normals to the surface of interest, their partial derivatives N_u and N_v are in the plane of the vectors $\mathbf{r}_u$ and

$\mathbf{r}_v$.[lxxxiii] Taking into account the definitions of the variables of the second fundamental form, e, f, and g, as well as the relations of $\mathbf{r}_u$, $\mathbf{r}_v$, N_u and N_v, it may be shown that:[194]

$$N_u \times N_v = \frac{eg - f^2}{EG - F^2} (\mathbf{r}_u \times \mathbf{r}_v). \qquad (6\text{-}25)$$

Furthermore, $K = \frac{eg-f^2}{EG-F^2}$ (equation (6-22), and therefore,

$$dA_s = K|\mathbf{r}_u \times \mathbf{r}_v|dudv = KdA. \qquad (6\text{-}26)$$

The derivation of equation (6-26) is therefore equivalent to the geometric interpretation of Gaussian curvature in equation (6-24, p. 184) and is consistent with the definition of K in terms of principal curvature, equation (6-20, p. 182) since, as discussed, equation, (6-22), used in the derivation above, may be derived from the product of principal curvatures.

Equation (6-26) indicates that the total or integral curvature of a figure $\int_\Delta KdA$, such as a triangle of area A_Δ on a specific surface, may be obtained by integrating the area of its image on the auxiliary sphere A_s. A particularly simple illustration, almost trivial, of a curved surface and the auxiliary sphere is that of the surface of a sphere of radius R. The auxiliary sphere of unit radius can be located concentrically, and the unit normals to both surfaces are co-linear. In this simple case, the image of a triangle on the surface is transformed into a triangle of area $A_{\Delta s} = A_\Delta/R^2$. As known by Lambert, the area of a triangle on a spherical surface is equal to the square of the radius times the excess of the sum of interior triangle angles (α, β, γ) above π in radians. Therefore, integrating equation (6-26),

---

[lxxxiii] As the partial derivatives like curve tangents are in the tangent plane of the surface, the partial derivatives may be written as $N_u = a\mathbf{r}_u + b\mathbf{r}_v$, and $N_v = c\mathbf{r}_u + d\mathbf{r}_v$. Details are given in Struik, p. 108.

$$\int_\Delta KdA = \int_{A_s} dA_s,$$

$$= A_\Delta/R^2,$$

$$= R^2(\alpha + \beta + \gamma - \pi)/R^2,$$

$$\int_\Delta KdA = (\alpha + \beta + \gamma - \pi) \qquad\qquad (6\text{-}27)$$

The consistency in this example of equation (6-27) with the definition of Gaussian curvature K given in equation (6-20, p. 182) is clear in that the radius of curvature ρ is R to all curves (great circles) formed from normal sections to a sphere of radius R. Thus, $\kappa_1 = 1/\rho_1 = \kappa_2 = 1/\rho_2 = 1/R$, and $K = \kappa_1 \cdot \kappa_2 = 1/R^2$. Then, if the Gaussian curvature K is substituted into equation (6-27), the area A_Δ is determined as expected.

Remarkably, equation (6-27) is valid on general surfaces for triangles formed by shortest lines (geodesic lines with $\mathbf{k} = \mathbf{k}_n$, or equivalently, $\mathbf{k}_g = 0$) between vertices. I will refer to the result shown in equation (6-27) as Gauss' Triangle. Theorem. The most general case involves closed figures, not necessarily formed by geodesic lines. The proof of the more general result involving $\mathbf{k}_g$ is typically demonstrated by integrating the geodesic curvature around the boundary of the figure and turning the integral curvature into a line integral. The general result known as the Gauss-Bonnet Theorem takes us beyond the mathematical scope for what is needed here; however, if the figures consist of lines along geodesics, then the Gauss-Bonnet Theorem reduces to equation (6-27). The interested reader may find details in Struik.[195]

There are numerous implications of equation (6-27). If for a surface, K > 0, as in convex cases like the sphere, then the sum of the interior angles of a triangle must be greater than π radians (180°). As discussed above, this is consistent with the Gaussian curvature of a sphere being equal to $1/R^2$ and with Lambert's recognition that the area of a triangle on a sphere of radius R is the product of R^2 and the angular excess of the interior angles of the triangle.

In the Euclidean geometry of a plane, the sum of a triangle's interior angles is π radians (180°), and since area A > 0, the Gaussian curvature

K = 0. This is consistent with the radius of curvature of a plane being infinite.[lxxxiv]

Of particular significance to us is the knowledge that the Hyperbolic Postulate of Lobachevsky and Bolyai requires that the sum of the interior angles of a triangle be less than π radians (180°). This requirement makes the right-hand side of equation (6-27) less than zero. Given that area is greater than zero, the Hyperbolic Postulate must apply to surfaces of negative curvature. This is the fulfillment of Lambert's speculation about the Hypothesis of the Acute Angle (HAA). The search for the realization of such a surface is the task of the next chapter.

[lxxxiv] You may wonder whether there are other surfaces for which the sum of the interior angles of triangles equals 180°. Drawing a triangle on a piece of paper and then forming a cylinder should easily convince you that the answer is yes, despite the curvature that you have introduced by making a cylinder of the paper. A cylindrical surface of radius R may be described by a position vector in Cartesian coordinates as $r = (x, \sqrt{R^2 - x^2}, z) = (R\cos\theta, R\sin\theta, z)$. In a plane perpendicular to the z-axis, the radius of curvature is R, and the normal curvature is $1/R$. However, a line parallel to the z-axis has a constant tangent vector in space. Like the case of the Euclidean plane, the radius of curvature in this direction is infinite, and therefore, the normal curvature is zero. As the Gaussian curvature is the product of the maximum and minimum normal curvatures, it is therefore zero. A similar analysis would apply to a cone that you could imagine forming with a piece of paper.

7 Models of Non-Euclidean Geometry

7.1 The Hypothesis of the Obtuse Angle (HOA) and the sphere revisited

Saccheri, like Lambert and Legendre after him, correctly rejected the Hypothesis of the Obtuse Angle (HOA) as being incompatible with Euclid's first four postulates because the HOA requires the sum of the interior angles of a triangle to be greater than 180°.[196] One proof that Euclid's first four postulates require that the sum be less than or equal to 180° is the Saccheri-Legendre Theorem (p. 57). Lambert, however, understood that the HOA also leads to the conclusion that the area of triangles depends only upon the excess of the sum of the triangle's interior angles above 180°. Furthermore, he was well aware that this result occurred for triangles formed on the surface of a sphere.[197] This had been known since at least the time of the Dutch mathematician Albert Girard's (1595-1626) publication of a treatise in 1626.[lxxxv] A proof was given by Euler in 1781.[198]

With spherical geometry being important to astronomy going back to the Babylonians and also quite capable of being visualized, it is, in hindsight, a bit surprising that there was no search for a set of postulates that would form its basis. The focus of Euclid's postulates held such sway over the imagination of mathematicians that this did not become clear until after Riemann's groundbreaking examination of the foundations

---

[lxxxv] Girard's treatise contained the earliest use of the abbreviations sin, tan, and sec for the trigonometric functions sine, tangent, and secant. The relation of the area of spherical triangles to their spherical excess was independently discovered by Cavalieri about the same time (Ball, p. 235).

of geometry in a 1854 lecture.[199] We will now examine those parts of Riemann's concepts related to spherical geometry. and the HOA.

As noted following the proof of the Saccheri-Legendre Theorem, the validity of the theorem depends on the unstated assumption that straight lines can be extended indefinitely to any length. This is implied by Euclid's second postulate E2, *"To produce a finite straight line continuously in a straight line,"* as no limit is given, and his fifth postulate, E5, which states the condition under which two straight lines will meet *"if produced indefinitely."*[lxxxvi] In addition to the Saccheri-Legendre Theorem, Proposition IE16 (p. 27) is also only valid if it can assured that straight lines can be indefinitely extended to any length. Hence, IE27 (p. 28) which establishes criteria for parallel lines using IE16 is similarly invalid. Other propositions that are affected are IE17, IE18, IE19, IE20, and IE21.[200]

Riemann recognized that a line may be boundless, but not infinite. As noted in the discussion of spherical coordinates, traveling around the equator, we may go as many times as we want, but the maximum length from the starting point is limited by the circumference of the earth. Thus, the straight line defined as a great circle is boundless, but not infinite. Riemann's insight implies that the HOA could be made compatible with a consistent geometry by modifying three of Euclid's five postulates, E1, E2, and E5, the Parallel Postulate. Regarding the modification to Euclid's E1 that establishes a straight line from one point to another ('' *To draw a straight line from any point to any point."*), Heath notes that Euclidean geometry implies that the straight line is unique.[201] The modified postulates in the geometry that became known as elliptic are listed below (and designated as E-R1, E-R2, and E-R5 in honor of Riemann).

Elliptic Geometry Postulates[202]
E-R1. Two distinct points determine at least one straight line.
E-R2. A straight line is boundless.

---

lxxxvi Heath, Vol. 1, pp. 154-155, 217. ; see also Postulates (p. 19).

E-R5. Any two straight lines in a plane intersect. (There are no parallel lines.)

I will not develop in detail all of the major implications of these postulates, but show informally how they relate to the geometry of the surface of a sphere and the HOA following the approach of Wolfe and of Eves.[203] Using figure 7-1 below, we wish first to prove that the postulates of Riemann given above lead to the conclusion, as in the HOA, that the sum of the interior angles of a triangle is greater than 180°

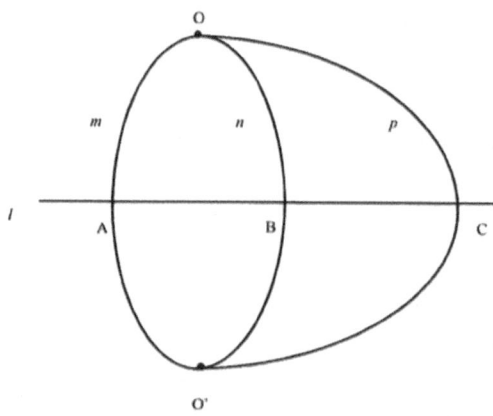

7-1 Intersecting lines in the elliptic plane

In the figure we start with the straight-line *l* in the elliptic plane governed by Euclid's postulates, as modified by Riemann. At an arbitrary point A, a straight-line *m* is extended upward, perpendicular to *l*. At another arbitrary point B on *l*, another straight-line *n* is also produced upward, perpendicular to l. By the modified postulate E-R5, the two lines must intersect at a point designated as O forming the triangle AOB. From proposition IE6 (see Appendix B), which is not affected by the modified postulates (see discussion below), the sides AO and OB are of equal length λ since ∠ABO = ∠BAO. Clearly, the sum of the interior angles of ΔAOB must be greater than 180°.

A point C is now selected on *l* such that the lengths AB = BC forming another triangle BOC. Because ∠ABO = ∠CBO, AB = BC = λ, and

BO is a common side, from Proposition IE4, $\triangle AOB \cong \triangle BOC$, and therefore, $CO = AO = \lambda$.[lxxxvii] Having proved that $AO = BO = CO = \lambda$, it is important to recall that points A and B on line l were selected arbitrarily. Hence, any line perpendicular to l will pass through point O and be of length λ. The point O is known as the pole of line l

The reasoning applied to the lines produced upward from line l may equally be applied to lines extended downward. OA is extended downward so that its length $O'A = \lambda$. Producing the line O'B results in $\triangle AO'B$. As $AO = AO' = \lambda$, $\angle OAB = \angle O'AB$, and AB is common to both triangles, $\triangle AOB \cong \triangle AO'B$. Therefore, the length of BO' also equals λ. Furthermore, at least informally, we have shown that the two points O and O' produce more than one line as allowed under Postulate E-R1. As noted below, it is possible that O and O' are the same point. With these results, we now wish to show that lines are boundless as required by Postulate E-R2.

In the following figure (7-2) two straight lines, m and n intersect at the pole O of line l. The length of the lines OA and OC, as proved above are each equal to λ. Line m and n are extended beyond l to the pole O'. Similarly, the length of lines AO' and CO' equal λ. Continuing from O'

[lxxxvii] Euclid proved that the sides and angles of figures are equal based on superposition (coincidence of figures) as stated in Common Notion E4. He used this in Proposition IE4 to prove that corresponding sides and angles of triangles are "equal" if the corresponding two sides of each triangle and the angles contained by the two sides are "equal." This concept of "equality" of figures is not specifically stated by Euclid, but means here that the figures may be made to coincide by placement of one onto the other – designated here as congruence (Heath, Vol. i, pp. 327-328). (Recall that elsewhere, Euclid used the term "equal" to refer to the equality of the areas of figures.) Euclid's approach to congruence relies upon unstated assumptions about the uniformity of space and on the rigid motion of figures allowing comparisons of figures through their superposition.(Bunt, Jones, and Bedient, p. 154.) In the specific case of the geometry of the surface of a sphere, the constant positive curvature is consistent with the establishment of congruence through the superposition of figures, that is one could imagines sliding one figure over another. In David Hibert's revision of Euclid's geometric system to address its deficiencies (Foundations of Geometry (1899), Proposition IE4 is taken as one of the postulates (Eves, pp. 82-85.)} Proposition IE4, is used in IE6 which established the equality of the lengths of AO and BO in the figure above.

through points B and D back to the pole O, the total length of each of lines *m* and *n* is 4λ.

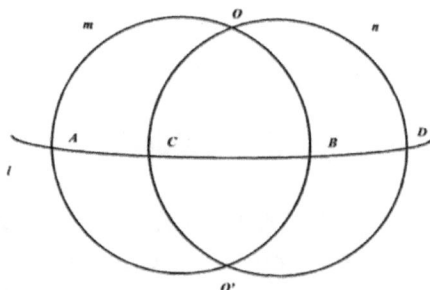

7-2 Boundless and finite lines in the elliptic plane

The lines *m* and *n* are finite, but boundless as required by Postulate E-R2. With these considerations, it should be clear that by considering straight lines to be great circles, the geometry of the surface of a sphere is consistent with elliptic geometry with two poles and the HOA. As will be discussed in detail in the next chapter, the great circles are geodesics along which the shortest distance between two points may be found.

Although, spherical geometry is consistent with the postulates of elliptic geometry, it is not synonymous with it. Rather it is one possible model. One way to see this is to consider the possibility that in the discussion above, the poles O and O' are the same points. In this case, there is only one surface, unlike that of the sphere which has an inner and outer surface.[204] I will not explore this here other than to note that perhaps the most well-known example of such a surface is the Möbius strip which can be visualized by twisting a strip of paper and pasting the ends together. A line can then be followed continuously on the previously separate sides of the paper.

One advantage of modeling elliptic geometry as a spherical surface is that it is embedded within a Euclidean space, and therefore one can be confident that its postulates form a logical system as consistent as that of Euclidean geometry. More concretely, the relationships of Euclidean geometry may be used to determine the geometric relations on this surface. As an example, the laws for spherical triangles, analogous to the Law of Cosines (see figure 4-3, p. 96) for Euclidean triangles, will now be developed. Let us look at a triangle $\triangle ABC$, with angles less than a right angle, on the surface of a sphere of radius r as shown in the figure below.[lxxxviii]

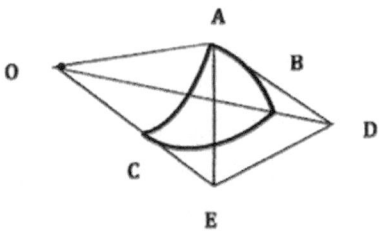

7-3 A spherical triangle

From the center of the sphere, O, radii of length r are drawn to points A, B, and C of the triangle. The straight-line AE is produced perpendicular to the radius OA in the plane OAC and intersects the extension of the straight-line OC. Similarly, the straight line AD is formed intersecting the extension of OB to OD. By their construction, $\angle OAE$ and $\angle OAD$ are right angles. The planes OAD and OAE cut through the sphere forming lines AB and AC, hence, $\angle EAD = \angle CAB$ as AE and AD are tangents at A to AC and AB. Note also that by construction arcs $BC = r\cdot\angle BOC = r\cdot\angle DOE$, $AC = r\cdot\angle AOC = r\cdot\angle AOE$, and $AB = r\cdot\angle AOB = r\cdot\angle AOD$.

From the Law of Cosines for $\triangle ODE$ and $\triangle ADE$:

$DE^2 = OD^2 + OE^2 - 2\cdot OD\cdot OE \cos(\angle DOE) = OD^2 + OE^2 - 2\cdot OD\cdot OE \cos(BC/r)$
$DE^2 = AD^2 + AE^2 - 2\cdot AD\cdot AE \cos(\angle EAD)$

Because $\angle OAD$ and $\angle OAE$ are right angles: $OD^2 = OA^2 + AD^2$ and $OE^2 = OA^2 + AE^2$. Substituting for OD^2 and OE^2 in the first equation for DE^2 and subtracting the second equation for DE^2, we obtain:

[lxxxviii] Todhunter shows that the derivation is valid even in such cases as triangles with angles greater than a right angle, (Todhunter, pp. 18-20).

$0 = 2 \cdot OA^2 + 2 \cdot AD \cdot AE \cos (EAD) - 2 \cdot OD \cdot OE \cos (BC/r)$; or
$\cos(BC/r) = (OA/OE) \cdot (OA/OD) + (AE/OE) \cdot (AD/OD) \cos(\angle EAD)$

From the definitions of sine and cosine:
$OA/OE = \cos (AC/r)$, $OA/OD = \cos (AB/r)$, $AE/OE = \sin (AC/r)$, $AD/OD = \sin (AB/r)$; therefore, after substitution,
$\cos(BC/r) = \cos(AC/r) \cdot \cos(AB/r) + \sin(AC/r) \cdot \sin(AB/r) \cdot \cos(EAD)$

As in the earlier proof of the Law of Cosines we will simplify our equation using Euler's notation: a = BC (opposite vertex A), b = AC (opposite vertex B), c = AB (opposite vertex C) and $\angle A = \angle CAB = \angle EAD$. This leads to the more easily recalled formula for cos (a/r). A similar derivation can be developed for cos (b/r) and cos (c/r) noting that $\angle B = \angle ABC$ and $\angle C = \angle ACB$. With these substitutions, we have the fundamental formulas for the sides and angles of triangles in spherical geometry:

$$\cos (a/r) = \cos(b/r) \cdot \cos(c/r) + \sin(b/r) \cdot \sin(c/r) \cdot \cos(A), \quad (7\text{-}1)$$

$$\cos (b/r) = \cos(c/r) \cdot \cos(a/r) + \sin(c/r) \cdot \sin(a/r) \cdot \cos(B),$$

$$\cos (c/r) = \cos(a/r) \cdot \cos(b/r) + \sin(a/r) \cdot \sin(b/r) \cdot \cos(C).$$

The equations given above can be recast in numerous ways to facilitate their use in different problems. The following equations each including the three angles and a single side are derived from the spherical law of cosines.[205]

$$\cos A = -\cos B \cos C + \sin B \sin C \cos a/r. \quad (7\text{-}2)$$

$$\cos B = - \cos A \cos C + \sin A \sin C \cos b/r.$$

$$\cos C = - \cos B \cos A + \sin B \sin A \cos c/r.$$

The consistency of the spherical law of cosines with the HOA may be illustrated by showing that the sum of the interior angles of an equilateral triangle (a = b = c) is greater than 180°. Solving for cos A in equation (7-1),

$$\cos A = \frac{\cos(^a/_r) - \cos^2(^a/_r)}{\sin^2(^a/_r)} \; (a \neq 0, \pi r, 2\pi r, \ldots),$$

$$= \frac{\cos(^a/_r)(1 - \cos(^a/_r))}{((1 + \cos(^a/_r)))(1 - \cos(^a/_r))},$$

$$= \frac{1}{(1/\cos(a/r) + 1)}$$

As 1/cos(a/r) >1 for a > 0, cos A < 1/2; hence, ∠A > 60°, and the sum of the equilateral triangle's interior angles is greater than 180°

The cosine laws for spherical geometry may be used to determine distances between two locations on the earth surface. I will illustrate their use by determining the distance along a geodesic (great circle path) between two cities approximately at the same latitude and compare it with the distance following instead the path of constant latitude. We so often see maps of the earth projected on a plane surface that it is easy to forget that the shortest distance between two locations of the same latitude does not follow the path of constant latitude. Rather the path for the crow and jets to fly is the great circle connecting the two points. Let us calculate the distance between Boston, Massachusetts and Rome, Italy which are both at a latitude of approximately 42°. The approximate longitude for Boston is 71.1° west of the Greenwich meridian and for Rome it is 12.5° east of Greenwich. Let us convert these coordinates to the spherical geometry coordinates r, θ, and ϕ shown in the figure below.

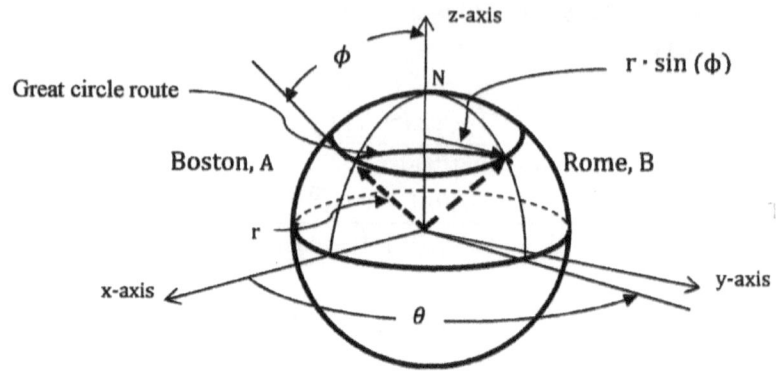

7-4 Great circle route from Boston to Rome

For our purpose, we only need the differences in longitude, so $\theta =$ 12.5° − (−71.1°) = 83.6°, or 83.6 $\left(\frac{\pi}{180}\right)$ = 1.46 radians. Latitude is measured from the plane of the equator, so $\phi = 90° − 42° = 48°$, or 48 $\left(\frac{\pi}{180}\right)$ =0.838 radians. Let us first calculate the distance between Boston and Rome along the constant latitude line of 42°. The path will be part of a circle of radius $r\cdot\sin\phi$, where r is the radius of the earth taken as approximately 4,000 miles. The distance along the path of constant latitude, $d_{lat.,}$ is therefore,

$$d_{lat.} = r \cdot \sin \phi \cdot \theta = 4000 \sin(48°) \cdot 1.46 = 4{,}340 \text{ miles.}$$

Now let us look at the great circle from Boston (point A) to Rome (point B) by forming the triangle ABN on the earth's surface with N being the North Pole. Let the arc AN = b, BN = a, and the arc between Boston and Rome, AB = c. Furthermore, let the spherical angle, ∠ANB = ∠C. Then from the law of cosines for spherical geometry:

$$\cos(c/r) = \cos(a/r) \cdot \cos(b/r) + \sin(a/r) \cdot \sin(b/r) \cdot \cos C.$$

Since Boston and Rome are at the same latitude, $a = b = r\phi$, and cos(c/r) $= \cos^2\phi + \sin^2\phi \cos C$ with $\phi = 48°$ and ∠C = θ = 83.6°. Therefore, cos(c/r) = 0.448 + 0.552(0.111), and cos(c/r) = 0.509. From a calculator with the inverse cosine function, the angle in radians for which the cosine equals 0.509 is 59.4°, or 1.037 radians. Thus, c/r =

1.037, and c = 1.037· 4000 = 4,184 miles. Indeed, the great circle route is shorter.

The Laws of Cosines for Euclidean geometry and spherical geometry are closely related in that for a large radius r, that is a/r, b/r, and c/r much less than 1, the spherical law of cosines is approximated by the Euclidean law. This can be shown explicitly by expressing in the spherical laws of cosines the sine and cosine functions as power series in a/r, b/r, and c/r as in equations (4-6) and (4-7, pp. 123,124). Retaining only terms no higher than second degree like $(a/r)^2$ (since higher degree terms are much smaller), the Euclidean Law of Cosines is recovered. This is consistent with the Gaussian curvature K for a sphere of radius r being $1/r^2$, As r increases the integral curvature approaches zero, and the angle excess approaches the Euclidean value of zero (equation (6-27, p. 186). For many practical purposes, the earth can be considered to be flat!

Another way to think of the relationship of Euclidean geometry to the concept of curvature is to relate the circumference of the circle on a sphere of radius r at the colatitude ϕ to the distance on the sphere's surface to the to the pole. For example, in figure 7-4 above, the circle could be formed by the latitude of Boston and Rome and the distance along the surface from the pole N is the arc $r\phi$ = NA = NB = R. Someone on the surface measuring the distance from the pole to such a circle and thinking that the surface is flat, would calculate the circumference of the circle as C' = 2πR. However, as we are aware that the surface is on a sphere, we know that the actual circumference is C = $2\pi r \cdot \sin \phi$. The sine function may be expressed as a power series as given in equation (4-6). Therefore,

$$C = 2\pi r \sin \phi = 2\pi r(\phi - \frac{\phi^3}{3!} + \frac{\phi^5}{5!} - \frac{\phi^7}{7!} + \cdots.$$

With $K = 1/_{r^2}$, and $R = r\phi$, this becomes:

$$C = C' - \frac{\pi}{3}KR^3 + \frac{\pi}{60}K^2R^5 + \cdots$$

So, for small values of the Gaussian curvature K, the apparent circumference, C' will approach the value, C expected for Euclidean geometry in which K = 0. Again, the Gaussian curvature may be considered to be a measure of a surface's deviation from Euclidean space. For K > 0, the actual circumference C is smaller than the circumference C'. We can see this also by rewriting the above equations in the following form:

$$\frac{3}{\pi}(C' - C)/R^3 = K - \frac{(KR)^2}{20} + \cdots.$$

For circles of small R, $K \approx \frac{3}{\pi}(C' - C)/R^3$ or in the limit,

$$K = \lim_{R \to 0} \frac{\frac{3}{\pi}(C' - C)}{R^3}.$$

We have derived this for the case of a sphere, but the definition and its geometric interpretation are general.

The identification of the HOA with a set of modified postulates consistent with a surface of constant positive Gaussian curvature, a sphere, recalls Lambert's speculation that the Hypothesis of the Acute Angle (HAA) would be met by a sphere of imaginary radius *ir.* Franz Taurinus (1794-1874) made just such a substitution into the trigonometric relationships of spherical geometry and derived many correct relationships for what would become known as hyperbolic geometry.[206] Despite the consistency of his results, like Saccheri and Lambert before him, Taurinus did not believe they represented a valid geometry. With our knowledge that the HAA is consistent with a surface of constant negative Gaussian curvature, it is time to look for models that satisfy the HAA.

7.2 Models of surfaces with negative Gaussian curvature

7.2.1 The sphere of imaginary radius

Lobachevsky in his development of a non-Euclidean geometry allowed for more than one parallel line to be formed through a given point P not on a given line l in the plane of P and l. I have designated that the Hyperbolic Postulate (see figure 3-1 (p. 71) and the discussion of the postulate and associated concepts such as the angle of parallelism). He went on to develop relations between collections of parallel lines and chords making equal angles to these lines forming what is known as a horocycle. From these relations, he was able to formulate a new trigonometry and a relationship for the angle of parallelism.[lxxxix] The results of Lobachevsky are equivalent to those provided by the speculations of Lambert and Taurinus of the geometry of surfaces of imaginary radius ir.[207] With the advantage of hindsight and the knowledge that a Gaussian curvature $K = 1(ir)^2 = -1/r^2$ is consistent with the requirement of the HAA that the interior angles of triangles sum to less than 180°, the results for hyperbolic trigonometry will be given below in the spirit of Taurinus.

In the previous section (see figure 7-3, p. 193) the law of cosines was derived for a spherical triangle with angles A, B, and C at the vertices and with sides a, b, and c opposite the respective angles on a sphere of radius r. Equation (7-1) is reproduced below for side a.

$$\cos(a/r) = \cos(b/r) \cdot \cos(c/r) + \sin(b/r) \cdot \sin(c/r) \cdot \cos(A).$$

Substituting the imaginary radius ir,

$$\cos(a/ir) = \cos(b/ir) \cdot \cos(c/ir) + \sin(b/ir) \cdot \sin(c/ir) \cdot \cos(A).$$

Hyperbolic functions were developed using the relations for $\cos(ix)$ and $\sin(ix)$ (equations (4-15) and (4-16, p. 125) The following equations result after simplification. I include those for the other two sides which can be similarly derived.

$$\cosh(a/r) = \cosh(b/r) \cosh(c/r) - \sinh(b/r) \sinh(c/r) \cos A. \quad (7\text{-}3)$$

$$\cosh(b/r) = \cosh(c/r) \cosh(a/r) - \sinh(c/r) \sinh(a/r) \cos B.$$

[lxxxix] Lobachevsky's development may be found in his *Theory of Parallels\rm* in Bonola (see also pp.88-89). The development following this approach is given in Wolfe, pp. 110-120, 131-152.

$$\cosh(c/r) = \cosh(a/r) \cosh(b/r) - \sinh(a/r) \sinh(b/r) \cos C.$$

Similar to the case of the spherical triangle in the previous section, we can illustrate with a hyperbolic equilateral triangle ($a = b = c$), that the sum of the interior triangles is less than 180°. Equation (7-3) becomes,

$$\cosh(a/r) = \cosh(a/r)^2 - \sinh(a/r)^2 \cos A.$$

Solving for cos A,

$$\cos A = [\cosh(a/r)^2 - \cosh(a/r)]/\sinh(a/r)^2 \ (a \neq 0),$$

$$= \frac{\cosh(a/r)(\cosh(a/r) - 1)}{(\cosh(a/r) + 1)(\cosh(a/r) - 1)}.$$

$$= \frac{1}{\left(1 + \dfrac{1}{\cosh(a/r)}\right)}.$$

As $1/\cosh(a/r) < 1$ for $a > 0$, $\cos A > 1/2$; hence, $\angle A < 60°$, and the sum of the equilateral triangle's interior angles is less than 180°.

A key characteristic of the hyperbolic geometry is the angle of parallelism, Π associated with limiting parallels from points along a perpendicular to a line (see figure 3-1, p. 71 and following discussion). As shown below, the angle of parallelism decreases with increasing distance along the perpendicular. This will be shown using the approach of Taurinus, but first I reproduce a simple proof of this given by Aleksandrov.[208]

Aleksandrov first proves without using the Parallel Postulate that if a line cutting two other straight lines forms equal alternate interior angles, a common perpendicular may be formed between the two straight lines. The proof that the angle of parallelism Π decreases, as the distance d between a point on the perpendicular to a line and the line increases, is outlined below using this proposition with the figure below.

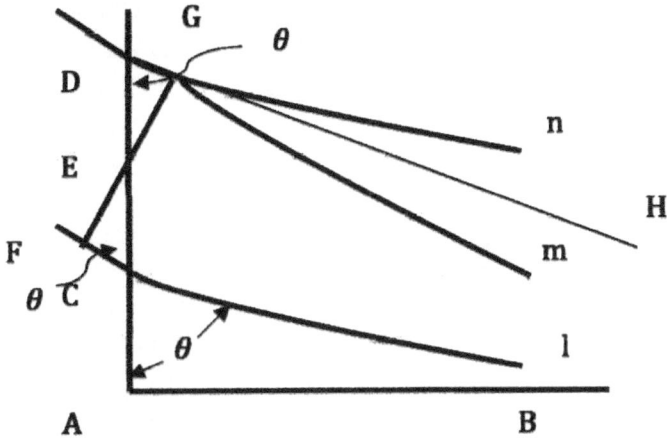

7-5 Proof for angle of parallelism Π

In figure 7-5, line AD is perpendicular to line AB. Line l is constructed as a limiting parallel to AB and intersects AD at point C at the angle of parallelism $\theta = \Pi(d_{AC})$. Line n is produced to intersects point D also at an angle θ. A common perpendicular FG, therefore, can be constructed between line l and n with $\angle GDE = \angle FCE = \theta$ because θ is the same for line l and n. At G, a line m can also be produced parallel to l at some angle less than 90° to FG since n is perpendicular to FG and n is parallel to l. A point H can be placed between the angle formed by lines n and m. The line DH must therefore also be parallel to l and AB, but $\angle ADH < \theta$. Therefore, the angle of parallelism at D is less than θ, or $\Pi(d_{AD}) < \Pi(d_{AC})$.

The qualitative description of the angle of parallelism may be quantified by once again following Taurinus's approach of analyzing the trigonometry of a sphere of imaginary radius. In this case we use the form of spherical law of cosines given in equation (7-2, p. 194) with the imaginary radius i r substituted for r.

$$\cos A = -\cos B \cos C + \sin B \sin C \cos(a/ir),$$

or

$$\cos A = -\cos B \cos C + \sin B \sin C \cosh(a/r). \qquad (7-4)$$

Imagine the hyperbolic triangle ABC with an extended line CA, a perpendicular to the line BC of length d_{BC}, and BA a limiting parallel to CA. The triangle is shown schematically in the figure below

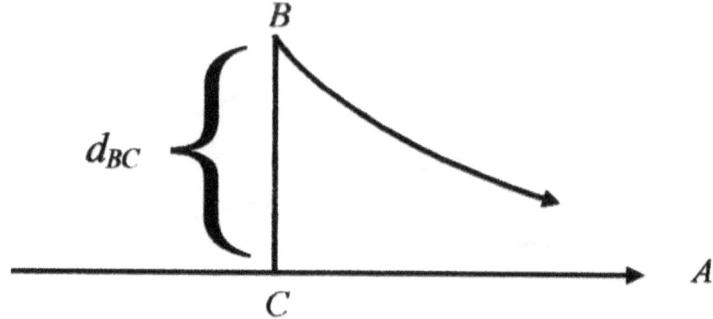

7-6 Triangle with one side a limiting parallel

Since line BA is a limiting parallel to CA, $\angle B = \Pi(d_{BC})$, $\angle C = 90°$, and $\angle A = 0°$ as it asymptotically approaches the line CA. The distance d_{BC} being opposite $\angle A$ corresponds in the triangle to a/r. Substituting these results into equation (7-4), $\cosh(a/r) = 1/\sin B$, and,

$$\sin\Pi(a/r) = 1/\cosh(a/r),\qquad (7\text{-}5)$$

Since in general, $\cos\theta = \sqrt{1 - \sin^2\theta}$, for $\theta = \Pi(a/r)$

$\cos\Pi(a/r) = \sqrt{1 - \sin^2\Pi(a/r)}$. Substituting from equation (7-5)

$$= \sqrt{(1 - 1/\cosh^2\Pi(a/r))},$$

$$= \sqrt{\frac{\cosh^2\Pi(a/r) - 1}{\cosh^2\Pi(a/r)}}$$

$$= \frac{\sinh(a/r)}{\cosh(a/r)}, \text{ or}$$

$$\cos\Pi(a/r) = \tanh(a/r)\qquad (7\text{-}6)$$

Lobachevsky developed an expression for the angle of parallelism in terms of tan $\Pi/2$, $\Pi/2$ being the half angle.[209] We will explore this using the trigonometric identity for a half angle, $\tan(\theta/2) = (1 - \cos\theta)/\sin\theta$.[xc] Therefore, also using equations (7-5) and (7-6),

$$\tan\left(\frac{\Pi(a/r)}{2}\right) = \frac{1 - \cos\Pi(a/r)}{\sin\Pi(a/r)},$$

$$= \frac{1 - \tanh(a/r)}{1/\cosh(a/r)},$$

$$= \cosh(a/r) - \tanh(a/r) \cdot \cosh(a/r),$$

$$= \cosh(a/r) - \sinh(a/r),$$

$$= \frac{e^{a/r} + e^{-a/r}}{2} - \frac{e^{a/r} - e^{-a/r}}{2}. \text{ Therefore,}$$

$$\tan\left(\frac{\Pi(a/r)}{2}\right) = e^{-a/r} \qquad (7\text{-}7)$$

Equation (7-7), discovered by Lobachevsky and Bolyai, concisely expresses one of the most important concepts of hyperbolic geometry. As a/r approaches 0, the angle of parallelism approaches the Euclidean value of $\pi/2$ since $\tan(\Pi(a/r)/2)$ approaches 1, and therefore the half angle $\Pi/2$ approaches $\pi/4$. This is consistent with the result that for increasingly large r, the Gaussian curvature ($K = 1/(ir)^2 = -1/r^2$) approaches the Euclidean value of zero, and the sum of the interior angles of the hyperbolic triangle approaches 180°. In contrast as a increases for fixed r, $\tan\left(\frac{\Pi(a/r)}{2}\right)$ approaches 0 and therefore so does the angle of parallelism.

Although hyperbolic geometry is consistent with that of a sphere of imaginary radius having negative Gaussian curvature, I would not be surprised if it still seems a bit contrived compared to the solidity of Euclidean geometry. After all, even Lobachevsky and Bolyai had

---

xc   The identity may be shown using the identities:

$\cos^2\frac{\theta}{2} = \frac{1}{2}(1 + \cos\theta)$ and $\sin^2\frac{\theta}{2} = \frac{1}{2}(1 - \cos\theta)$ (see footnote lxiv, p. 147). Then,

$\tan^2\frac{\theta}{2} = \frac{\sin^2\frac{\theta}{2}}{\cos^2\frac{\theta}{2}} = \frac{1 - \cos\theta}{1 + \cos\theta}$, and $\tan\frac{\theta}{2} = \sqrt{\frac{1 - \cos\theta}{1 + \cos\theta}} = \sqrt{\frac{1 - \cos\theta}{1 - \cos\theta}} = \frac{1 - \cos\theta}{\sin\theta}$.

questions concerning its meaning. We will continue in the next section to look at other models with negative Gaussian curvature.

7.2.2 Poincaré's half-plane and disk models

While Taurinus's interpretation of the HAA in terms of the sphere of imaginary radius may have been inspired, ultimately it was unsatisfying to him even though he recognized the consistency of the resulting conclusions. However, after Bolyai and Lobachevsky proclaimed their new geometry, compatible with the HAA and replacing the Parallel Postulate, a number of explicit models were developed. These models, discussed briefly in Section 3.3 *Modeling the hyperbolic plane – a first look*, complied with the postulates of the new hyperbolic geometry and the requirement that they represent surfaces of negative Gaussian curvature. Beltrami's groundbreaking model of 1868, called the pseudosphere, mapped a portion of the hyperbolic plane on a surface in Euclidean space. However, the first model which was complete in the sense that it covered the entire hyperbolic plane was a disk model of Beltrami in which the plane is mapped into the interior of a circle. The model was part of an analysis of geodesics undertaken between 1865 and 1869.[210] Beltrami's disk model was independently developed by Klein in 1871, based upon a paper by Arthur Cayley (1821-1891) on a related subject and published in 1859. As McCleary notes, the model is most often referred to as the Klein model, but also the Klein-Beltrami model, or even the Cayley-Klein-Beltrami model."[211] Whatever the name, the flurry of related investigations attests to the understanding that the new geometry was vital to understanding the meaning of mathematics.

The model of Beltrami of the entire hyperbolic plane has the disadvantage that although straight lines are represented as rays, the measure of angles of intersecting lines is not the same as represented in Euclidean geometry by two intersecting straight lines. Poincaré developed two models in which the angles are represented as in a Euclidean space. These are qualitatively described in Section 3.3. One is a half-plane model representing the upper half of the hyperbolic plane; the other is another disk model in which the entire hyperbolic plane is mapped into the interior of a circle. With the background provided on representing a differential line segment with the First Form (equation (6-8, p. 171) and its use in calculating Gaussian curvature (equation (6-23, p.183), we can analyze these models in greater detail. I will focus on the Half-plane model because of the relatively simpler analytical expressions that describe the model compared to the disk model.

The differential length ds of the Poincaré Half-plane model may be specified as,[212]

$$ds^2 = \frac{dx^2 + dy^2}{y^2}. \qquad (7\text{-}8)$$

Rewriting as the First Form, $ds^2 = Edx^2 + 2Fdxdy + Gdy^2$ then $E = 1/y^2$, $G = 1/y^2$, $F = 0$.

From equation (6-23), the Gaussian curvature may be calculated as:

$$K = -\frac{1}{\sqrt{EG}} \left[\frac{\partial}{\partial x} \left(\frac{1}{\sqrt{E}} \frac{\partial \sqrt{G}}{\partial x} \right) + \frac{\partial}{\partial y} \left(\frac{1}{\sqrt{G}} \frac{\partial \sqrt{E}}{\partial y} \right) \right]$$

$$K = -y^2 \left[0 + \frac{\partial}{\partial y} \left(y \frac{\partial(1/y)}{\partial y} \right) \right],$$

$$= -y^2 \frac{\partial}{\partial y} (y(-1/y^2)),$$

$$= -1.$$

Thus, the Poincaré Half-plane model represents a surface of negative curvature. I will not prove that it satisfies all of the requirements of hyperbolic geometry in the upper half plane; however, I will illustrate some of its characteristics. The figure below shows two points with Euclidean coordinates $(x_1, y_1) = (-\sqrt{2}, \sqrt{2})$ and $(x_2, y_2) = (\sqrt{2}, \sqrt{2})$. The two points are joined by a Euclidean straight line and by a circle with its center at the origin having a Euclidean radius of 2. In this model, with the methods of the next chapter, the shortest line between two points (a geodesic) can be shown to be a circle with its center on the x-axis.

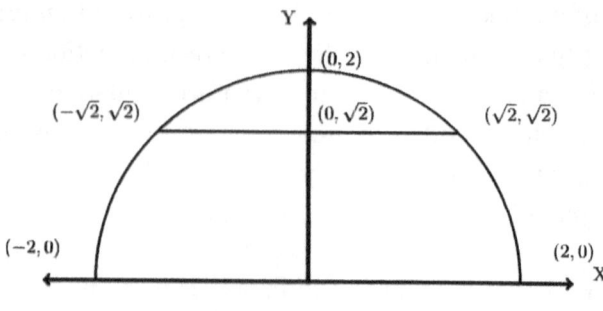

7-7 Half-plane model

The distance represented by the Euclidean straight line between the two points is simple to evaluate since the coordinate y is constant $(y = \sqrt{2})$. In the Half-plane model, however, this is not a straight line.

$$S_{y=\sqrt{2}} = \int_{(x_1,y_1)}^{(x_2,y_2)} \sqrt{\frac{dx^2 + dy^2}{y^2}} = \int_{(x_1,y_1)}^{(x_2,y_2)} \frac{dx}{\sqrt{2}} = \frac{(\sqrt{2} - (-\sqrt{2}))}{\sqrt{2}} = 2.$$

The circular path between the two points is expressed as $x^2 + y^2 = 4$. As has been shown previously (see figure 5-2, p. 136) and discussion), along the circular path, $dy/dx = -x/y$. Because similar calculations will be required, it will be generalized to a circle of radius r centered at a point c on the x-axis, that is $(x-c)^2 + (y-c)^2 = r^2$ with $dy/dx = -(x - c)/y$.

$$S_{geodesic} = \int_{(x_1,y_1)}^{(x_2,y_2)} \frac{\sqrt{dx^2 + dy^2}}{y},$$

$$= \int_{(x_1,y_1)}^{(x_2,y_2)} \frac{\sqrt{1 + (dy/dx)^2}}{y} dx,$$

$$= \int_{(x_1,y_1)}^{(x_2,y_2)} \frac{\sqrt{1 + (-(x-c)/y)^2}}{y} \, dx,$$

$$= \int_{(x_1,y_1)}^{(x_2,y_2)} \frac{\sqrt{y^2 + (x-c)^2}}{y^2} \, dx,$$

$$= \int_{x_1}^{x_2} \frac{r \, dx}{(r^2 - (x-c)^2)} \, dx.$$

Proceeding in a manner similar to previous integrations, let $x - c = r \cos \theta$. It follows that, $dx = -r \sin\theta d\theta$.. (It also follows that $\sin \theta = y/r$, $\tan \theta = \frac{y}{x-c}$, and $\cot \theta = 1/\tan \theta = \frac{x-c}{y}$..) Making the substitution for x,

$$S_{geodesic} = - \int \frac{r^2 \sin \theta d\theta}{r^2(1 - \cos^2\theta)},^{xci}$$

$$= - \int \frac{d\theta}{\sin \theta},$$

$$= - \ln \left[\frac{1 - \cos \theta}{\sin \theta} \right].$$

As $\sin \theta = y/r$, and $\cos \theta = (x-c).r$, then

$$S_{geodesic} = - \ln \left[\frac{r}{y} \left(1 - \frac{(x - c_-)}{r} \right) \right] \Big|_{x_1,y_1}^{x_2,y_2}.$$

Recalling that $a - \ln b = \ln \frac{a}{b}$, after simplification,

---

207

$$S_{\text{geodesic}} = \left| \ln \left(\frac{(r - (x_1 - c))y_2}{(r - (x_2 - c))y_1} \right) \right|. \tag{7-9}$$

Substituting into the above equation the coordinates of figure 7-7,

$$S_{\text{geodesic}} = \ln \left(\frac{\left(2 - (-\sqrt{2} - 0) \right) \sqrt{2}}{\left(2 - (\sqrt{2} - 0) \right) \sqrt{2}} \right) \approx 1.763.$$

Indeed, the geodesic is shorter than the "Euclidean straight line" with its length of 2. Notice in equation (7-9) that if $x_1 = x_2$, then $S_{\text{geodesic}} = \ln y_2 / y_1$. Then, starting at y_1 and going towards the x-axis ($y = 0$), the distance increases beyond any limit as y_2 approaches 0, that is, the x-axis is at infinity.

If this analysis has obscured the meaning of the calculation of hyperbolic length in the Half-plane model, then a numerical calculation should help to clarify it. The length along the geodesic from $(-\sqrt{2}, \sqrt{2})$ to $(\sqrt{2}, \sqrt{2})$ may be estimated as a sum,

$$S = \sum_{k=1}^{N} \frac{\sqrt{\Delta x_k^2 + \Delta y_k^2}}{y_k}.$$

with $\Delta x_k = 2\sqrt{2}/N$, $y_k = \sqrt{4 - (k\Delta x_k)^2}$, $\Delta y_k = y_k - y_{k-1}$. To begin the calculation of the sum, $y_0 = \sqrt{2}$. For $N = 50$, $S = 1.763$ as expected.

As a further illustration of the Half-plane model, a triangle may be formed using the geodesic formed between two points and the geodesics to a third point. In the figure below, points B and C are the points used in the discussion above. A third point A on the y-axis has the coordinates $(0, 2\sqrt{2})$. The geodesics forming the triangle sides BA and CA must be circles with centers on the x-axis. These circles are shown in the figure with the circle forming BA having a center c' at $(\sqrt{2}, 0)$ and the other geodesic with center c'' at $(-\sqrt{2}, 0)$.

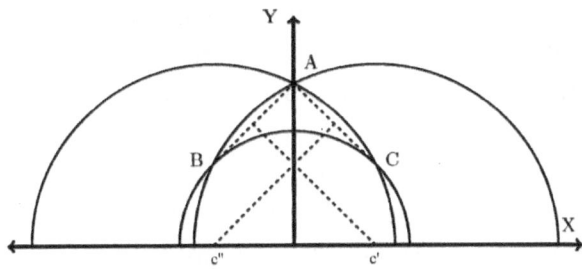

7-8 Half-plane triangle

The locations of the centers c' and c" may be easily found as they lie on the lines perpendicular to the midpoints of the Euclidean lines joining BA and CA. The lines joining the points and the perpendiculars to the midpoints are shown by dashed lines in the figure. For example, the midpoint of BA is $((-\sqrt{2}+0)/2,(\sqrt{2}+2\sqrt{2})/2) = (-\sqrt{2}/2,3\sqrt{2}/2)$. The slope of BA is $\frac{2\sqrt{2}-\sqrt{2}}{0-(-\sqrt{2})} = 1$. The slope of the perpendicular is therefore -1, and the line through the midpoint is $\frac{y-3\sqrt{2}/2}{x-(-\sqrt{2}/2)} = -1$. At $y = 0$, the center $x = c' = \sqrt{2}$. A similar calculation gives, $c" = -\sqrt{2}$. The radii of both circles is the Euclidean distance from c' to B and from c" to A. In both cases the radius r is $\sqrt{10}$. Equation (7-9) can now be used to calculate the lengths of the geodesics from B to A, $S_{geodesicBA}$ and from A to C, $S_{geodesicAC}$.[xcii]

$$S_{geodesic} = \left| \ln\left(\frac{(r-(x_1-c))y_2}{(r-(x_2-c))y_1}\right) \right|.$$

$$S_{geodesicBA} = \left| \ln\left(\frac{\left(\sqrt{10}-(-\sqrt{2}-\sqrt{2})\right)2\sqrt{2}}{\left(\sqrt{10}-(0-\sqrt{2})\right)\sqrt{2}}\right) \right| \approx 0.962.$$

$$S_{geodesicAC} = \left| \ln\left(\frac{\left(\sqrt{10}-(0-(-\sqrt{2}))\right)\sqrt{2}}{\left(\sqrt{10}-(\sqrt{2}-(-\sqrt{2}))\right)2\sqrt{2}}\right) \right| \approx 0.962.$$

$S_{geodesicBC} \approx 1.763$ (as previously calculated). With the side lengths of triangle ABC known, angles A, B, and C can be calculated from equation (7-3, p.199) for angle A and the analogous forms for angles

---

[xcii] The geodesics BA and AC are equal by symmetry; however, the calculations are given to make this explicit in this unfamiliar geometry.

B and C. In the equations below, we associate a/r with BC, b/r with AC, and c/r with BA. Furthermore, equation (7-3) may be evaluated in the spirit of Taurinus's sphere of imaginary radius ir. As the Gaussian curvature $K = -1 = \dfrac{1}{(ir)^2} = -\dfrac{1}{r^2}$, then $r = 1$.

$$\begin{aligned}
\cos A &= (\cosh(b/r) \cdot \cosh(c/r) - \cosh(a/r))/(\sinh(b/r) \cdot \sinh(c/r)), \\
&\approx (\cosh^2(0.962) - \cosh(1.763)/(\sinh^2(0.962)), \\
&\approx -0.602, \\
\angle A &\approx 127°.
\end{aligned}$$

$$\begin{aligned}
\cos B &= (\cosh(a/r) \cdot \cosh(c/r) - \cosh(b/r))/(\sinh(a/r) \cdot \sinh(c/r)), \\
&\approx (\cosh 1.763 \cosh(0.962) - \cosh(0.962)/(\sinh 1.763 \sinh(0.962)), \\
&\approx 0.9487, \\
\angle B &\approx 18.4° = \angle C.
\end{aligned}$$

The sum of the interior angles of triangle ABC is approximately 164° satisfying the hyperbolic geometry requirement that the sum be less than 180.

As another illustration of the consistency of the Half-plane model, the angles may also be determined from the inner product of the tangent vectors to the geodesics at the vertices. This is also a check of the claim that angles in the model are the same as Euclidean angles.

The tangent vectors may be determined from the slopes of the geodesics. The slope at a point is $-(x - c)/y$, using the coordinates of the appropriate geodesic. For example, at A, the slope along the geodesic AB is $-(0 - \sqrt{2})/2\sqrt{2} = 1/2$. A tangent vector with that slope is $\mathbf{t}_{AB} = (2,1)$. Similarly, $\mathbf{t}_{AC} = (-2,1)$. From the discussion and definition of the inner vector product (p. 156):

$$\begin{aligned}
\cos A &= \mathbf{t}_{AB} \bullet \mathbf{t}_{AC}/(|\mathbf{t}_{AB}||\mathbf{t}_{AC}|), \\
&= \frac{(2,1)) \bullet (-2,1)}{(\sqrt{5} \cdot \sqrt{5})}, \\
&= -3/5. \\
\angle A &\approx 127°.
\end{aligned}$$

Similar calculations could be made for angles B and C further illustrating the consistency of the model.

As a final illustration of the Half-plane model, we will look at the angle of parallelism. In the figure below, three limiting parallels, m, n, and p are shown. The lines are parallel to the line l represented as a circle of radius 2, centered at the origin. The centers for the limiting parallels are on the x-axis with x-coordinates of -0.4, -2, and -4. Their radii are indicated by dashed lines. All of the lines meet on the x-axis at $x = 2$, that is they meet at infinity in the model.

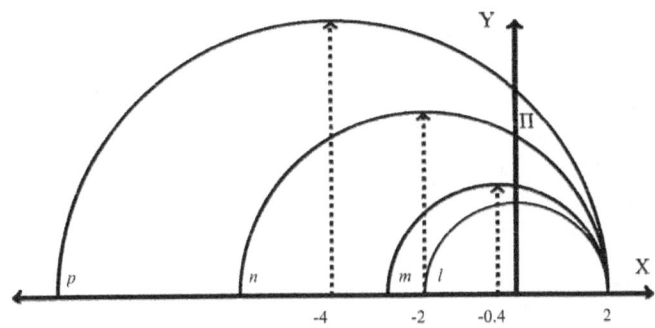

7-9 Half-plane angles of parallelism

Note that the y-axis is perpendicular to line l; hence the y intercepts for lines m, n, and p, being points on the perpendicular to line l, are analogous to point P in figure 3-1 (p. 71) The angle of parallelism is the angle formed by the y-axis and its intersection with the limiting parallel. For the parallel line p represented by the circle centered at $x = -4$ the angle is designated as Π, the symbol used by Lobachevsky. The y-intercept of line p, y_{incpt}. and the associated angle of parallelism is easily calculated. The equation of the line p is $(x - c)^2 + y^2 = (x + 4)^2 + y^2 = 6^2$. Then, the line p intersects the y-axis at $y_{incpt}.(p) = 2\sqrt{5}$. The angle Π may be obtained by noting that the slope of the tangent line of p at $x = 0$ is $-(x-c)/y = c/y_{incpt}.$ $(p) = -4/2\sqrt{5} \approx -0.8944$. The tangent line meets the x-axis at an angle θ, and $\tan\theta = 0.8944$. Using a calculator for the inverse tangent, $\theta = -41.8°$. As the tangent line forms a right triangle with the x-axis, the angle of $\Pi \approx 90° - 41.8° = 48.2°$.

By equation (7-7, p. 203), $\tan(\Pi/2) = e^{-d}$ in which d corresponds to the distance from $y_{\text{incpt.}}(p)$ to the point of the interception of the y-axis with line l, $y_{\text{incpt.}}(l)$ (also see figure 7-6 (p. 202). For line p, $\Pi/2 = 24.1°$ and $\tan\frac{\Pi}{2} \approx 0.4472$. Now the hyperbolic distance along a line of fixed x in the Half-plane model $d = \ln(y_{\text{incpt.}}(p)/y_{\text{incpt.}}(l))$. Focusing on the right hand side of equation (7-7), $e^{-d} = e^{-(\ln(y_{\text{incpt.}}(p)/y_{\text{incpt.}}(l)))} = y_{\text{incpt.}}(l)/y_{\text{incpt.}}(p) = 2/2\sqrt{5} \approx 0.4472$ satisfying equation (7-7). Poincaré's model has once again consistently modelled the hyperbolic geometry of Lobachevsky and Bolyai. Results for all of the limiting parallels in figure 7-9 are given in the table below.

Table 7-1 Half-plane calculations of Π

Line	c	$y_{\text{incpt.}}$	$\tan\theta = c/y_{\text{incpt.}}$	θ	Π	$\tan\Pi/2$	$d = \ln(y_{\text{incpt.}}/2)$	$e^{-d} = 2/y_{\text{incpt.}}$
p	-4	$2\sqrt{5}$	−0.8944	−41.8°	48.2°	0.4472	2.236	0.4472
m	-2	$2\sqrt{3}$	−0.5774	−30°	60°	0.5774	0.5493	0.5774
n	-0.4	2.3664	−0.1690	−9.59°	80.4°	0.8452	0.1690	0.8452
l	0	2	0	0	90°	1	0	1

The table above illustrates one of the key previously discussed characteristics of hyperbolic geometry. The 6th column shows that the angle of parallelism Π increases with decreasing distance from the line l, approaching 90° the Euclidean angle of parallelism. The 7th column, $\tan(\Pi/2)$ is in agreement with the 9th column, e^{-d} confirming the geometry of the geodesics defined as circles centered on the x-axis, and the definition of distance defined by the differential length in equation (7-8, p. 205). It should not be hard to imagine ever increasing circles centered at increasing distances from the origin on the negative x-axis forming limiting parallels such that a vertical line at x =2 can be considered to be a limiting parallel of infinite radius. Of course, a similar discussion could be used to establish limiting parallels approaching x = −2 by geodesics on the positive x-axis.

It is a remarkable result that all of the results of the Half-plane model flow from the equation of differential length. These results include: the Gaussian curvature K = −1, the relationship of triangle sides and angles, the sum of interior angles of triangles being less than 180°, as well as the relationship between distance and the angles of parallelism illustrated in the table. It is, however, important to remember that model of hyperbolic geometry is just one model consistent with the postulates of hyperbolic geometry. One vital element that has not been discussed is how the equation of the differential length defines the geodesic lines. Here I have simply defined straight lines as the segments of Euclidean circles centered on the x-axis. This question will be answered in the next chapter.

I will conclude this section with the Poincaré disk model which has the advantage of covering the entire hyperbolic plane. The equation of the differential length of this model is given below.[213]

$$ds^2 = \frac{dx^2 + dy^2}{\left(1 - \frac{x^2 + y^2}{4}\right)}.$$

Introducing polar coordinates, $x = r\cos\theta$, $y = r\sin\theta$, the differential may also be written as,

$$ds^2 = \frac{dr^2 + r^2 d\theta^2}{\left(1 - \frac{r^2}{4}\right)}. \tag{7-10}$$

In either case, the differential is of the form, $ds^2 = Edu^2 + Gdv^2$, and from equation (6-23, p. 183), the Gaussian curvature $K = -1$. The aptly named Disk model maps the entire hyperbolic plane into a circle of radius 2. The boundary $r = 2$ is at infinity, like the x-axis in the Half-plane model. Also, like the Half-plane model, geodesics are circles intersecting the disk at right angles with angles having their Euclidean measure. A detailed analysis similar to that given for the Half-plane model could also be performed. For qualitative details, I refer you back to the discussion in Section 3.3 *Modeling the hyperbolic plane – a first look.*"

From this section, it should be clear that mathematical models exist which satisfy the postulates of hyperbolic geometric; however, unlike the modeling of the elliptic geometry by a spherical surface, the hyperbolic models are not clearly connected to Euclidean space and may appear to be abstract mathematical structures without connection to the familiar space of our physical world. This concern would be resolved with a model of hyperbolic geometry in Euclidean space that would also profoundly change mathematicians' understanding of geometry and the axiomatic method.

7.2.3   The pseudosphere – the model that changed mathematics

Despite Bolyai's and Lobachevsky's extraordinary discovery of a new geometry starting in the 1820s, the significance of the geometry remained unclear to its discoverers and mostly unknown to other mathematicians for more than thirty years. Bolyai remained uncertain about whether the Parallel Postulate could be proved by Euclid's preceding postulates which if possible would put into question the consistency of his entire work. While he continued to investigate his new geometry, as indicated by his papers, he never published a resolution to his concerns of consistency. Bolyai did not hear of Lobachevsky's discovery until 1848 when he was informed of it by Gauss.[214] Lobachevsky, in contrast to Bolyai, published a number of papers and books on his work,[xciii] Lobachevsky appears to have been focused on the physical reality of his new geometry rather than its implications for axiomatic systems. Lobachevsky concluded that experimentally it would be impossible to determine whether the geometry of the universe followed the Euclidean

---

[xciii]   *On The Principles of Geometry (1830), Imaginary Geometry (1835), New Principles of Geometry, with a Complete Theory of Parallels* (1836), *The Application of the Imaginary Geometry to Some Integrals* (1836), and a small summary of his work in 1840 (Bonola, pp. 85-86). Gauss became aware of Lobachevsky's work about 1841. As with the work of Bolyai, in a letter to a colleague in 1846 Gauss claimed to have already arrived at Lobachevsky's results, discovering them fifty-four years earlier. However, Gauss never publicly associated himself with the new geometry (see Section 3.1 *Gauss' insight*). Bolyai believed that his work had been transmitted to Lobachevsky by Gauss. It appears, however, that Lobachevsky was unaware of Bolyai (Ibid., p. 56).

postulates or his system, based on calculations involving the radius of the earth's orbit and the parallax of a fixed star.[215]

In the 1860s, a number of translations of the work of Lobachevsky in French and German began to appear, including in 1866 one with some extracts of Gauss's letters to his colleagues, adding approval to the investigations of Lobachevsky.[216] (Gauss died in 1855. Lobachevsky died in 1856.) During this period, it was still an open question as to whether the new geometry was fully consistent.[217] Then in 1868, Eugenio Beltrami made the connection between a surface of negative curvature in Euclidean space and the geometry of Lobachevsky and Bolyai. Beltrami proved that taking lines to be geodesics on a surface of negative curvature, Saccheri's Hypothesis of the Actuate Angle (HAA) held, hence all of Lobachevsky's and Bolyai's geometry.[218]

The surface known as a pseudosphere has constant negative curvature and is formed by the revolution of a curve known as a tractrix about an axis. The tractrix has the property that the line formed by the tangent to the curve and its intersection with the ordinate axis is of constant length. This axis of revolution is the curve's asymptote. The form of the curve is illustrated in the figure below in which the axis of revolution is taken as the z-axis, and the tractrix can be described as $z = f(r)$.

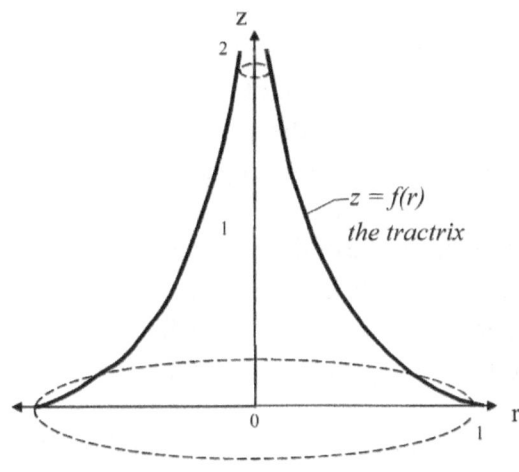

7-10 The pseudosphere

Taking the length of the tangent to the tractrix curve f(r) at the point (r, z) as a, then the following relation holds:

$$\frac{dz}{dr} = f'(r) = -\frac{\sqrt{a^2 - r^2}}{r}. \tag{7-11}$$

This should be clear as in the plane with ordinate z and abscissa r, the tangent line forms a right triangle with a hypotenuse of length a, side of length r, and the third side formed on the z-axis. Note that for r and z positive, the slope is negative. The surface of the pseudosphere $\mathbf{r}(x, y, z)$ in Cartesian coordinates may be expressed in terms of two polar parameters r and θ as,

$$\mathbf{r} = (r \cos \theta, r \sin \theta, f(r)).$$

Using the methods leading to the expression for the First Form, equation (6-8, p. 171), the differential line segment may therefore be expressed as,

$$ds^2 = (1++(f'(r))^2)dr^2 + r^2 d\theta^2 = Edr^2 + Gd\theta^2. \text{ Therefore,}$$

$$E = 1 + (f'(r))^2 = 1 + \frac{a^2 - r^2}{r^2} = \left(\frac{a}{r}\right)^2, G = r^2.$$

Substituting the results for E and G into equation (6-23, p. 183), $K = -\frac{1}{a^2}$.

With this result, we know that all of the results of hyperbolic geometry are valid on the surface of the pseudosphere. The curve with its constant length tangent was known by Jacques Bernoulli (1759-1789).[219] Ferdinand Minding (1806-1885), a student of Gauss, was aware of the constant negative curvature of the pseudosphere and even that spherical trigonometry applied to its surface when interpreted as relations of a sphere of imaginary radius. However, Minding was not aware of the connections to the concepts of Bolyai and Lobachevsky.[220]

What is the equation of the tractrix, z = f(r)? The function f(r) may be obtained by integrating equation (7-11).

$$z = -\int \frac{\sqrt{a^2 - r^2}}{r}\, dr.$$

Let $r = a\cos \phi$, then $dr = -a\sin \phi \, d\phi$, $\sin\phi = \frac{\sqrt{a^2 - r^2}}{a}$.

$$z = -a \int \frac{\sqrt{1 - \cos^2\phi}}{a\cos \phi}(-a\sin \phi)d\phi,$$

$$= a \int \frac{(1 - \cos^2\phi)}{\cos \phi}\, d\phi,$$

$$= a \int \frac{d\phi}{\cos \phi} - a \int \cos \phi \, d\phi.$$

Similar to integration for the length of a geodesic in Poincaré's Half-plane model,[xciv]

$$z = a\ln \left(\frac{1 + \sin \phi}{\cos \phi}\right) - a\sin \phi,$$

$$= a\ln \left(\frac{1}{\cos \phi} + \tan \phi\right) - a\sin \phi,$$

$$= a\ln \left(\frac{a}{r} + \frac{\sqrt{a^2 - r^2}}{r}\right) - \frac{a\sqrt{a^2 - r^2}}{a}.$$

$$= a\ln \left(\frac{a + \sqrt{a^2 - r^2}}{r}\right) - \sqrt{a^2 - r^2}.$$

As in figure 7-10 above, $a = 1$, $K = -1$, and $z(1) = 0$

$$z = \ln \left(\frac{1 + \sqrt{1 - r^2}}{r}\right) - \sqrt{1 - r^2}.$$

The significance of the identification by Beltrami of the pseudosphere with the geometry of Bolyai and Lobachevsky was enormous. As

---

[xciv] See footnote xci, p. 207 and related integration. Here let $u = \frac{1+\sin \phi}{\cos \phi}$. and note that $\frac{d\ln u}{d\phi} = \frac{1}{u}\cdot\frac{du}{d\phi} = 1/\cos\phi.$

Aleksandrov wrote, "Beltrami's discovery at once changed the attitude of mathematicians to Lobačhevskiĭ geometry; from being `fictious ` it became real."[221] As the pseudosphere is embedded within three-dimensional Euclidean geometry, it can be considered to be as consistent a model as the Euclidean geometry that underlies it. The discovery of such a model without the use of the Parallel Postulate meant that, after 2000 years, the search for a proof of the Parallel Postulate could end, knowing that such a proof is impossible. The Parallel Postulate is independent of the Euclid's other postulates. That a non-Euclidean geometry could be physically realized, as the result of the replacement of Euclid's fifth postulate with another, meant that geometry could no longer be considered to spring from either the self-evident truths of the ancient Greeks or from Kant's *a priori* intuition of the human mind. If there remained any doubt about this, then it was removed by Einstein`s discovery that space and time are inextricably interlinked in his Special and General Relativity.

Beltrami's discovery can be considered to be the beginning of the modern abstract axiomatic approach in which a mathematical system is based upon a set of postulates and primitive undefined terms with its validity depending on the observed internal consistency of results. As shall be discussed in the final chapter, the search for proofs of consistency would also lead in the twentieth century to another explosion in the understanding of mathematics. Finally, the new geometry was a "plane" geometry of two coordinate parameters on a surface of constant negative Gaussian curvature. Riemann's famous lecture on the foundations of geometry, given in 1854, but not published until 1867, provided the intellectual basis to extend this to abstract geometries of an n-dimensional space with surfaces on which curvature varied.[222] This extension to the concept of geometry, explored in next chapter, would be crucial to the formulation of Einstein's General Relativity.

8 Riemann's New View of Geometry

8.1 The answer to "what are straight lines?" - geodesics

In the previous chapter, I have defined straight lines as the shortest distance between two points and asserted that this definition can be extended to non-Euclidean geometries by forming these lines along geodesics. Geodesics can be defined in terms of the curvature vector, that is, curves for which the geodesic curvature vector $\mathbf{k}_g = 0$ (equation (6-14, p. 177). In the case of the surface of a sphere, the geodesics are identified with great circles; in Poincaré's Half-Plane and disk models of hyperbolic geometry, they are identified with circular arcs. It is time now to demonstrate the validity of these assertions, tying together these various concepts.

The ancient Greeks recognized a distinction between straight and curved lines; however, Euclid did not make an explicit statement concerning this distinction.[223] Although Euclid in his first postulate states the ability *"To draw a straight line from any point to any point,"* nowhere does he state that it is unique.[224] In contrast to straight lines on the Euclidean plane, an infinite number of great circles connect two poles in spherical geometry. That a straight line is the shortest between two points is reflected in Euclid's Proposition 20: "In any triangle two sides taken together in any manner are greater than the remaining one." Nevertheless, it appears that Euclid's concept of a straight line involves omissions as part of his unstated assumptions, either unrecognized or thought unnecessary.[225] In the context of our concerns here, the concept of the straight line is implicit in Gauss's explicit use of triangles *"formed by shortest lines"* in his proof that the integral curvature is equal to the triangle's angular excess (equation (6-27, p. 186).[226]

Some insight into the problem of finding the shortest line between two points can be obtained by looking at the familiar example of a straight line in the Euclidean plan. As we have seen, the distance S along a line between two points (x_1, y_1) and (x_2, y_2) is (equation 5-7, p. 151),

$$S = \int_{(x_1,y_1)}^{(x_2,y_2)} \sqrt{1 + \left(\frac{dy}{dx}\right)^2}\, dx$$

Suppose we did not know the function $y(x)$ between the two points that provides the shortest line. Then we would need a procedure that would minimize the integral for S. The problem of finding the function minimizing (or possibly maximizing) an integral is known as the calculus of variations and was studied extensively by Jean Bernoulli and later Euler and Lagrange.[227]

The problem can be looked at more generally than the specific problem of finding the line of shortest distance. In general, one seeks to find the function $y(x)$ that, in our case, minimizes the integral I between $y_1 = y(x_1)$, and $y_2 = y(x_2)$ of a function F dependent on the variables x, y, and $dy/dx = y'$.

$$I = \int_1^2 F(x, y, y')\, dx.$$

Introducing a family of curves $\tilde{y} = y + \varepsilon\eta(x)$ with $\eta(x) = 0$ at the end points 1 and 2. The term $\varepsilon\eta(x)$ provides variation from the solution y that minimizes the integral. Using this formulation, the conditions to minimize the integral I may be obtained. Details of this approach are given in Aleksandrov et al.[228] The minimum (or possibly maximum) value of the integral occurs when F satisfies the Euler-Lagrange equation:

$$\frac{d}{dx}\left(\frac{\partial F}{\partial y'}\right) = \frac{\partial F}{\partial y}. \qquad (8\text{-}1)$$

In the case of the integral for Euclidean length, $F = \sqrt{1+y'^2}$. Therefore, since $\frac{\partial F}{\partial y} = 0$,

$$\frac{d}{dx}\left(\frac{\partial F}{\partial y'}\right) = \frac{d}{dx}\left(\partial\left(\sqrt{1+y'^2}\right)/\partial y'\right) = 0,$$

$$= \frac{d}{dx}\left(1/2\frac{2y'}{(1+y'^2)^{\frac{1}{2}}}\right),$$

$$= \frac{y''}{(1+y'^2)^{1/2}} - 1/2\frac{2y'^2y''}{(1+y'^2)^{\frac{3}{2}}},$$

$$= \frac{y''(1+y'^2) - y'^2y''}{(1+y'^2)^{3/2}},$$

$$= \frac{y''}{(1+y'^2)^{3/2}} = 0. \text{ Therefore,}$$

$y'' = 0$, and $y = Ax + B$ (with A and B constants of integration).

Indeed, the result of the Euler-Lagrange equation is consistent with a straight line being a line of constant slope and with this line being the shortest distance between two points just as Euclid assumed. Using the Euler-Lagrange equation, the shortest line between two points on the surface of a sphere will now be shown to be formed on a great circle. We shall first derive the general equation of a great circle in terms of spherical coordinates θ and ϕ (see figure 6-1, p. 167) and then show that this satisfies the Euler-Lagrange equation, hence forms the shortest line.

A great circle is formed by the intersection with a sphere's surface of a plane containing the origin of the sphere. The equation of such a plane in Cartesian coordinates is $ax + by + cz = 0$ (equation (5-10, p. 157). The Cartesian coordinates on the surface of the sphere may be expressed as $x = R\sin\phi\cos\theta$, $y = R\sin\phi\sin\theta$, $z = R\cos\phi$ where R is the radius of the sphere, ϕ is the colatitude, and θ the angle in the plane of the equator as previously given. Substituting these coordinates into the equation of the plane and simplifying results gives the great circle in terms of the relation of ϕ to θ.

$$\cot \phi = A\cos \theta + B\sin \theta. \qquad (8\text{-}2)$$

A and B are constants related to the equation of the plane. The constants A and B may be determined for a particular great circle between two points as the solution of two equations in two unknowns from the specification of each point's meridian and colatitude.

Now the next task is to show from the Euler-Lagrange equation that the great circle provides the shortest path between two points on a sphere's surface. From the differential length ds in spherical coordinates (equation (6-4, p. 169), the integral over a path $\phi = \phi(\theta)$ may be written (with $\phi' = \frac{d\phi}{d\theta}$),

$$S = \int_{\theta_1,\phi_1}^{\theta_2,\phi_2} R\sqrt{\sin^2\phi + \phi'^2}\, d\theta.$$

The conditions from the Euler-Lagrange equation to minimize the integral given above follow.

$\frac{d}{d\theta}\left(\frac{\partial F}{\partial \phi'}\right) = \frac{\partial F}{\partial \phi}$ with $F = R\sqrt{\sin^2\phi + \phi'^2}$.

$$\frac{\partial F}{\partial \phi} = R/2\,\frac{2\sin\phi\cos\phi}{\sqrt{\sin^2\phi + \phi'^2}},$$

$$\frac{\partial F}{\partial \phi'} = R/2\,\frac{2\phi'}{\sqrt{\sin^2\phi + \phi'^2}}$$

$$\frac{d}{d\theta}\left(\frac{\partial F}{\partial \phi'}\right) = R\frac{d}{d\theta}\left(\frac{\phi'}{\sqrt{\sin^2\phi + \phi'^2}}\right)$$

$$= R\frac{\phi''}{(\sin^2\phi + \phi'^2)^{1/2}} - R/2\,\frac{\phi'(2\sin\phi\cos\phi\,\phi' + 2\phi'\phi'')}{(\sin^2\phi + \phi'^2)^{3/2}}$$

Substituting into the Euler-Lagrange equation and simplifying,

$$\frac{\phi''}{\sin^2\phi} - \frac{2\phi'^2 \cos\phi}{\sin^3\phi} = \cot\phi.$$

It may be shown[xcv] that the last equation is equivalent to:

$$\frac{d^2}{d\theta^2}(\cot\phi) = -\cot\phi.$$

Notice that in the last equation, the cotangent function is differentiated twice with respect to θ, returning the negative of the cotangent function. The sine and cosine functions both have this behavior, thus the general solution to the Euler-Lagrange equation is,

$$\cot\phi = A\cos\theta + B\sin\theta.$$

Here A and B are again constants that depend on the particular points that the path passes through. The result is the same as equation (8-2) that describes a great circle. Thus, the great circle has been shown to provide the path of shortest length. A similar procedure could be used to show that the arcs used as straight lines for Poincaré's Half-plane and Disk models of hyperbolic geometry are shortest lines.

As previously noted, Gauss used the term "shortest lines" as applied to triangles on curved surfaces in his work in which he introduced his ideas of local and total curvature. However, it was replaced by the term geodesic curve after its use in an influential work in 1844 by Joseph Liouville (1809-1882). The term geodesic originally was used in regard to measurements of the Earth's surface as part of the subject

---

[xcv]

$$\frac{d}{d\theta}(\cot\phi) = -\phi'\csc^2\phi.$$

$$\frac{d}{d\theta}\left(\frac{d}{d\theta}(\cot\phi)\right) = \frac{d}{d\theta}(-\phi'\csc^2\phi) = -\phi''(\csc^2\phi) - \phi'\frac{d}{d\theta}(\csc^2\phi).$$

$$-\phi''(\csc^2\phi) - \phi'\frac{d}{d\theta}(\csc^2\phi) = -\phi''(\csc^2\phi) - 2\phi'\csc\phi(-\csc\phi\cot\phi(\phi')). \text{ Therefore,}$$

$$\frac{d^2}{d\theta^2}(\cot\phi) = -\left(\frac{\phi''}{\sin^2\phi} - \frac{2\phi'^2 \cos\phi}{\sin^3\phi}\right).$$

223

of geodesy. Liouville took the term from C. Jacobi (1804-1851) who studied "shortest" curves on an ellipsoid of rotation referring to them as "geodesic curves."[229] The lines formed on great circles may be termed geodesics, however, noting that the distance between two points on the great circle depends on the direction of movement along the path. Thus, a geodesic curve may not be the shortest distance. This ambiguity may be removed by defining a geodesic as any curve for which the geodesic curvature vector $\mathbf{k}_g = 0$.[230]

In the example above, the equation for geodesic as a portion of a great circle on the surface of a sphere is given as a function of the spherical coordinates, that is $\phi = \phi(\theta)$ The determination of this function using the Euler-Lagrange equation can be shown to be consistent with the definition of a geodesic as a curve for which $\mathbf{k}_g = 0$.[231]

Although the Euler-Lagrange equation (8-1) provides a method to determine geodesics on two-dimensional surfaces, a generalization of this approach can be extended to n-dimensional geometries and leads to the capability of establishing a differential equation for geodesics which can be determined solely on the basis of the coefficients of Gauss's First Form. This latter capability would be crucial for Einstein's formulation of General Relativity. The extension of equation (8-1) is developed below.

Recall that a curve is parametrized by a single parameter, say t, and the surface by two parameters, u and v, so that the curve is given by u = u(t) and v = v(t) (see equation (6-9, p. 171)). Then, it may be shown, in a manner analogous to that previously described, that a set of Euler-Lagrange equations for each variable u and v will determine extremes (in our case the minimum) of the integral I with $\dot{u} = \frac{du}{dt}$, and $\dot{v} = \frac{dv}{dt}$.[232]

$$I = I_{min.} = \int F(t, u, \dot{u}, v, \dot{v})dt, \text{ if}$$

$$\frac{d}{dt}\left(\frac{\partial F}{\partial \dot{u}}\right) = \frac{\partial F}{\partial u}, \text{ and} \qquad (8\text{-}3)$$

224

$$\frac{d}{dt}\left(\frac{\partial F}{\partial \dot{v}}\right) = \frac{\partial F}{\partial v}. \tag{8-4}$$

For comparison with the use of equation (8-1), we apply equation (8-3) to the case of a spherical surface of unit radius with $u = \theta$, and $v = \phi$.

$$ds = \sqrt{\dot{\phi}^2 + \sin^2\phi(\dot{\theta})^2}\, dt,$$

$$F = \sqrt{\dot{\phi}^2 + \sin^2\phi(\dot{\theta})^2},$$

$$\frac{d}{dt}\left(\frac{\partial F}{\partial \dot{u}}\right) = \frac{d}{dt}\left(\frac{\partial F}{\partial \dot{\theta}}\right) = \frac{\partial F}{\partial u} = \frac{\partial F}{\partial \theta}.$$

$$\frac{\partial F}{\partial \theta} = 0, \frac{\partial F}{\partial \dot{\theta}} = 1/2 \frac{\sin^2\phi(2\dot{\theta})}{\sqrt{\dot{\phi}^2 + \sin^2\phi(\dot{\theta})^2}}.$$

Now, $dt/ds = 1/\sqrt{\dot{\phi}^2 + \sin^2\phi(\dot{\theta})^2}$, therefore,

$$\frac{\partial F}{\partial \dot{\theta}} = \sin^2\phi\left(\dot{\theta}\cdot\frac{dt}{ds}\right) = \sin^2\phi\left(\frac{d\theta}{ds}\right).$$

$$\frac{d}{dt}\left(\frac{\partial F}{\partial \dot{\theta}}\right) = \frac{ds}{dt}\left\{\frac{d}{ds}\left(\frac{\partial F}{\partial \dot{\theta}}\right)\right\} = \frac{\partial F}{\partial \theta} = 0. \text{ Therefore,}$$

$$\frac{ds}{dt}\left\{\frac{d}{ds}\left(\sin^2\phi\left(\frac{d\theta}{ds}\right)\right)\right\} = 0, \text{ and } \frac{d}{ds}\left(\sin^2\phi\left(\frac{d\theta}{ds}\right)\right) = 0.$$

Taking the derivative with respect to s and simplifying, the geodesic equation for θ can be expressed as,

$$\frac{d^2\theta}{ds^2} + 2\cot\phi\,\frac{d\phi}{ds}\frac{d\theta}{ds} = 0. \tag{8-5}$$

Following the same approach, the geodesic equation for ϕ is,

$$\frac{d^2\phi}{ds^2} - \sin\phi\cos\phi\frac{d\phi}{ds}\frac{d\phi}{ds} = 0. \tag{8-6}$$

When the equations for the coordinates are expressed in these forms, the coefficients of the terms $\frac{d\phi}{ds}\frac{d\theta}{ds}$ and $\frac{d\phi}{ds}\frac{d\phi}{ds}$ can be identified as what are known as Christoffel symbols, named after the mathematician E. B. Christoffel (1853-1925).[233] These symbols may be computed directly from a generalization of Gauss' First Form, the metric tensor. The metric tensor, central to Riemann's expanded view of geometry, is introduced in the next section and their relation to the Christoffel symbols in Section 8.3 A general form of the geodesic equation may be expressed using notation such as x^i to represent the two parametric coordinates of the surface, that is, i takes on the values of 1 or 2 (θ or ϕ in the case of the sphere and Γ^i_{jk} being Christoffel symbols).

$$\frac{d^2x^i}{ds^2} + \sum_{j=1}^{2}\sum_{k=1}^{2}\Gamma^i_{jk}\frac{dx^j}{ds}\frac{dx^k}{ds} = 0. \tag{8-7}$$

If a curve satisfies the geodesic equations, then it may be shown that this is equivalent to the condition that its geodesic curvature vector $\mathbf{k}_g$ = 0 or alternatively the result of the Euler Lagrange equations.[234] In the case of the spherical surface, the only non-zero Γ s are:

$$\Gamma^1_{12} = \Gamma^1_{21} = \Gamma^\theta_{\theta\phi} = \cot\phi, \text{ and } \Gamma^2_{11} = \Gamma^\phi_{\theta\theta} = -\sin\phi\cos\phi.$$

With these values for the Christoffel symbols in equation (8-7), the geodesics of a sphere may be determined. The crucial point of equation (8-7) is that the geodesics may be determined for a general surface based solely on the expression that the First Form takes for the line element ds^2. The formulation extends to n-dimensional spaces and is a key element of General Relativity with its four-dimensional space-time. Its expression as part of the language of the coordinate independent entities, known as tensors, was needed by Einstein to express the laws of physics in reference frames with arbitrary motion. However, to gain insight into the source of the standard geodesic equations, we must enter into the extended view of geometry that followed from Riemann's vision.

8.2   The foundation of geometry: the metric tensor

For two thousand years, Euclid's geometry was considered to be the only true description of space. The geometric relations of space so described formed the absolute space in which Newton formulated his Laws of Motion and Law of Universal Gravitation. Thought by the ancient Greeks to be derived by deduction from self-evident truths, it was seen by the influential philosopher Immanuel Kant in his *Critique of Pure Reason*, published in 1781, to be an *a priori* truth of our consciousness that precedes and is necessary for our perception.[xcvi] That there could be other geometries and that the truth needed to be sorted out empirically was antithetical to Kant's views. As I discussed in Section 3.1 *Gauss' insight*, such was the prestige of this view that Gauss never published his ideas on non-Euclidean geometry. Yet in a period of about forty years, the non-Euclidean geometry of Bolyai and Lobachevsky would be discovered and found to exist embedded in the Euclidean world on the surface of constant negative curvature, the pseudosphere. Although Gauss would not publicly disclose his thoughts on non-Euclidean geometry, his work analyzing the characteristics of surfaces with his First Form and his undoubted influence on his student Riemann would lead to a generalization of geometry. The generalization, now known as Riemannian geometry, extended the meaning of non-Euclidean geometry beyond hyperbolic and elliptic geometries to n-dimensional spaces. with non-constant curvature. Like Gauss's First Form, the differential length ds is central to the definition of a Riemannian geometry; however, it is expressed using the mathematical entity called the metric tensor. The use of tensors, which includes vectors, allows geometric properties to be expressed independently of the coordinate systems. This property of tensors was critical to Einstein's formulation of General Relativity. Einstein sought a description incorporating gravity that would be valid for any reference frame in any state of relative motion to other reference

---

[xcvi]   See footnote xxxiii(p.67) and footnoted text.

frames. His General Relativity, ruling out Newton's absolute space and time, expresses the gravitational dynamics of the universe on a geometrical foundation in four-dimensional space-time. In this section and the following two sections, the metric tensor and properties of tensors will be introduced as an entrée to Riemannian geometry and to provide sufficient basics to give meaning to the Einstein's Field Equations and their geometric and physical implications. With that goal, the metric tensor and its properties fundamental to General Relativity will now be introduced.

Gauss's First Form is given in equations (6-7) and (6-8, p. 171).

$$ds^2 = \mathbf{dr} \cdot \mathbf{dr} = (\mathbf{r}_u du + \mathbf{r}_v dv) \cdot (\mathbf{r}_u du + \mathbf{r}_v dv).$$
$$ds^2 = Edu^2 + 2Fdudv + Gdv^2$$

Simply changing notation, it can also be expressed as,

$$ds^2 = g_{11}(dx^1)^2 + g_{12}dx^1 dx^2 + g_{21}dx^2 dx^1 + g_{22}(dx^2)^2.$$

Here, Gauss's notation with E, F, and G is replaced by $g_{11} = E.g_{22} = G$, and if $g_{12} = g_{21}$, then $g_{12} = F$. Also, the differentials of the parametric coordinates du and dv are replaced by the symbols dx^1 and dx^2. You may have noticed the use of superscripts with dx^1 and dx^2. This notation has not been casually chosen and will turn out to have a fundamental meaning; however, the significance of the notation will be discussed later after the interpretation of a vector as a form of the mathematical entity called a tensor. The convenience of the new notation may be further illustrated below using summation notation.

$$ds^2 = \sum_{i=1}^{2} \sum_{j=1}^{2} \frac{\partial \mathbf{r}}{\partial x^i} \frac{\partial \mathbf{r}}{\partial x^j} dx^i dx^j, \tag{8-8}$$

$$= \sum_{i=1}^{2} \sum_{j=1}^{2} g_{ij} dx^i dx^j. \tag{8-9}$$

Finally, as a further simplification in notation, the convention is used that repeated indices indicate summation over the repeated indices. Einstein introduced this convention.

$$ds^2 = g_{ij}dx^idx^j. \qquad (8\text{-}10)$$

We will take this last equation as the definition of the metric tensor g_{ij} in a Riemannian geometry extending its meaning to n-dimensional space. Note that repeated indices are often called dummy indices since changing them does not change the result. For example, in the equation above for the metric tensor, j could be changed to k, that is, $ds^2 = g_{ik}dx^idx^k$. The metric tensor is taken as symmetric, that is $g_{ij} = g_{ji}$.[235] With ds^2 being non-negative, the metric tensor is restricted to forms producing positive results in equation (8-10). Furthermore, endowed with the properties of a tensor discussed below, the form of the metric tensor may be assumed and "the [resulting] structure of Riemannian space is built up around it by definitions that are based on surface or three-space analogies".[236]

In Euclidean geometry using Cartesian coordinates, the metric tensor takes the simple form $g_{ij} = 1$, if $i = j$, and $g_{ij} = 0$, if $i \neq j$ since in three-dimensions, $ds^2 = (dx^1)^2 + (dx^2)^2 + (dx^3)^2$. This simple form of the metric tensor is often known as the Kronecker delta, designated as δ_{ij}.[xcvii] It is sometimes helpful to visualize the metric in its matrix form.[xcviii] Below, I have shown the metric tensor for the case of a spherical surface with $ds^2 = R^2\sin^2\phi\,d\theta^2 + R^2d\phi^2$.

$$g_{ij} = \begin{pmatrix} g_{11} & g_{12} \\ g_{21} & g_{22} \end{pmatrix} = \begin{pmatrix} R^2\sin^2\phi & 0 \\ 0 & R^2 \end{pmatrix}.$$

Here the x^1 and x^2 coordinates corespond to θ and ϕ, respectiviely.

[xcvii] The designation is after the mathematician Leopold Kronecker (1823-1891); (Merzbach and Boyer, pp. 542-543).

[xcviii] An extensive algebra is associated with matrices; however, it will not be employed in this book. An introduction to the algebra of matrices is available in Wrede (pp. 114-121).

The identification of g_{ji} as a tensor will greatly increase our analytical capabilities; however, we must first introduce the properties that define a tensor. The metric tensor g_{ji} is an example of a tensor of rank (or order) 2, that is, there are two indices. In the case of the metric tensor of a surface in two dimensions, as indicated above there are 4 components. In an n-dimensional space, it has n^2 components. A vector is a tensor of order one. In n-dimensional space, it has n components. Vectors may take on two forms, called contravariant and covariant. A contravariant vector is described by the notation v^i going from 1 to n, whereas the covariant form is designated as v_i. (The names contravariant and covariant are older designations that are frequently replaced by designating contravariant vectors as vectors and covariant vectors as one-forms. A covariant tensor of second order is a two-form, etc. I shall use the older names consistent with the use in my references) As we shall see the location of the indices is indicative of the way that the components of a vector, and more generally a tensor, transform from one coordinate system to another. I will illustrate such a transformation for a vector in the Euclidean plane by deriving how components of one Cartesian rectangular coordinate system transform to another Cartesian rectangular system rotated with respect to the original.

In the figure below, the vector **V** is shown relative to the rectangular Cartesian coordinate system (x^1, x^2). The vector **V** makes an angle α with respect to the axis of the x^1 coordinate. Another coordinate system, $(\bar{x}^1, \bar{x}^2)$ is shown with axes indicated by dashed lines. The axes of the $(\bar{x}^1, \bar{x}^2)$ coordinate system are rotated an angle θ with respect to the axes of the (x^1, x^2) coordinate system. The vector **V** originating at the origin, may be represented by coordinates of the point (x^1, x^2). or $(\bar{x}^1, \bar{x}^2)$.

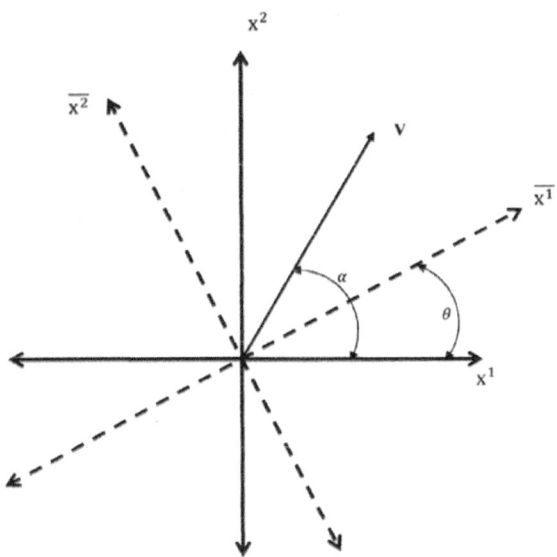

8-1 Vector representation in rotated Cartesian axes

The coordinates of the barred coordinate system are easily interpreted in terms of the angles α, θ, and taking the vector magnitudes as $|V|$.

$$\bar{x}^1 = |V| \cos(\alpha - \theta),$$

$$\bar{x}^2 = |V| \sin(\alpha - \theta)$$

$$x^1 = |V| \cos \alpha,$$

$$x^2 = |V| \sin(\alpha).$$

By using the trigonometric identities for the sine and cosine of the difference of two angles, the barred coordinates may be related to the unbarred.[xcix]

[xcix] From equation (4-9) (p.124): $e^{i(\alpha - \theta)} = \cos(\alpha - \theta) + i\sin(\alpha - \theta)$. Also, $e^{i(\alpha - \theta)} = e^{i\alpha}e^{-i\theta} = (\cos \alpha + i\sin \alpha)(\cos(-\theta) + i\sin(-\theta)) = (\cos \alpha + i\sin \alpha)(\cos(\theta) - i\sin(\theta))$. Multiplying and equating the real part with $\cos(\alpha - \theta)$ and the imaginary part with $\sin(\alpha - \theta)$ gives the identities for differences in angles.

$$\bar{x}^1 = |V|(\cos\alpha\cos\theta + \sin\alpha\sin\theta),$$

$$= x^1\cos\theta + x^2\sin\theta,$$

$$\bar{x}^2 = |V|(\sin\alpha\cos\theta - \sin\theta\cos\alpha),$$

$$= x^2\cos\theta - x^1\sin\theta.$$

With this transformation from the unbarred to the barred coordinate system,

$$(\bar{x}^1)^2 + (\bar{x}^2)^2 + (\bar{x}^3)^2 = (x^1)^2 + (x^2)^2 + (x^3)^2 = |V|.$$

It is common to view a vector as defined as a geometric entity having magnitude and direction. The above equation illustrates that it may also be defined in terms of components satisfying a specific transformation rule between coordinate systems.

The coefficients in the above transformation are equal to the various partial derivatives formed by the transformation of the coordinate system from the unbarred to the barred frame, that is $\partial\bar{x}^i/\partial x^j$ as,

$$\frac{\partial\bar{x}^1}{\partial x^1} = \cos\theta, \frac{\partial\bar{x}^1}{\partial x^2} = \sin\theta, \frac{\partial\bar{x}^2}{\partial x^1} = -\sin\theta, \frac{\partial\bar{x}^2}{\partial x^2} = \cos\theta.$$

Using our indicial notation and the Einstein convention the coordinate transformation can be written as,

$$\bar{x}^i = \frac{\partial\bar{x}^i}{\partial x^j} \cdot x^j, \text{that is,}$$

$$\bar{V}^i = \frac{\partial\bar{x}^i}{\partial x^j} \cdot V^j \tag{8-11}$$

Satisfying the transformation from the unbarred to barred coordinate systems in equation (8-11), V^j and $\bar{V}^i$ are the components of contravariant vectors.[237] The example given above involves only two components on the Euclidean plane; however, the definition for a contravariant vector is general for coordinate systems in n-dimensional space. (A helpful memory device is that the summed indices are in an upper and lower position and the unsummed index is with a barred quantity on both sides of the equation.)

Another example of a contravariant vector is the tangent vector **t** to a curve discussed in Section 6.1 *Curves on the Euclidean Plane*. If **r**(t) is a position vector of a curve in three-dimensional space then using vector and indicial notation, $\mathbf{r} = x^j(t)\mathbf{i}_j$ with x^j being rectangular Cartesian coordinates and the traditional unit vectors **i**, **j**, and **k** designated as $\mathbf{i}_j$, then

$$\mathbf{t} = \frac{d\mathbf{r}}{dt} = \frac{dx^j}{dt}\mathbf{i}_j, \text{ or}$$

$$t^j = \frac{dx^j}{dt}.$$

Let the curve be defined in other coordinates $\bar{x}^j$ and $\bar{\bar{x}}^j$, then in component form, the tangent vector is related to the Cartesian and barred systems by the rules governing total derivatives and partial differentiation (see equation (5-1, p. 138).[238]

$$\frac{dx^j}{dt} = \frac{\partial x^j}{\partial \bar{x}^k}\frac{d\bar{x}^k}{dt}, \tag{8-12}$$

$$\frac{dx^j}{dt} = \frac{\partial x^j}{\partial \bar{\bar{x}}^k}\frac{d\bar{\bar{x}}^k}{dt}, \tag{8-13}$$

$$\frac{d\bar{x}^j}{dt} = \frac{\partial \bar{x}^j}{\partial \bar{x}^k}\frac{d\bar{x}^k}{dt}. \tag{8-14}$$

In each case above, the transformation from one coordinate system to another follows the form of equation (8-11). The first two transformation are with the Euclidean coordinate transformations

$x^j = x^j(\bar{x}^k)$, and $x^j = x^j(\bar{\bar{x}}^k)$; however, the transformation is the same for the general transformation $\bar{x}^j = \bar{x}^j(\bar{\bar{x}}^k)$.

As an alternative way to look at the transformation equations, the first two of the above equations can be explicitly transformed into the third equation (and in the process, give some insight into partial differentiation). A form of the Kronecker delta $\bar{\delta}^l_k = \frac{\partial \bar{x}^l}{\partial \bar{x}^k}$ may be defined recalling its required characteristic is that $\bar{\delta}^l_j = 1$, if $l = k$, otherwise equals 0 (see e.g., p. 229). Furthermore, using the chain rule for partial derivatives (see equations (5-3) and (5-4, p. 139), $\bar{\delta}^l_k = \frac{\partial \bar{x}^l}{\partial x^j}\frac{\partial x^j}{\partial \bar{x}^k}$. Setting the right-hand sides of equations (8-12) and (8-13) equal, and multiplying by $\frac{\partial \bar{x}^l}{\partial x^j}$ leads to:.

$$\frac{\partial \bar{x}^l}{\partial x^j}\frac{\partial x^j}{\partial \bar{x}^k}\frac{d\bar{x}^k}{dt} = \bar{\delta}^l_k \frac{d\bar{x}^k}{dt} = \frac{d\bar{x}^l}{dt} = \frac{\partial \bar{x}^l}{\partial x^j}\frac{\partial x^j}{\partial \bar{x}^k}\frac{d\bar{x}^k}{dt} = \frac{\partial \bar{x}^l}{\partial \bar{x}^k}\frac{d\bar{x}^k}{dt},$$ that is, as in equation (8-14):

$$\frac{d\bar{x}^l}{dt} = \frac{\partial \bar{x}^l}{\partial \bar{x}^k}\frac{d\bar{x}^k}{dt}.$$

In rectangular Cartesian coordinate systems, vector components can be expressed explicitly using base vectors tangential to the coordinate axes, that is, **i**, **j**, and **k**. Analogous base vectors exist for contravariant vectors and have already been defined for two-dimensional curved surfaces. Recall that for the surface **r**(u, v), tangential vectors $\partial \mathbf{r}/\partial u$ and $\partial \mathbf{r}/\partial v$ form base vectors (see figure 6-2, p. 170)). This can be extended to n-dimensional space by defining the base vector as:

$$e_j = \frac{\partial \mathbf{r}}{\partial x_j}.$$

Furthermore, it follows from equations (8-8) and (8-9, p. 228) that for coordinate systems x^i and $\bar{x}^i$,

$$e_i \bullet e_j = g_{ij}, \text{ and } \bar{e}_i \bullet \bar{e}_j = \bar{g}_{ij}. \tag{8-15}$$

The index for the base vectors is in the lower position as it transforms differently from contravariant vectors. Suppose that we wish to transform the vectors $\mathbf{e}_i$ from the unbarred coordinate system x^i to the system $\bar{x}^i$, then,

$$\bar{\mathbf{e}}_i = \frac{\partial \mathbf{r}}{\partial \bar{x}^i}. \text{ Using the chain rule for partial derivatives,}$$

$$= \frac{\partial x^j}{\partial \bar{x}^i} \frac{\partial \mathbf{r}}{\partial x^j},$$

$$= \frac{\partial x^j}{\partial \bar{x}^i} \mathbf{e}_j.$$

This last result differs from equation (8-11) and defines the base vector $\mathbf{e}_i$ as covariant. In general, a vector V_i is defined as a covariant vector, if

$$\bar{V}_i = \frac{\partial x^j}{\partial \bar{x}^i} V_j. \tag{8-16}$$

Just as we can express in Cartesian coordinates the vector $\mathbf{V} = V_1\mathbf{i} + V_2\mathbf{j} + V_3\mathbf{k}$, we can express the general contravariant vector $\mathbf{V}=V^i\mathbf{e}_i$ with covariant base vectors. This form is consistent with the transformation rules for contravariant vectors as shown below.

$$\mathbf{V} = \bar{V}^i \bar{\mathbf{e}}_i = \left(\frac{\partial \bar{x}^i}{\partial x^j} V^j\right)\left(\frac{\partial x^l}{\partial \bar{x}^i} \mathbf{e}_l\right) = \left(\frac{\partial \bar{x}^i}{\partial x^j}\frac{\partial x^l}{\partial \bar{x}^i}\right)V^j\mathbf{e}_l == \delta^l_j V^j \mathbf{e}_l = V^l\mathbf{e}_l = \mathbf{V}.$$

An example of a covariant form of a vector that occurs frequently in physics is the gradient of a scalar. If the scalar is, for example temperature T, then in rectangular Cartesian coordinates x^i the gradient is defined as $\frac{\partial T}{\partial x^i}$. Transforming to another coordinate system $\bar{x}^j$ using the usual rules for partial differentiation,

$$\frac{\partial T}{\partial \bar{x}^j} = \frac{\partial x^i}{\partial \bar{x}^j}\frac{\partial T}{\partial x^i}. \tag{8-17}$$

Clearly, the gradient transforms as a covariant vector. Since the gradient of any scalar is a vector, the operation of forming a gradient is often expressed by the symbol ∇, an example of a vector operator. Suppose we wish to determine the rate of change of temperature with path length s along a path given in rectangular coordinates $\mathbf{r}(x^i(s))$, then forming the total derivative as in equation (5-1, p. 138),

$$\frac{dT}{ds} = \frac{\partial T}{\partial x^1}\frac{dx^1}{ds} + \frac{\partial T}{\partial x^2}\frac{dx^2}{ds} + \frac{\partial T}{\partial x^3}\frac{dx^3}{ds}. \tag{8-18}$$

The gradient operator may therefore be interpreted as:

$$d\mathbf{T} = \nabla \cdot d\mathbf{r} \tag{8-19}$$

The gradient of a scalar being a covariant vector may be transformed to other coordinate systems retaining this meaning. We will encounter it again in Newton's and Einstein's description of gravity.

The covariant base vectors $\mathbf{e}_j$ are not necessarily orthogonal like the familiar rectangular Cartesian base vectors $\mathbf{i}, \mathbf{j}$, and $\mathbf{k}$. Thus, in general, they imply the existence of another set of base vectors orthogonal to them. This set, a contravariant base $\mathbf{e}^j$ is defined by its orthogonality to the covariant base vectors. Because of this relationship, the covariant and contravariant bases are said to be reciprocal.[239]

$$\mathbf{e}^p \cdot \mathbf{e}_q = \delta_q^p. \tag{8-20}$$

A contravariant metric tensor g^{jk} may be defined as the inverse of the contravariant metric g_{jk}, that is,

$$g_{jk}g^{jp} = \delta_k^p. \tag{8-21}$$

For example, if g_{ij} is defined for spherical coordinates by $ds^2 = dr^2 + r^2\sin^2\phi\, d\theta^2 + r^2 d\phi^2$, then

$$g_{ij} = \begin{pmatrix} 1 & 0 & 0 \\ 0 & r^2\sin^2\phi & 0 \\ 0 & 0 & r^2 \end{pmatrix},$$

and

$$g^{ij} = \begin{pmatrix} 1 & 0 & 0 \\ 0 & \dfrac{1}{r^2\sin^2\phi} & 0 \\ 0 & 0 & \dfrac{1}{r^2} \end{pmatrix}$$

At this point, it should be stated that for rectangular Cartesian vectors, no distinction need be made between covariant and contravariant vectors as an orthogonal set of base vectors would be the same orthogonal ones.

From equations (8-15, p. 234), (8-20), and (8-21, p. 236), the following relations between covariant and contravariant forms and their transformed coordinate systems $\bar{x}^i$ can be proved. The processes shown are often referred to as raising or lowering indices.[240]

Raising and lowering of tensor indices   (8-22)

$$\bar{e}_i = \bar{g}_{ij}\bar{e}^j,$$
$$\bar{e}^i = \bar{g}^{ij}\bar{e}_j,$$
$$\bar{V}_i = \bar{g}_{ij}\bar{V}^j,$$
$$\bar{V}^i = \bar{g}^{ij}\bar{V}_j,$$

These relations may also be used to express the contravariant form of the metric tensor in terms of the contravariant base vectors,

$$\bar{g}^{ij} = \bar{e}^i \cdot \bar{e}^j.$$

The metric tensors are second order tensors and are transformed to other coordinate systems by extending the transformations for covariant and contravariant vectors as shown below.

$$\bar{g}_{ij} = \frac{\partial x^p}{\partial \bar{x}^i}\frac{\partial x^q}{\partial \bar{x}^j} g_{pq},$$

$$\bar{g}^{ij} = \frac{\partial \bar{x}^i}{\partial x^p}\frac{\partial \bar{x}^j}{\partial x^q} g^{pq}.$$

This approach can be also applied to tensors of higher orders with both covariant and contravariant indices as should be clear by the pattern indicated in the following example.

$$\overline{T}^{ijk\cdots}_{pqr\cdots} = \frac{\partial \overline{x}^i}{\partial x^a} \frac{\partial \overline{x}^j}{\partial x^b} \frac{\partial \overline{x}^k}{\partial x^c} \cdots \frac{\partial x^u}{\partial \overline{x}^p} \frac{\partial x^v}{\partial \overline{x}^q} \frac{\partial x^w}{\partial \overline{x}^r} \cdots T^{abc\cdots}_{uvw\cdots}.$$

Although the mathematical structure of tensors is complex, it is important to step back and remember that covariant and contravariant forms are just different representations of the same geometric object. This is illustrated below by the different representations of a vector.

$$\mathbf{V} = V^i \mathbf{e}_i = (g^{ij}V_j)(g_{ik}\mathbf{e}^k) = (g^{ij}g_{ik})V_j\mathbf{e}^k = \delta^j_k V_j\mathbf{e}^k = V_j\mathbf{e}^j = \mathbf{V}.$$

Furthermore, the mathematical structure of tensors allows geometric properties such as length, area, and angles to be represented independent of coordinate systems. In developing geometric definitions in Riemannian spaces, a key characteristic is that the coordinate independent definitions should also be valid in Euclidean space. This is illustrated below by the definitions of vector magnitude and the angle between vectors.

Riemannian definition of vector magnitude:

$$|V| = \sqrt{g_{ij}V^iV^j} = \sqrt{V^iV_i} = \sqrt{g^{ij}V_iV_j}.$$

Riemannian definition of angle between vectors V and W:

$$\cos\theta = \frac{g_{ij}V^iW^j}{\sqrt{g_{pq}V^pV^q}\sqrt{g_{rs}W^rW^s}}.$$

In Euclidean geometry with Cartesian bases, $g_{ij} = \delta_{ij}$, and there is no distinction between contravariant and covariant vectors. The above equations for vector magnitude and angles are easily seen to be consistent with the Euclidean definitions. For example, the orthogonality condition for vectors V^i and W^j becomes $V^iW^j = 0$, the same condition provided by the Cartesian inner product.

Much of the machinery of tensor analysis has now been introduced; however, in order to define geodesics and curvature in Riemannian geometries, a coordinate independent description of derivatives of tensors must be developed.

8.3 Tensors and a universal geodesic equation for straight lines and the orbits of planets

Geodesics on the surface of a sphere were identified as great circles using the Euler-Lagrange equation. The analysis was quite specific to the form that the differential length ds takes on a spherical surface. However, it was noted at the end of the section that the results fit into the form of a general equation for geodesics applicable to the geometry of any surface. The geodesic equation (8-7, p. 226)) is reproduced below, with the as yet undefined Christoffel symbols, Γ^i_{jk}.

$$\frac{d^2x^i}{ds^2} + \sum_{j=1}^{2}\sum_{k=1}^{2}\Gamma^i_{jk}\frac{dx^j}{ds}\frac{dx^k}{ds} = 0.$$

With the Einstein convention used in the previous section, this takes on the even simpler looking form,

$$\frac{d^2x^i}{ds^2} + \Gamma^i_{jk}\frac{dx^j}{ds}\frac{dx^k}{ds} = 0. \tag{8-23}$$

Our goal in introducing tensors is to develop a mathematical structure to express geometric relations in a manner independent of the coordinate system. Equation (8-23) looks like it would advance us towards this goal if we knew how the Christoffel symbols were related to different coordinate systems. The key to obtaining this relationship is uncovered by developing a method to take the derivative of a tensor in a manner that will transform appropriately. Some of issues are made apparent by looking at the derivative of a vector V^j in a rectangular Cartesian system x^j, and the implications for the associated contravariant vector $\bar{V}^k$ in the coordinate system $\bar{x}^k$.[241] As $\bar{V}^k$ is a contravariant vector,

$$V^j = \frac{\partial x^j}{\partial \bar{x}^k} \bar{V}^k.$$

Now taking the partial derivative with respect to Cartesian coordinates x^p,

$$\frac{\partial V^j}{\partial x^p} = \frac{\partial}{\partial x^p}\left(\frac{\partial x^j}{\partial \bar{x}^k}\bar{V}^k\right),$$

$$= \frac{\partial^2 x^j}{\partial x^p \partial \bar{x}^k}\bar{V}^k + \frac{\partial x^j}{\partial \bar{x}^k}\frac{\partial \bar{V}^k}{\partial x^p}.$$

The terms on the right-hand side can be modified using the chain rule of partial differentiation so that a comparison with the transformation rules of tensors may be more easily made. Using the chain rule,

$$\frac{\partial}{\partial x^p} = \frac{\partial \bar{x}^q}{\partial x^p}\frac{\partial}{\partial \bar{x}^q}, \text{and}$$

$$\frac{\partial \bar{V}^k}{\partial x^p} = \frac{\partial \bar{x}^q}{\partial x^p}\frac{\partial \bar{V}^k}{\partial \bar{x}^q}, \text{Therefore}$$

$$\frac{\partial V^j}{\partial x^p} = \frac{\partial \bar{x}^q}{\partial x^p}\frac{\partial^2 x^j}{\partial \bar{x}^q \partial \bar{x}^k}\bar{V}^k + \frac{\partial \bar{x}^q}{\partial x^p}\frac{\partial x^j}{\partial \bar{x}^k}\frac{\partial \bar{V}^k}{\partial \bar{x}^q}, \tag{8-24}$$

and[c]

$$\frac{\partial V^j}{\partial x^p} = \frac{\partial \bar{x}^q}{\partial x^p}\frac{\partial x^j}{\partial \bar{x}^k}\left(\frac{\partial \bar{x}^k}{\partial x^t}\frac{\partial^2 x^t}{\partial \bar{x}^q \partial \bar{x}^s}\bar{V}^s + \frac{\partial \bar{V}^k}{\partial \bar{x}^q}\right). \tag{8-25}$$

---

[c]   Focusing on the second partial derivative in equation (8-24), $\frac{\partial^2 x^j}{\partial \bar{x}^q \partial \bar{x}^k}\bar{V}^k = \delta^j_t \frac{\partial^2 x^t}{\partial \bar{x}^q \partial \bar{x}^k}\bar{V}^k$. Now, $\delta^j_t = \frac{\partial x^j}{\partial x^t} = \frac{\partial x^j}{\partial \bar{x}^s}\frac{\partial \bar{x}^s}{\partial x^t}$. Therefore, $\frac{\partial^2 x^j}{\partial \bar{x}^q \partial \bar{x}^k}\bar{V}^k = \frac{\partial x^j}{\partial \bar{x}^s}\frac{\partial \bar{x}^s}{\partial x^t}\frac{\partial^2 x^t}{\partial \bar{x}^q \partial \bar{x}^k}\bar{V}^k$. Dummy indices k and s can be switched and the results then substituted into equation (8-24) leading to equation (8-25). Note also that partial derivatives commute: $\frac{\partial^2 x^j}{\partial \bar{x}^q \partial \bar{x}^k} = \frac{\partial^2 x^j}{\partial \bar{x}^k \partial \bar{x}^q}$.

Notice if the first term in the parenthesis on the right-hand side of equation (8-25) equals zero, then the partial derivative of the vector transforms as a normal tensor with one index covariant and the other contravariant This will not be true, however, unless the coordinate transformations is linear, $x^i = a^i_j x^j + b^i$ with a^i_j and b^i being constants as in the transformation of a Cartesian coordinate system. Thus, a vector or more generally a tensor is not transformed into a tensor by simply taking the partial derivative. However, the term within the parentheses of equation (8-25) transforms like a tensor with one contravariant index and one covariant index. and, as we shall see, provides a definition of the Christoffel symbol $\bar{\Gamma}^k_{qs}$. The entity within the parenthesis of equation (8-25) is known as the covariant derivative of a contravariant vector and is defined below with the semi-colon notation used by Einstein.[242] ci

$$\bar{V}^k_{;q} = \frac{\partial \bar{x}^k}{\partial x^t} \frac{\partial^2 x^t}{\partial \bar{x}^q \partial \bar{x}^s} \bar{V}^s + \frac{\partial \bar{V}^k}{\partial \bar{x}^q}, \tag{8-26}$$

$$= \bar{\Gamma}^k_{sq} \bar{V}^s + \frac{\partial \bar{V}^k}{\partial \bar{x}^q}, \text{ with} \tag{8-27}$$

$$\bar{\Gamma}^k_{sq} = \frac{\partial \bar{x}^k}{\partial x^t} \frac{\partial^2 x^t}{\partial \bar{x}^q \partial \bar{x}^s}. \text{ Therefore,} \tag{8-28}$$

$$\frac{\partial V^j}{\partial x^p} = \frac{\partial \bar{x}^q}{\partial x^p} \frac{\partial x^j}{\partial \bar{x}^k} \bar{V}^k_{;q}. \tag{8-29}$$

Equation (8-29) makes explicit that the covariant derivative transforms as a second order tensor. The significance of the covariant derivative should be clear in that transforming to rectangular Cartesian coordinates, it becomes simply the partial derivative of the vector. Note that it may be shown that the result is general and does not require the involvement of a rectangular Cartesian coordinate system. The steps leading to the covariant derivative were developed following Riemann by a number

---

ci   Other notations exist, for example Wrede (p. 334) expresses this symbolically as $\nabla_q V^k$.

of mathematicians including Beltrami, Christoffel, Rudolph Lipschitz (1831-1904), and particularly, Gregorio Ricci Curastro (1853-1925) and Tullio Levi-Civita (1873-1945).[243]

Einstein, in his description of the use of tensors to formulate the reference-frame-independent theory of General Relativity, developed the covariant derivative and Christoffel symbols in terms of the concept of the parallel displacement of a vector rather than focusing on tensor transformation relations as in the discussion above.[244] I will informally summarize Einstein's approach to give further insight into the covariant derivative.(Another approach is to start with the dependence of base vectors on position as discussed in Appendix C.)

Suppose within some space under consideration, a vector $\mathbf{V}$ (as a contravariant vector V^k) is defined at all points of a coordinate system x^k. We wish to know how $V^k(x^k)$ differs at a point, say P_1, from that at a point P_2 an infinitesimal distance dx^q away. Because in general the base vectors of a general coordinate system are a function of position, even if V^k at P_2 is the same vector as at P_1, its components will change. We wish to evaluate the changes in the vector that are not simply due to changes in the base vectors. To do this we imagine that V^k at P_1 is displaced in a "parallel" manner to P_2, that is V^k is displaced, unchanged in magnitude and direction, an infinitesimal distance dx^q to a position P_2. Then, due to the change of base vectors at P_2, the components of the parallel displaced vector at P_2 change to $V^k + \delta V^k$. However, the vector $\mathbf{V}$ at P_2 is $V^k + dV^k$ with,

$$dV^k = \frac{\partial V^k}{\partial x^q} dx^q.$$

Therefore, the change in $\mathbf{V}$ over the displacement dx^q. is $(V^k + dV^k) - (V^k + \delta V^k) = dV^k - \delta V^k$. The change in components solely due to the coordinate system is given the following expression in the presentation of Einstein in which he introduces the Christoffel symbol, Γ_{qj}^k.[245] Keep in mind that repetition of indices indicates summation,[cii]

$$\delta V^k = -\Gamma_{jq}^k V^j dx^q. \tag{8-30}$$

Thus, the change of the vector **V** over the distance dx^q is,

$$dV^k - \delta V^k = \left(\frac{\partial V^k}{\partial x^q} + \Gamma_{qj}^k V^j\right) dx^q.$$

This change in the vector **V** over the distance dx^q is the basis for the the tensor formulation for the covariant derivative of a contravariant vector and as shown in equation (8-29) transforms as a tensor. As defined in equation (8-28), the Christoffel symbols can be obtained from the transformation from a rectangular Cartesian (unbarred) system to a general coordinate (barred) system, $\bar{x}^j = \bar{x}^j(x^i)$. The equation for the Christoffel symbols is reproduced below and can equally be derived starting with equation (8-30) as in Einstein (end note 245) and Lawden (pp. 95-102).

$$\bar{\Gamma}_{sq}^k = \frac{\partial \bar{x}^k}{\partial x^t} \frac{\partial^2 x^t}{\partial \bar{x}^q \partial \bar{x}^s}.$$

However, as Einstein reminds us, *"Since the quantities $g_{\mu\nu}$ determine all the metrical properties of the continuum* [or geometry], *they must also determine the $\Gamma_{\alpha\beta}^\nu$."*[246]

Recalling that for Euclidean geometry, $g_{pq} = \delta_{pq}$ and $g^{rt} = \delta^{rt}$, we have by covariant and contravariant transformations,

$$\bar{g}_{ij} = \frac{\partial x^p}{\partial \bar{x}^i} \frac{\partial x^q}{\partial \bar{x}^j} \delta_{pq},$$

$$\bar{g}^{bc} = \frac{\partial \bar{x}^b}{\partial x^r} \frac{\partial \bar{x}^c}{\partial x^s} \delta^{rs}.$$

[cii]   The Christoffel symbols turn out to be symmetric in the lower indices; $\Gamma_{qj}^k = \Gamma_{jq}^k$ (Einstein, p. 70).

Taking partial derivatives in the above expressions and forming relationships to replace the coordinate-based definition, equation (8-28), with one based upon metric tensors, we eventually arrive at,[247]

$$\Gamma_{ij}^{k} = \frac{1}{2} g^{kq} \left(\frac{\partial g_{jq}}{\partial x^{i}} + \frac{\partial g_{qi}}{\partial x^{j}} - \frac{\partial g_{ij}}{\partial x^{q}} \right). \tag{8-31}$$

As the metric tensor is symmetric ($g_{ij} = g_{ji}$), Γ_{ij}^{k} is symmetric in the lower indices, i, j. Note that the Christoffel symbols are not tensors, having their own transformation rule;[248] however, this is not a key part of the story here.

The covariant derivative of a covariant vector V_i may be determined in a manner similar to that of the covariant derivative of a contravariant vector, equation (8-27, p. 241), by starting with the parallel displacement of a covariant vector.[249]

$$V_{i;j} = \frac{\partial V_{i}}{\partial x^{j}} - \Gamma_{ij}^{k} V_{k}. \tag{8-32}$$

For comparison the covariant derivative of a contravariant vector, equation (8-27), is reproduced below:

$$V_{;j}^{i} = \frac{\partial V^{i}}{\partial x^{j}} + \Gamma_{sj}^{i} V^{s}.$$

The covariant derivatives of vectors may be extended to tensors of higher order of covariant and contravariant indices in a straightforward manner with the standard transformation of tensors in regard to upper and lower indices.[250]

$$T_{a\cdots;k}^{p\cdots} = \frac{\partial T_{a\cdots}^{p\cdots}}{\partial x^{k}} + \Gamma_{lk}^{p} T_{a\cdots}^{l\cdots} - \Gamma_{ak}^{l} T_{l\cdots}^{p\cdots} + \cdots$$

The preceding analyses have shown how the structure of tensors may be used to define derivatives to accommodate the dependence of base vectors and coordinates on position. The newly defined covariant derivatives retain the transformation properties of tensors while giving the expected results in a rectangular coordinate system. The goal now is to use this approach to define the derivative of a vector along a defined curve in a Riemannian space and, thereby, to define a geodesic curve in that space. As the result must reduce to the Euclidean result, we will use the definition of a straight line in rectangular Cartesian coordinates for motivation. In rectangular Cartesian coordinates in Euclidean space, a straight line can be defined as a curve with a tangent vectors $t = \frac{dx^i}{ds}$ which is the same at any point along the curve. If s is the distance along the straight line then $\frac{dt}{ds} = \frac{d^2x^i}{ds^2} = 0$. We will look for analogous result in Riemannian space.

Let us first look at the contravariant vector V^i. From the discussion on partial derivatives in general coordinates x^i, we know that parallel displacement of the vector from x^i to $x^i + dx^i$ along the path $s = s(x^i)$ will change its components simply due to the dependence of the base vectors on position. As in equation (8-30), this change by displacement ds is described for a contravariant representation of the vector as,

$$V^i + \delta V^i = V^i - \Gamma^i_{jk} V^j dx^k; \text{ however,} \qquad (8\text{-}33)$$

$$V^i_{(x^i + dx^i)} = V^i + \frac{dV^i}{ds} ds. \qquad (8\text{-}34)$$

The components of the parallel displaced vector defined by equation (8-33) equal those defined by equation (8-34) if

$$\frac{dV^i}{ds} ds + \Gamma^i_{jk} V^j dx^k = 0, \text{ or}$$
$$\frac{dV^i}{ds} + \Gamma^i_{jk} V^j \frac{dx^k}{ds} = 0. \qquad (8\text{-}35)$$

Therefore, in general, vectors V^i satisfying equation (8-35) are parallel vectors.

Equation (8-35) may simplified by defining the absolute derivative of a vector V^i as

$$\frac{DV^i}{ds} \equiv V^i_{;k} \frac{dx^k}{ds} = \left(\frac{\partial V^i}{\partial x^k} + \Gamma^i_{jk} V^j\right) \frac{dx^k}{ds}.$$

Noting that the total derivative $\frac{dv^i}{ds} = \frac{\partial v^i}{\partial x^k} \frac{dx^k}{ds}$, equation (8-35) can also be expressed equivalently as,

$$\frac{DV^i}{ds} = V^i_{;k} \frac{dx^k}{ds} = \frac{dV^i}{ds} + \Gamma^i_{jk} V^j \frac{dx^k}{ds} = 0. \qquad (8\text{-}36)$$

If the absolute derivative of a vector along a path equals zero, then in n-dimensional Riemannian space the vectors are parallel. The parallel vectors along the curve form what is known as a vector field. It may be proved that the vectors of the field are constant in magnitude. Furthermore, if each of two vectors forms a vector field along the curve, then the angles between the vectors are constant along the curve. These characteristics apply on the Euclidean plane and are extended to Riemannian geometries through the absolute derivative. [251]

Now consider the vector $v^i = \frac{dx^i}{ds} = \mathfrak{t}$, the tangent vector along a curved path. A curve for which the tangent vectors are parallel as in equation (8-36) is the generalization of the straight line of Euclidean space. Substituting the tangent vector into equation (8-36) (or equivalently, equation (8-35) gives the equation of a "straight line," the geodesic of Riemannian space.

$$\frac{D\mathfrak{t}}{ds} = \frac{D(dx^i/ds)}{ds} = \frac{d^2x^i}{ds^2} + \Gamma^i_{jk} \frac{dx^j}{ds} \frac{dx^k}{ds} = 0. \qquad (8\text{-}37)$$

Clearly, the equation holds for Euclidean straight lines as the Christoffel symbols are zero. On a curved surface, the curve can be shown to satisfy the Euler-Lagrange equations and the condition that the geodesic curvature $\mathbf{k}_g = 0$. or equivalently the curvature vector $\mathbf{k}$ is perpendicular to the tangent plane.[252]

Let us apply equation (8-37) to curves on a surface of a sphere of radius R. Recall then that $ds^2 = R^2 \sin^2\phi\, d\theta^2 + R^2 d\phi^2$. For convenience

I reproduce the metric tensor in its matrix form. Let coordinate $x^1 = \theta$, and $x^2 = \phi$.

$$g_{ij} = \begin{pmatrix} g_{11} & g_{12} \\ g_{21} & g_{22} \end{pmatrix} = \begin{pmatrix} R^2\sin^2\phi & 0 \\ 0 & R^2 \end{pmatrix}.$$

From equation (8-21, p. 236), $g^{11} = 1/(R^2\sin^2\theta)$, and $g^{22} = 1/R^2$. The relationship of the Christoffel symbols to the metric tensors is given in equation (8-31, p. 244) and reproduced below,

$$\Gamma_{ij}^k = \frac{1}{2}g^{kq}\left(\frac{\partial g_{jq}}{\partial x^i} + \frac{\partial g_{qi}}{\partial x^j} - \frac{\partial g_{ij}}{\partial x^q}\right).$$

All of the Christoffel symbols are zero except Γ_{11}^2 and $\Gamma_{12}^1 = \Gamma_{21}^1$.

With $i = k = 1, j = 2$,

$$\Gamma_{ij}^k = \Gamma_{12}^1 = \frac{1}{2}g^{11}\left(\frac{\partial g_{21}}{\partial x^1} + \frac{\partial g_{11}}{\partial x^2} - \frac{\partial g_{12}}{\partial x^1}\right) = \frac{1}{2}g^{11}\frac{\partial g_{11}}{\partial x^2}. \text{ Therefore,}$$

$$\Gamma_{\theta\phi}^\theta = \frac{1}{2R^2\sin^2\phi}\left(\frac{\partial}{\partial\phi}(R^2\sin^2\phi)\right),$$

$$= \frac{1}{2R^2\sin^2\phi}(2R^2\sin\phi\cos\phi),$$

$$= \cot\phi.$$

With $i = j = 1, k = 2$,

$$\Gamma_{ij}^k = \Gamma_{11}^2 = \frac{1}{2}g^{22}\left(\frac{\partial g_{12}}{\partial x^1} + \frac{\partial g_{21}}{\partial x^1} - \frac{\partial g_{11}}{\partial x^2}\right) = -\frac{1}{2}g^{22}\frac{\partial g_{11}}{\partial x^2}. \text{ Therefore,}$$

$$\Gamma_{\theta\theta}^\phi = -\frac{1}{2R^2}\left(\frac{\partial}{\partial\phi}(R^2\sin^2\phi)\right),$$

$$= -\frac{1}{2R^2}(2R^2\sin\phi\cos\phi),$$

$$= -\sin\phi\cos\phi.$$

Substituting these values of the Christoffel symbols into equation (8-37), we have reestablished the equations for the geodesics on the surface of a sphere determined by the Euler-Lagrange equations (8-5) and (8-6, pp. 225, 226).

$$\frac{d^2\theta}{ds^2} + \Gamma^\theta_{\theta\phi} \frac{d\theta}{ds} \frac{d\phi}{ds} = \frac{d^2\theta}{ds^2} + \cot\phi \frac{d\theta}{ds} \frac{d\phi}{ds} = 0.$$

$$\frac{d^2\phi}{ds^2} + \Gamma^\phi_{\theta\theta} \frac{d\theta}{ds} \frac{d\theta}{ds} = \frac{d^2\phi}{ds^2} - \sin\phi \cos\phi \frac{d\theta}{ds} \frac{d\theta}{ds} = 0.$$

Equation (8-37) is the governing equation for geodesics described by their metric. The metric tensor, considered as another description of the First Form, provides as Gauss knew the key to the description of the geometry of two-dimensional surfaces. We have seen how it is used to prescribe length, angles, and shortest lines. However, the First Form when realized in Riemann's vision as a metric extends the concept of geometry to spaces beyond two-dimensions. The development of the metric by mathematicians after Riemann, such as Christoffel, made it possible not only to determine the geodesics of two-dimensional surfaces, but eventually in the hands of Einstein, the geodesics of a four-dimensional space-time that describes the movement of planets, stars, galaxies and even the bending of light passing a star.

Another concept that can be extended to n-dimensional space, like metrics and geodesics, is curvature. Gaussian curvature K is decisive in distinguishing the characteristics of hyperbolic, elliptic, and Euclidean geometries, but it is only defined for two-dimensional surfaces. The curvature of four-dimensional space-time was discovered by Einstein to be central to the interaction between the space-time geometry of the universe and its distribution of mass and energy. Not only does this interaction govern the geodesics of space-time, but the evolution of the universe itself. The final section of this chapter will therefore introduce the curvature tensor.

8.4   Leaving Gauss' surface – the curvature of space

In the previous section, the concept of the parallel displacement of vectors was central to defining Christoffel symbols, parallel vectors in Riemannian space, the absolute derivative, and the equation of a geodesic. This development began with the understanding that a parallel vector V^i displaced an infinitesimal distance along a curve would have its component change by δV^i due to the dependence of the base vectors on position. If we consider the continuous parallel displacement of the vector around a closed curve in Riemannian space, another phenomenon becomes evident that does not occur on the Euclidean plane - the vector on returning to its starting point may not point in its original direction.

As with so many of the concepts that have been introduced, this may be illustrated on the surface of a sphere. Consider our familiar sphere of radius R with coordinates (θ, ϕ) representing respectively, the angle of the meridian and the colatitude expressed in radians. Let us imagine a closed curve formed by a triangle with the three vertices A, B, and C at $(0, \pi/2)$, $(0, 0)$ and $(\pi/2, \pi/2)$ as shown in the figure below.

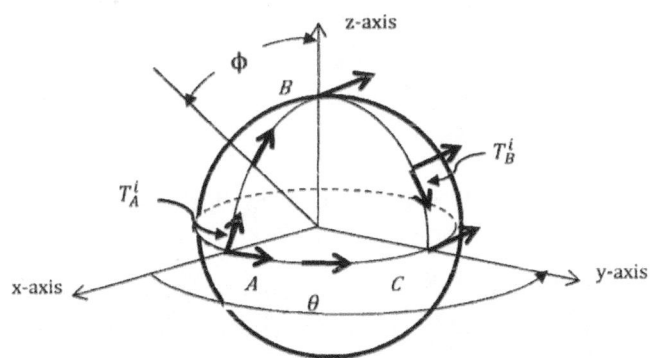

8-2 Parallel displacement of a vector around a closed curve

Starting at point A on the equatorial plane, the tangent vector T_A^i is displaced in a parallel manner along the geodesic line AB until it reaches point B. This parallel displacement is often referred to as parallel transport. Now as noted at the end of the previous section, two parallel vectors maintain a constant angle as they are parallel displaced along a curve. At vertex B, we identify a second tangent vector T_B^i to be the second vector. The vector T_A^i is perpendicular to the geodesic line BC, hence perpendicular to T_B^i as the planes forming AB and BC on the surface of the sphere are perpendicular to each other. The angle between the vectors remains constant as they undergo parallel displacement along BC.

At vertex C, the vector T_A^i is parallel with the equatorial line CA as the meridian BC is perpendicular to CA. Thus, on returning to vertex A, the vector T_A^i has rotated in a clockwise direction an angle of $\pi/2$ radians. This angle is the same as the excess angle that Lambert identified as being associated with the area of the triangle under the Hypothesis of the Obtuse Angle (HOA) in figure 2-9 (p. 61), and more specifically, the excess angles associated with Gauss' total (integral) curvature $\int_A K dA$ in equation (6-27, p. 186). Without going into more detail, we can say that the rotation of the vector in completing the parallel displacement around the close curve is intimately related to the curvature of the surface.

Einstein provides a general discussion of parallel displacement of a vector around a closed curve C resulting in a change from the vector's initial direction.[253] The total change in direction ΔV^i is the integral of the changes over the curve C caused by parallel displacement in the Riemannian space, that is,

$$\Delta V^i = \oint_C \delta V^i = -\oint_C \Gamma_{jk}^i V^j dx^k.$$

The change in direction of the vector V^j undergoing parallel displacement around a closed curve can be used to define a fourth order tensor called the Riemann curvature tensor, R_{jkl}^i. The curvature tensor is a complex function of Christoffel symbols, hence ultimately the metric tensor.[ciii]

[ciii] Additional details from those provided by Einstein are given in Lawden with the curvature tensor designated as B_{jkl}^i (pp. 102-105). Wrede's development of the Riemann curvature tensors follows a different approach (pp. 354-361).

$$\Delta V^i = -\frac{1}{2} R^i_{jkl} V^j f^{kl}, \text{with}$$

$$R^i_{jkl} = \Gamma^i_{rk}\Gamma^r_{jl} - \Gamma^i_{rl}\Gamma^r_{jk} + \frac{\partial \Gamma^i_{jl}}{\partial x^k} - \frac{\partial \Gamma^i_{jk}}{\partial x^l}, \text{and} \qquad (8\text{-}38)$$

f^{kl}, a tensor related to the coordinates of the closed curve.

From the definition of the Christoffel symbols in equation (8-31, p. 244) it should be clear that $R^i_{jkl} = 0$, if the components of the metric tensor are not a function of position. This is the case for Euclidean space with Cartesian coordinates. If $R^i_{jkl} = 0$, then the space is described as flat.[254] This may be true on surfaces that are not Euclidean planes. For example, the surface of a right circular cylinder provides such a case. The position vector on the cylindrical surface of radius R may be taken as $r(\theta, z) = (R\cos\theta, R\sin\theta, z)$ in which case $ds^2 = R^2 d\theta^2 + dz^2$, and the metric components, $g_{11} = R^2$, $g_{22} = 1$, $g_{12} = g_{21} = 0$, are constants. That this space is also considered to be flat from the point of view of the Riemann curvature tensor is consistent with our previous discussion that the sum of the interior angles of a triangle on such a surface is 180° and the Gaussian curvature $K = 0$ (see footnote lxxxiv and footnoted text, p. 187). The identification of "flat" surfaces with the condition $R^i_{jkl} = 0$, along with the tensor's connection to the Gaussian curvature K, discussed below, accounts for it being known as the Riemann curvature tensor.[255]

Using the index lowering property of the metric tensor g_{im} (equation (8-22, p. 237), the fully covariant tensor of the associated Riemann curvature tensor may be formed.

$$R_{ijkl} = g_{im} R^m_{jkl}, \qquad (8\text{-}39)$$

For a two-dimensional surface, the indices will take on the values of either 1 or 2, and the following simple relation results, providing a link with Gaussian curvature K: [256]

$$K = \frac{R_{1212}}{(g_{11}g_{22} - g_{12}^2)}. \tag{8-40}$$

This result may be obtained noting that the Christoffel symbols are functions of the metric tensor, and the Gaussian curvature K is also a function of the metric tensor (see equation (6-23, p. 183) with footnote lxxxii) recalling that in Gauss's First Form $E = g_{11}$, $F = g_{12} = g_{21}$ and $G = g_{22}$. Let us apply equation (8-40) to the familiar case of a spherical surface of radius R with $g_{11} = R^2\sin^2\phi$, $g_{22} = R^2$, $g_{12} = g_{21} = 0$. At the end of the previous section the Christoffel symbols were calculated as

$$\Gamma_{12}^1 = \Gamma_{\theta\phi}^\theta = \cot\phi,$$
$$\Gamma_{11}^2 = \Gamma_{\theta\theta}^\phi = -\sin\phi\cos\phi.$$

All other Christoffel symbols equal zero. $R_{1212} = g_{22}R_{121}^2$. From equation (8-38) with i=2, j=1, k=2, l=1:

$$R_{jkl}^i = R_{121}^2 = \Gamma_{r2}^2\Gamma_{11}^r - \Gamma_{r1}^2\Gamma_{12}^r + \frac{\partial\Gamma_{11}^2}{\partial x^2} - \frac{\partial\Gamma_{12}^2}{\partial x^2},$$

$$= \Gamma_{12}^2\Gamma_{11}^1 + \Gamma_{22}^2\Gamma_{11}^2 - \Gamma_{11}^2\Gamma_{12}^1 - \Gamma_{21}^2\Gamma_{12}^2 + \frac{\partial\Gamma_{11}^2}{\partial x^2} - \frac{\partial\Gamma_{12}^2}{\partial x^2},$$

$$= -\Gamma_{11}^2\Gamma_{12}^1 + \frac{\partial\Gamma_{11}^2}{\partial x^2}$$

$$= -\Gamma_{\theta\theta}^\phi\Gamma_{\theta\phi}^\theta + \frac{\partial\Gamma_{\theta\theta}^\phi}{\partial x^\phi},$$

$$= -\left(-\sin\phi\cos\phi\frac{\cos\phi}{\sin\phi}\right) + \frac{\partial(-\sin\phi\cos\phi)}{\partial\phi},$$

$$= \cos^2\phi + (-\cos^2\phi + \sin^2\phi) = \sin^2\phi.$$

As stated above,

$R_{1212} = g_{22}R_{121}^2 = R^2\sin^2\phi$. Therefore, in equation (8-40),

$$K = R_{1212}/(g_{11}g_{22}) = R^2\sin^2\phi/(R^2\sin^2\phi \cdot R^2) = 1/R^2.$$

The above calculation for a spherical surface gives the same result as that obtained from the Gaussians curvature defined in terms of the radii of curvature, that is $K = 1/(\rho_1\rho_2)$ in equation (6-20, p. 182). This illustrates the close connection between the Riemann curvature tensor and Gaussian curvature. However, the Riemann curvature tensor can be also be applied to n-dimensional spaces. Einstein searched for mathematical forms that would meet his objectives for General Relativity in four-dimensional space-time. Among these objectives were the requirements that all frames of references could equally be used for specifying physical laws, maintenance of features of Special Relativity based upon the constancy of the velocity of light, and consistency with Newton's laws of motion and gravitation as an approximation. He eventually selected the Ricci tensor R_{ij} and its invariant $\mathcal{R}$ called the curvature scalar which are formed from the Riemann curvature tensor,[257] As described by Einstein,[258] the Ricci tensor is formed by contracting (summing) the i and l indices in equation (8-38, p. 251). The Ricci scalar can then be formed by a further contraction with the metric tensor.

$$R_{jk} = R^i_{jki\cdot} = \Gamma^i_{rk}\Gamma^r_{ji} - \Gamma^i_{ri}\Gamma^r_{jk} + \frac{\partial\Gamma^i_{ji}}{\partial x^k} - \frac{\partial\Gamma^i_{jk}}{\partial x^i}. \qquad (8\text{-}41)$$

$$\mathcal{R} = g^{jk}R_{jk}. \qquad (8\text{-}42)$$

For two-dimensional surfaces, the connection between Riemannian and Gaussian curvature shown in equation (8-40) may also be expressed as $\mathcal{R} = 2K$.[259]

Our task will be to develop an understanding of the role these tensors play in Einstein's geometric formulation of the laws that govern matter and gravity in the universe. However, before that, we will need to review Newton's ideas of motion in time in a Euclidean and absolute space, as well as the implications for the concept of energy developed following Newton. These ideas would be overturned by Einstein's discovery of space-time.

PART III

THE GEOMETRIC UNIVERSE

9 Newton

9.1 On the shoulders of giants

"Join me in singing the praises of Newton, who reveals all this,
Who opens the treasure chest of hidden truth,...
No closer to the gods can any mortal be."

From the ode by Edmund Halley
prefacing the *Principia*

In a letter Newton wrote, *"If I have seen farther than Descartes, it is because I have stood on the shoulders of giants."*[260] Indeed, Newton was a beneficiary of a scientific revolution that had been initiated by three developments: the theory of Nicolaus Copernicus (1473 – 1543) which shifted the sun to the center of the solar system, the three laws of Kepler describing the planetary motions,[261] and the astronomical observations with a telescope by Galileo Galilei (1564 – 1642). Galileo's observations supported Copernicus' system through the example of a mini–solar system in the moons revolving about Jupiter, and the observation of phases of Venus, similar to those of the moon, which supported the description of Venus revolving around the sun as an inner planet.[civ]

Newton explained the planetary motions, the orbit of the moon, the tides, as well as the motion of falling objects and projectiles at the earth's surface through his Laws of Motion and Universal Law of Gravitation in

---

[civ] Drake, S., *Discoveries and Opinions of Galileo*, translated by Stillman Drake; a translation by Drake of Galileo's letters describing his observations of Jupiter and Venus, among others, see pp. 51 – 58, pp. 93 – 94. A general discussion of Galileo's scientific and mathematical contributions is given in Kline, pp. 46-52.

the *Principia*[cv] published in 1687.[cvi] To do this he developed the calculus, which was independently discovered in this period by Gottfried Leibniz with important differences in notation. The discovery of the calculus by Newton and Leibniz must be numbered among the great turning points in civilization. Not only did it inspire an explosive growth in mathematics, but it also provided the beginnings of the language that would describe the physics of Mechanics, Electromagnetism, General Relativity, and Quantum Mechanics. Moreover, the mathematical language, brought to life in the discovery of the calculus, remains central to science and engineering.

The calculus was used by Newton as an analytical method to determine the changes in motion of particles in response to forces, in particular his universal gravity. The motion of bodies accelerated by gravity had been studied by Galileo. In addition to Galileo's contributions to the ongoing revolution in astronomy, Galileo also developed the relationships of velocity and distance with time of a uniformly accelerated object under the force of gravity. He verified the relationships experimentally through observations of the time of descent of bodies sliding down inclined planes.[262] These relationships and Kepler's three laws would fall like Newton's legendary apple from Newton's Laws of Motion and his Universal Law of Gravitation. The Laws of Motion as expressed in the *Principia* are given below.[263]

Law 1: *Every body perseveres in its state of being at rest or moving uniformly straight forward, except as it is compelled to change its state by forces impressed.*

Law 2: *A change in motion is proportional to the motive force impressed and takes place along the straight line in which that force is impressed.*

[cv]   *Philosophiæ Naturalis Principia Mathematica (Mathematical Principles of Natural Philosophy.)*

[cvi]   The translation of Newton's *Principia* by I. B. Cohen and A. Whitman is preceded by a very helpful introduction including a history of its publication, many insightful details to aid the reader through dense passages, and overviews of the structures of the three books of the Principia and its concluding section.

Law 3: *To any action there is always an opposite and equal reaction; in other words, the actions of two bodies upon each other are always equal and always opposite in direction.*

The Laws of Motion should be familiar from their introduction in high school physics; however a word needs to be said about its expression in Law 2. Here, by *"motion,"* Newton is referring to a body's momentum **p**, the product of its mass m and velocity **v**, that is, **p** = m**v**, or as Newton expressed it, *"Quantity of motion is a measure of motion that arises from the velocity and the quantity of matter jointly,"*[264]. In addition to this clarification, by *"motive force"* Newton means the product of the impressed force **F** over a period of time dt. This is made clear through numerous examples.[cvii] Finally, the vectorial nature of the relationships of the Laws of Motion is made clear by Newton's statement that *"A body acted on by the* [two] *forces acting jointly describes the diagonal of a parallelogram in the same time in which it would describe the sides if the forces were acting separately"* (see figure 5-5, p. 154). With these clarifications, Law 2 may be written in the notation of the calculus in the form introduced by Leibniz (Newton proved most of his results using geometrical forms).

$$\mathbf{F}dt = d\mathbf{p}, \text{ or } \mathbf{F} = \frac{d\mathbf{p}}{dt} = \frac{d(m\mathbf{v})}{dt}. \qquad (9\text{-}1)$$

With the mass of the body being constant,

$$\mathbf{F} = m\frac{d\mathbf{v}}{dt}, \qquad (9\text{-}2)$$

or in probably the most frequently encountered form,

$$\mathbf{F} = m\mathbf{a}, \mathbf{a} \text{ being the body's acceleration.} \qquad (9\text{-}3)$$

As Newton noted, from the first two of the Laws of Motion, Galileo's result for falling bodies can be obtained, that is *"the descent of heavy*

[cvii] A discussion of the various ways Newton defined force is given in I. B. Cohen's *A Guide to Newton's Principia*, pp. 103-106, preceding the translation of the *Principia*.

bodies is in the square ratio of time...as experiment confirms, except insofar as these motions are somewhat retarded by the resistance of the air.'[265] Using the calculus and taking the force of gravity mg as constant (of which more will be said below), the conclusion is almost trivial.

Let the body's descent be taken as one-dimensional in the vertical coordinate z with the origin at the location where the body begins to fall at time zero. As such only the z component in Law 2 is indicated.

$$F_z = ma_z,$$

$$mg = ma_z, \text{ or } a_z = dv_z/dt = g, \text{ Therefore,}$$

$$v_z = \frac{dz}{dt} = \int_0^t g \, dt = gt, \text{ and}$$

$$z = 1/2gt^2$$

These results are shown schematically in the figure below emphasizing the derivative as a slope and the integral as the area under a curve. Starting on the right with a = g, the distance z = $1/2gt^2$ may be obtained taking the integral as the area under the curves for acceleration and velocity. Or starting on the left with the result for z(t) and interpreting the derivative as a slope, the acceleration a = g is recovered.

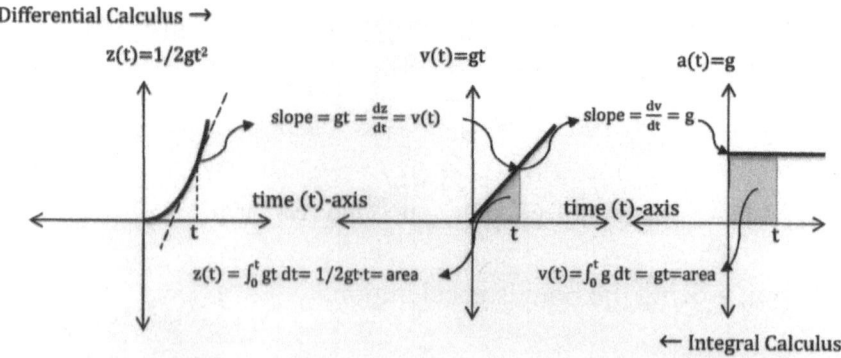

9-1 Calculation of the descent of a falling body under gravity

Of particular interest to our themes are the use of the Laws of Motion and the Law of Universal Gravitation to describe the orbits of the planets and their satellites. Newton's picture of the planets mechanically orbiting in their courses under the influence of force acting at a distance was accepted for over two hundred years. Einstein's General Relativity, discussed in Chapters 11 and 12, replaced Newton's gravitational force with a revolutionary geometrical picture in which the motion of the planets followed geodesics in space-time. As an illustration of the application of Newton's laws of motion and gravitation, the circular orbit of a hypothetical planet around its sun will be used. Other aspects of the laws such as their implication for the meaning of space, time and energy will be explored in the following section and chapter, to make clear the contrast with the radical revisions required by Einstein's theories.

Newton began in Book 1 of his *Principia* with a determination of the motion of bodies under the influence of radially directed forces, what he termed centripetal forces.[266] He proved that such forces result in motion in which a body, such as our hypothetical planet, sweeps out equal areas from the center of motion in equal times.[267] An elliptical orbit results if the force varies as the inverse square of the distance from the source of gravitational attraction, its sun, located at the focus of the elliptical orbit. Furthermore, the cube of the major axis of the ellipse is proportional to the square of the orbital period.[268] These are the laws that Kepler discovered empirically from observations; however, Newton only referred to Kepler in regard to the last of these laws.[269]

Analyzing the reported astronomical observations of planets and finding them consistent with his results for an inverse square law centripetal force, Newton identified his centripetal force with a universal gravity:

"Hitherto we have called 'centripetal' that force by which celestial bodies are kept in their orbits. It is now established that the force is gravity, and therefore, we shall call it gravity from now on.'[270]

The universality of Newton's Law of Gravitation is essentially stated for the inverse square centripetal force in the proposition:

"Gravity exists in all bodies universally and is proportional to the quantity of matter in each.'[271]

Now let us illustrate these laws with the example of a hypothetical planet of mass m moving steadily in a circular orbit about its sun of mass M. Assuming the magnitude of the planet's velocity is constant, V, the acceleration **a** of the planet is solely due to the change in its direction. The circular motion can be conveniently analyzed in a plane using the methods of Section 6.1 *Curves on the Euclidean Plane*." The planet's position about its sun can be expressed by the position vector in Cartesian coordinates $\mathbf{r}(x, y) = (R \cos \theta, R \sin \theta)$ with R being the radius of the orbit. The motion can be described as a function of time t with the curve parameter $\theta = \theta(t)$.

The velocity vector of the planet $\mathbf{V} = \frac{d\mathbf{r}}{dt} = \left(\frac{d\mathbf{r}}{d\theta}\right)\left(\frac{d\theta}{dt}\right)$ $= R(-\sin\theta, \cos\theta)\frac{d\theta}{dt}$. This result may written more simply explicitly in polar coordinates noting that in terms of their Cartesian components the unit radial vector $\mathbf{e}_r = (\cos\theta, \sin\theta)$, and the unit azimuthal vector $\mathbf{e}_\theta = (-\sin\theta, \cos\theta)$ following the discussion of the position vector and tangent to a circle (pp. 164,165). Therefore, $\mathbf{r} = R\mathbf{e}_r$, and $\mathbf{V} = R\frac{d\theta}{dt}\mathbf{e}_\theta = V\mathbf{e}_\theta$. Note that $\frac{d\theta}{dt}$ is the angular velocity of the planet, and the velocity magnitude $V = R \cdot \frac{d\theta}{dt}$.

The acceleration **a** is found by taking the next derivative,

$$\mathbf{a} = \frac{d\mathbf{V}}{dt} = \frac{d^2\mathbf{r}}{dt^2},$$

$$= \frac{d}{dt}(V(-\sin\theta, \cos\theta)),$$

$$= V\frac{d}{d\theta}(-\sin\theta, \cos\theta)\frac{d\theta}{dt},$$

$$= -\frac{V^2}{R}(\cos\theta, \sin\theta), \text{ and}$$

262

$$\mathbf{a} = -\frac{V^2}{R}\mathbf{e}_r. \quad \text{cviii}$$

The force of gravity (a centripetal force, that is a radially directed force) must balance the acceleration, that is $F_g^r = -m\,V^2/R$, (the negative sign indicating that it is a force of attraction). The magnitude of the planet's velocity $V = (2\pi R)/T$, with T being the time of one revolution of our hypothetical planet. From these considerations, the form of Newton's Gravitational Law follows.

$$F_g^r = -m\frac{V^2}{R},$$

$$= -m\frac{\left(\frac{2\pi R}{T}\right)^2}{R}$$

$$= -\frac{4m\pi^2 R}{T^2}.$$

---

cviii The acceleration may also be determined with the absolute derivative defined in equations (8-36) and (8-37, p. 246) and used to determine geodesics. Polar coordinates may be used with $(x^1, x^2) = (r, \theta)$ and $ds^2 = dr^2 + r^2 d\theta^2$. Therefore, $g_{11} = g_{rr} = 1$; $g_{22} = g_{\theta\theta}$ $= r^2$; $g_{12} = g_{r\theta} = g_{21} = g_{\theta r} = 0$. For constant circular motion, $\frac{dx^i}{dt} = \frac{dx^i}{ds}\cdot\frac{ds}{dt}$, and $\frac{ds}{dt} = r\frac{d\theta}{dt} = dx^r/dt =$ constant. As the force is centripetal, the only component of the acceleration a^r is in the radial direction and the acceleration may be expressed equally as a derivative in time or path length with $\frac{ds}{dt}$ being constant.

$$a^r = \frac{D(dx^r/dt)}{dt} = \frac{d^2 x^r}{dt^2} + \Gamma^r_{jk}\frac{dx^j}{dt}\frac{dx^k}{dt}.$$

From the definition of the Christoffel symbols in equation (8-31, p. 244), $\Gamma^1_{22} = \Gamma^r_{\theta\theta} = -r$. Therefore,

$$a^r = \frac{D(dx^r/dt)}{dt} = 0 + \Gamma^r_{\theta\theta}\frac{d\theta}{dt}\frac{d\theta}{dt},$$

$$= -r\frac{V}{r}\frac{V}{r},$$

$$= -\frac{V^2}{r}.$$

As a bit of a complication, the contravariant components are not necessarily the same as their physical counterparts associated with physical displacements $ds^2 = g_{ij}dx^i dx^j$. However, the physical representation can be obtained as $V^i = V^i/g_{ii}$ (no sum on i; Wrede, pp. 234-236). In the present case $g_{rr} = 1$, and there is no difference between the physical and contravariant components.

From Kepler's Laws, the cube of the orbit's radius R^3 is proportional to the square of the orbit's period T^2. Taking into account this proportionality, and the mutual attraction of bodies of masses M and m, the Universal Law of Gravitation can be expressed as

$$\mathbf{F_g} = -\frac{GMm}{R^2}\,\mathbf{e_r}. \tag{9-4}$$

Here G is a constant including the constant of proportionality between R^3 and T^2. It has been illustrated for the hypothetical case of a planet in circular motion about its sun, but it is general for particles of masses m and M separated by a distance R.[cix]

For a body of mass m at a distance d above the earth's surface with R the radius of earth and $d/R \ll 1$:

$$\mathbf{F_g} = -\frac{GmM_{earth}}{(R+d)^2} \approx -GmM_{earth}/R^2 = -mg, \tag{9-5}$$

Therefore, close to the earth's surface the gravitational force is approximately constant. With the vertical coordinate z being an approximation to the radial direction (measured downward), a, the mass's acceleration, and neglecting air resistance and other forces,

$$F_g = mg = ma, \text{ and}$$
$$a = g, \text{ independent of the mass of the falling body.}$$

In general, given the form of the Universal Law of Gravitation in equation (9-4) and the Second Law of Motion in equation (9-3, p. 259), the acceleration due to the gravitational force is the same for all bodies, independent of its mass. That is to say, the mass in the Second Law of Motion, inertial mass, and in the Universal Law of Gravitation, gravitational mass, are the same. This would be one of starting points for Einstein in his development of General Relativity.

[cix] A major conclusion proved by Newton in the *Principia* is that a planet's gravitational force beyond its surface can be calculated as if the planet's entire mass is located at its center (Newton, pp. 806-809).

Newton was well aware that the cause of gravitational force with its remarkable action at distance remained unknown. Despite the extraordinary breadth of the phenomena described by his Laws of Motion and Universal Law of Gravity, he was aware that the lack of an explanation for gravity could be a source of criticism. In defending his work, Newton made clear a point of view that that remains at the heart of science - the fundamental laws of science are inductive and depend on experimental support. In his own words,[272]

*"Thus far I have explained the phenomena of the heavens and of our sea by the force of gravity, but I have not yet assigned a cause to gravity. indeed, this force arises from some cause that penetrates as far as the centers of the sun and planets without any diminution of its power to act, and that acts not in quantity of the **surfaces** of the particles on which it acts (as mechanical causes are wont to do) but in proportion to the quantity of **solid** matter, and whose action is extended everywhere to immense distances, always decreasing as the squares of the distances....I have not as yet been able to deduce from the phenomena the reason for these properties of gravity, and I do not feign hypotheses. For whatever is not deduced from phenomena must be called hypothesis; and hypothesis whether metaphysical or physical, or based on occult qualities, or mechanical, have no place in experimental philosophy. In this experimental philosophy, propositions are deduced from the phenomena and are made general by induction...the laws of motion and the law of gravity has been found by this method. And it is enough that gravity really exists and acts according to the laws that we have set forth and is sufficient to explain all the motions of the heavenly bodies and the sea,"*

Newton's emphasis on experimental support makes clear a fundamental difference between geometry and physics. Both disciplines may be thought of as being founded on postulates with additional propositions being deduced from them. In the case of geometry, the mathematician declares these propositions to be true if the postulates form a consistent system. In contrast the physicist must ask if the postulates are true in the light of experience and in the same spirit can ask is Euclidean geometry true? That Newton understood the limitations of truths formed by induction is clear from his Rule 4 of his *Rules for the Study of Natural Philosophy.*"[273] There he noted that

"In experimental philosophy, propositions gathered from phenomena by induction should be considered either exactly or very nearly true notwithstanding any contrary hypotheses, until yet other phenomena make such propositions either more exact or liable to exceptions."

Such was the case with Newton's Laws for over two hundred years, until new phenomena, such as electromagnetism, would require changes in the Newtonian system. Among them were changes to the Newtonian understanding of time and space which will be described in the next section.

9.2   The stage of Newton's universe: time and space

In the overview of the Newton's Laws in the previous section, it may have occurred to you that nothing was said either of the nature of space, the coordinate systems used to record the motion and location of bodies, or time as the fundamental parameter of motion. Newton had this to say about time.

"Absolute, true, and mathematical time, in and of itself and of its own nature, without reference to anything external, flows uniformly and by another name is called duration. Relative, apparent, and common time is any sensible and external measure (precise or imprecise) of duration by means of motion; such a measure - for example, an hour, a day, a month, a year - is commonly used instead of true time."[274]

Time remains an elusive concept to this day. [cx] Einstein has noted that *"In order to give physical significance of the process of time, processes of some kind are required which enable relations to be established between different places."* As we shall see in the discussion of Special Relativity, the velocity of light is crucial to Einstein's assignment of time in different reference frames.[275] One of Newton's assumptions about time is clear, being absolute; the passage of time is the same for all observers. Such a common sense assumption seems almost trivial, but Einstein discovered in his theory of Special Relativity that this assumption must be incorrect.

Of space, Newton also assumed the existence of an absolute.

"Absolute space, of its own nature without reference to anything external, always remains homogeneous and immovable. Relative space is any moveable or dimension of this absolute space; such a measure or dimension is determined by our senses from the situation of the space with respect to bodies and is popularly used for immovable space, as in the case of space under the earth or in the air or in the heaven, where the dimension is determined from the situation of the space with respect to the earth...Place is the part of space that a body occupies, and it is, depending on the space either absolute or relative...Absolute motion is the change of position of a body from one absolute place to another; relative motion is change of position from one relative place to another."[276]

As an example of absolute motion, Newton states that *"...the fixed stars are also at rest...."*[277]

[cx] Even the basis for the direction of time is in question. We recognize the increasing direction of time in the expectation from experience that a bottle falls and is broken into many pieces rather than reassembling to a whole from its fragments. Thus, in a common scientific view the direction of time is the direction of increasing disorder, that is the thermodynamic property known as entropy. Nevertheless, the physicist Richard Muller questioned this in *The Physics of Time* noting examples of evolving order in the universe. Muller believes that the creation of the universe in the Big Bang, predicted by or at least consistent with General Relativity, as we shall see, is responsible for the flow of time. Time and its direction are created by an expanding universe (Muller, pp. 292-293.)

Einstein summarizes Newton's assumptions of time and space by noting that in pre-relativity physics, time is absolute and length is absolute.[278] In regard to length, "*if an interval, at rest relatively to* [a reference frame] K *has a length s, then it has the same length s, relative to a system* K' *which is in motion relatively to* K. Or using rectangular Cartesian coordinates in Euclidean space, and Δx^i being the differences in the interval's coordinate endpoints, then $s^2 = (\Delta x^1)^2 + (\Delta x^2)^2 + (\Delta x^3)^2 = (\Delta x'^1)^2 + (\Delta x'^2)^2 + (\Delta x'^3)^2$.

Assuming Newton's Laws of Motion are valid in absolute space then they are valid in any reference frame moving with constant velocity with respect to absolute space. These reference frames are known as inertial systems. This can be made clear by supposing that reference frames K' and K" are inertial frames moving with constant velocity v^i with respect to each other. Then if for example, at relative time 0, the origins of rectangular Cartesian reference frames coincide, the position of a particle in the K" frame at time t is,

$$x''^i = x'^i - v^i t. \tag{9-6}$$

From equation (9-6), if a particle in K' is moving at velocity $u'^j = \dfrac{dx'^j}{dt'}$ then,

$$u''^j = \frac{dx''^j}{dt} = \frac{dx'^j}{dt} - v^i = u'^i - v^i ^{\text{cxi}} \tag{9-7}$$

Moreover, if either frame is inertial, then both frames satisfy Newton's Second Law:

$$F''^i = m\frac{d^2 x''^i}{dt^2} = m\frac{d^2 x'^i}{dt^2} = F'^i. \tag{9-8}$$

[cxi] The relations that follow from equation (9-6) are referred to by Einstein as the Galilean transformation (Einstein, p.26).

The implication of this transformation is that the velocity depends on the reference frame. Experimental observations and theoretical considerations of electromagnetic waves led Einstein to theorize that the velocity of light is constant in all reference frames and to reject Newton's absolutes of space and time.

10 Beyond Newton: Conservation of Energy

Embedded in Newton's Laws of Motion are two types of energy: kinetic and potential. Both arise from the analysis of the displacement of an object by a force acting in the direction of displacement. Although, Newton employed in his calculations force acting through a distance and determined its impact on an object's velocity, he did not explicitly identify it as energy, nor did his contemporaries understand it as such.[279]

Let us look at $\int_A^B \mathbf{F} \cdot d\mathbf{r}$ with the force $\mathbf{F}$ and displacement $d\mathbf{r}$. The inner product $\mathbf{F} \cdot d\mathbf{r}$ projects the component of the force along the direction of the path of displacement from location A to B. The force is assumed to act on an object of mass m causing the object to move with velocity $\frac{d\mathbf{r}}{dt} = \mathbf{V}$, magnitude V, and acceleration $\frac{d\mathbf{r}^2}{dt^2}$ following Newton's Second Law, $\mathbf{F} = m\frac{d\mathbf{r}^2}{dt^2}$.

$$\int_A^B \mathbf{F} \cdot d\mathbf{r} = \int_A^B m\frac{d\mathbf{r}^2}{dt^2} \cdot d\mathbf{r},$$

$$= \int_A^B m\frac{d}{dt}\left(\frac{d\mathbf{r}}{dt}\right) \cdot d\mathbf{r},$$

$$= \int_A^B m\frac{d}{dt}\left(\frac{d\mathbf{r}}{dt}\right) \cdot \frac{d\mathbf{r}}{dt} dt,$$

$$= \frac{1}{2}\int_A^B m\frac{d}{dt}\left(\frac{d\mathbf{r}}{dt} \cdot \frac{d\mathbf{r}}{dt}\right) dt,$$

$$= \frac{1}{2}\int_A^B m\frac{d}{dt}(dV^2). \quad \text{Therefore,}$$

$$\int_A^B \mathbf{F} \cdot d\mathbf{r} = \frac{1}{2}mV_B^2 - \frac{1}{2}mV_A^2. \tag{10-1}$$

The right-hand side of equation (10-1) is the difference in the kinetic energy of the mass at positions B and A.

As a simple example, let a ball of mass m be dropped from a height h above the ground. Let z be the vertical coordinate with the origin at the ground (that, is gravity is in the negative z direction, see equation (9-5, p. 264). Neglecting all other forces such as air resistance,

$$\int_h^0 F_z dz = \int_h^0 -mg\,dz = -mgz\Big|_h^0 = -mg(0 - h) = mgh.$$

From equation (10-1),

$$\int_h^0 F_z dz = \int_h^0 -mg\,dz = \frac{1}{2}mV_{z=0}^2 - \frac{1}{2}mV_{z=h}^2 = \frac{1}{2}mV_{z=0}^2, \text{ and therefore,}$$

$$mgh = \frac{1}{2}mV_{z=0}^2.$$

The term mgh is the potential energy of gravity Φ. It is the energy necessary to raise a body of mass m a height h against the downward force of gravity. The ability to define potential energy depends on the energy formed by the force being independent of the path between the beginning or end point. This is known as a conservative force. Since the force is independent of the path, the potential energy may be expressed as a total differential, $d\Phi = -\mathbf{F} \cdot d\mathbf{r}$, or perhaps more clearly,

$$-\int_A^B \mathbf{F} \cdot d\mathbf{r} = \int_A^B d\Phi = \Phi_B - \Phi_A. \tag{10-2}$$

Combining this result with equation (10-1),

$$\frac{1}{2}mV_A^2 + \Phi_A = \frac{1}{2}mV_B^2 + \Phi_B. \tag{10-3}$$

This is the law of conservation of mechanical energy. It only applies when all forces are conservative and potential energies can be defined. In the presence of such forces as friction, it cannot strictly be applied, but is often used as an approximation.

Notice in equations (10-1) and (10-3), only differences in energy occur. We can therefore arbitrarily take any datum as a reference point of zero potential energy. In our example let $z = z_0$, be the reference datum. Then the potential energy at any height $\Phi(z) = - \int_{z_0}^{z} \mathbf{F} \cdot d\mathbf{r} = - \int_{z_0}^{z} (-mg) dz = mg(z - z_0)$. With the reference point for potential energy being z_0, the total mechanical energy at the height h is the potential energy $mg(h - z_0)$. When the ball hits the ground, $\Phi(z = 0) = -mgz_0$. Thus,

$$\tfrac{1}{2} mV_{z=0}^2 + \Phi_{z=0} = \tfrac{1}{2} mV_{z=0}^2 + (-mgz_0) = \tfrac{1}{2} mV_{z=h}^2 + \Phi_{z=h} = 0 + mg(h - z_0), \text{ or as before}$$

$$\frac{1}{2} mV_{z=0}^2 = mgh.$$

In the general case of Newton's Law of Gravity with $F = -(GmM/r^2) \mathbf{e}_r$, the potential energy may be similarly defined. Here, the datum for zero potential is selected as a distance r infinitely far away ($r \to \infty$).

$$\Phi(r) = - \int_{\infty}^{r} \mathbf{F} \cdot d\mathbf{r},$$

$$= - \int_{\infty}^{r} (-GmM/r^2) \mathbf{e}_r \cdot d\mathbf{r},$$

$$= \int_{\infty}^{r} (GmM/r^2) dr,$$

$$\Phi(r) = - GmM/r. \tag{10-4}$$

As an example, conservation of mechanical energy can now be used to determine the velocity V_0 that an object needs to just escape earth's gravity. From equation (10-3) with R being the earth's radius, M the mass of the earth, and m the mass of the object,

$$\frac{1}{2}mV_0^2 + \Phi(R) = \frac{1}{2}mV_\infty^2 + \Phi(\infty),$$

$$\frac{1}{2}mV_0^2 + (-GmM/R) = 0 + 0.$$

$$V_0 = \sqrt{2GM/R}. \tag{10-5}$$

With $G = 6.673 \times 10^{-11}\,m^3kg^{-1}s^{-2}$, $M = 5.979 \times 10^{24}kg$, and $R = 6.371 \times 10^3$ km, $V_0 = 11.2$ km/s.

The use of the gravitational potential instead of the vectors of gravitational force is often advantageous because the potential, being a scalar, does not involve the complication of direction. However, this is not the reason we will revisit it when we turn to General Relativity in the next chapter. Rather, Newton's Universal Law of Gravity, as expressed in terms of the potential, provided Einstein with the insight into the way to formulate his equations. Furthermore, given the success of Newton's Laws, it is necessary that the equations of General Relativity reproduce Newton's experimentally verified results while providing verifiable predictions for phenomena beyond the scope of Newton's laws. These considerations in addition to Einstein's major insight, that an accelerating frame of reference is equivalent to one in which gravity acts, led to the equations of General Relativity[280] (often referred to as the Einstein Field Equations).

The concept of energy as a conserved quantity did not emerge until the nineteenth century. Numerous contributors made important advances before conservation of energy became a recognized principle, as the First Law of Thermodynamics, most notably in the age of steam. I shall mention just two of the early contributors of this period.

Heat was thought in the eighteenth century to be a discrete fluid-like substance that was conserved. Count Rumford with his observation that heat was generated continuously during the operation of the boring of a canon, a mechanical action, made this understanding of heat untenable. Rumford noted that *"...the Heat generated by friction... appeared to be inexhaustible."*[281] This was an early step towards the understanding that energy could be transformed from one form to another. James Prescott Joule recognized that work could be transformed into heat and quantified the transformation by measuring the increase in temperature of water agitated mechanically. In Joule's experiments, the agitation was caused by the falling of a weight which generated motion of a water wheel which in turn raised the temperature of the water.[282] The total energy in such an experiment remains constant and is only transformed from one type to another. Heat would eventually be defined as the transfer of energy by a temperature difference.

Considering the simple example of the ball dropped from a height h, we know from experience it will bounce back up, but to a lesser height and continue the process until the ball comes to rest. Where did the energy go? It was transformed into heat by the action of the ball's compression and expansion on impact with the ground and effects of air resistance.

Eventually, conservation of energy would encompass many forms of energy such as electromagnetic, chemical, and internal energy. Furthermore, the description of energy as motion on a molecular level explained the concepts of heat and internal energy.[cxii] With these ideas in place, Einstein would add the most famous equation of energy, $E = mc^2$, one of many radical consequences of the velocity of light being the same in all reference frames. We now turn to Special Relativity where this principle was first announced.

[cxii] Until the beginning of the twentieth century, the reality of atoms and molecules was still in question. Einstein contributed significantly to the acceptance of that reality through his investigation of Brownian motion connecting diffusion, and thus heat, to molecular motion (Pais, pp. 90 - 100).

11 Einstein

11.1 Special Relativity: the merging of time and space

11.1.1 The Lorentz transformation

With the publication of the *Principia* in 1687, Newton created a paradigm that would dominate scientific thought in physics for more than two hundred years. The universe of Newton was driven mechanically and deterministically by forces between particles. The mathematician and scientist Pierre Simon Laplace (1749 – 1827) was so confident that all could be explained by this approach that he stated,

> *"An intellect which at any given moment knew all the forces that animate nature and the mutual positions of the beings that compose it,...for such an intellect nothing could be uncertain; and the future just like the past would be present before its eyes."*[283]

However throughout the nineteenth century, new physical phenomena were being discovered that would eventually upend this mechanical view of the universe. Of particular importance in initiating a new phase in physics were the discoveries of electromagnetic phenomena by, among others, Hans Christian Oersted (1777 – 1851) and Michael Faraday (1791-1867).[284] Oersted observed that an electric current creates a magnet field, while Faraday observed the reciprocal phenomenon that an electric current is induced by a changing magnetic field.[cxiii] These phenomena were initially analyzed from a Newtonian point of view. Faraday viewed electric and magnetic phenomena as stresses

---

[cxiii] A discussion of these phenomena is given by Einstein and Infeld, pp.125-144.

and motions of a material medium. The observed electromagnetic phenomena were comprehensively explained and generalized in a mathematical theory by James Clerk Maxwell (1831-1879) with similar material interpretations. [285] In the theory, however, phenomena were controlled by equations with the central role for magnetic and electric fields rather than the electrically charged particles and magnetic materials of the experiments.

The electromagnetic fields of the Maxwell Equations propagate as waves, unlike Newton's gravitational force which acts instantaneously at a distance. At first, the fields were considered to be expedient methods of calculation, but their reality soon became evident. Maxwell concluded from his theory that light was an electromagnetic wave. This was confirmed experimentally by Heinrich Hertz (1857 - 1894).[286] Still trying to preserve the connection with Newton's mechanical universe, it was theorized that light propagated in a medium called the ether in a manner similar to material waves such as sound or water waves. The search was therefore on for the medium in which the electromagnetic waves propagated and for the material characteristics of the medium. If light propagated in the ether, one would expect that velocity of light to change depending on the motion of the light source with respect to the ether. The optical experiments of Albert Michelson (1852 - 1931) and Edward Morley (1838 - 1923) are the most famous of those seeking evidence of the changes in the speed of light due to its propagation in the ether. No such change was detected by Michelson and Morley or has one ever been detected.[287] The problem of the ether was solved by Einstein in 1905 with his theory of Special Relativity by dismissing its existence as a viable physical hypothesis and replacing it by two principles:[288]

Special Relativity Principle 1: *"The velocity of light* [c] *in vacuo is the same in all CS* [Coordinate Systems] *moving uniformly relative to each other."*

Special Relativity Principle 2:} *"All laws of nature are the same in all CS moving uniformly relatively to each other."*

I will refer to Special Relativity's first principle as the constancy of the velocity of light. This principle has a profound impact on our understanding of space and time, and requires a new formulation of the laws of motion. To begin to see this, let us look at the propagation of light as seen by two coordinate systems moving uniformly relative to each other. For context, recall that constancy of the velocity of light contradicts our normal perception. For example, if two people are running at the same velocity, but one is on the ground and the other gets on a moving train, their velocities will be different with the runner on the train having a velocity augmented by the addition of the train's velocity. This is just the Galilean transformation of velocities derived from $x''^i = x'^i - v^i t$ (equation (9-6, p. 268) for coordinate systems moving uniformly with respect to each other.

Assume that a light wave is initiated by a point source of light at the origin of each of two Cartesian coordinate systems at time zero as they pass each other. In one of the coordinate systems, the location of the spherical light wave is such that $x^2 + y^2 + z^2 = (ct)^2$, or $x^2 + y^2 + z^2 - (ct)^2 = 0$. This is an application of nothing more than distance $\sqrt{x^2 + y^2 + z^2}$ equals the product of the speed c and time t. The second system moves at a velocity $\mathbf{v}$ with respect to the first. What is shocking in Einstein's theory is that in the other coordinate system (x', y', z'), it is also true that, $x'^2 + y'^2 + z'^2 - (ct')^2 = 0$ with t' being time measured in the moving frame. The velocity of the primed frame has no impact on the velocity of light! Because the invariant includes time, we must have a space – time geometry rather than separate realms for an absolute space and absolute time. Recall that in Newton's mechanics, this was the backdrop for the inertial coordinate systems, where his Laws of Motion are valid.

The relationship between coordinate systems moving uniformly with respect to each other that maintains space – time invariants was determined by Hendrik A. Lorentz (1853 – 1928) years before Einstein published his theory of Special Relativity in 1905. However, Lorentz derived his transformation in the context of issues related to the search for the ether.[cxiv] Einstein recognized their fundamental meaning in terms of the constancy of the velocity of light.[289]

The Lorentz transformation given below is for the simplified case in which one Cartesian reference frame (the primed coordinate system) moves with a positive velocity, v parallel to another Cartesian reference frame's x-axis (the unprimed system), and the coordinate frames' origins coincide at $t = t' = 0$. The simplified case makes the radical implications of Special Relativity more readily apparent.

$$x' = (x - v\,t)/\sqrt{1 - v^2/c^2}. \qquad (11\text{-}1)$$

$$y' = y. \qquad (11\text{-}2)$$

$$z' = z. \qquad (11\text{-}3)$$

$$ct' = \left(c\,t - v\frac{x}{c}\right)\Big/\sqrt{1 - v^2/c^2}. \qquad (11\text{-}4)$$

You should check that the space-time invariant, $x^2 + y^2 + z^2 - (ct)^2 = 0$, is maintained in the primed coordinate system. Also note that if two events occur at (x_1, y_1, z_1, t_1) and (x_2, y_2, z_2, t_2) in the unprimed frame with $x_2 - x_1 = \Delta x$, $y_2 - y_1 = \Delta y$, $z_2 - z_1 = \Delta z$, and $t_2 - t_1 = \Delta t$, then using the Lorentz transformation,

$$\Delta x' = (\Delta x - v\,\Delta t)/\sqrt{1 - v^2/c^2}. \qquad (11\text{-}5)$$

$$\Delta y' = \Delta y\,'. \qquad (11\text{-}6)$$

$$\Delta z' = \Delta z. \qquad (11\text{-}7)$$

[cxiv] Lorentz's transformation followed the proposal in 1893 of George Francis Fitzgerald (1851 - 1901) to explain the inability of experiments to detect the ether by relating length contraction to relative motion (Asimov, pp. 336-340; Pais, pp. 122-123).

$$c \, \Delta t' = (c \, \Delta t - v \, \Delta x/c)/ \sqrt{1 - v^2/c^2}. \qquad (11\text{-}8)$$

Also, the event interval, ΔS is invariant in all frames:

$$\Delta S^2 = \Delta x^2 + \Delta y^2 + \Delta z^2 - c^2 \Delta t^2 = \Delta x'^2 + \Delta y'^2 + \Delta z'^2 - c^2 \Delta t'^2.$$

A quick look at Lorentz transformation equation (11-8) tells us that if someone in the primed frame measures a time change $\Delta t'$ while observing you in the unprimed frame, then your wristwatch ($\Delta x = 0$) will see only a time change of, $\Delta t = \Delta t' \sqrt{1 - v^2/c^2}$. In other words your time has slowed down. As the velocity v approaches the speed of light, time for you will become slower and slower. Experimentally we see this in the longer lifetimes of accelerated decaying atomic particles.

Other fascinating consequences of the transformation are that lengths contract in moving systems and that events are not simultaneous in all systems. To see how the phenomenon of length contraction is predicted by the Lorentz transformation, let there be a stationary bar in the moving frame of length $\ell' = x'_2 - x'_1$. In the unprimed frame, the bar length is measured by marking simultaneously the location of its ends at a time t as it moves by the unprimed frame and $\ell = x_2 - x_1$. Using Lorentz transformation equation (11-5), $\ell' = \ell/\sqrt{1 - v^2/c^2}$, or from the point of view of the unprimed frame, the bar is contracted compared to the measurement in the moving frame, $\ell = \ell' \sqrt{1 - v^2/c^2}$. However, the transformation equations indicate that the contraction only occurs in the direction of motion. No contraction occurs in the directions perpendicular to motion.

In regard to events that are simultaneous in one frame, notice that when the bar is marked simultaneously at time t, at its ends, x_1 and x_2 the Lorentz transformation for time tells us that an observer in the moving frame will tell us the marks were made at two difference times, $t'_1 = (t - vx_1/c^2)/\sqrt{1 - v^2/c^2}$, and $t'_2 = (t - vx_2/c^2)/\sqrt{1 - v^2/c^2}$. Einstein was aware of criticism of his theory for giving a central theoretical role of the propagation of light in the definition of time. To this criticism Einstein responded.

"In order to give physical significance to the concept of time, processes of some kind are required which enable relations to be established between different places. It is immaterial what kind of processes one chooses for such a definition of time. It is advantageous, however, for the theory, to choose only those processes concerning which we know something certain. This holds for the propagation of light in vacuo to a higher degree than for any other process which could be considered."[290]

The profound impact on the descriptions of the physical universe made by the principle of the constancy of the velocity of light and the Lorentz transformation that is a consequence of that principle cannot be overstated. Absolute time and space disappear. Along with them, the Euclidean invariant $\ell^2 = x^2 + y^2 + z^2$ has been replaced by a new space-time invariant $S^2 = x^2 + y^2 + z^2 - (ct)^2$. The transformation from one inertial system to another, so straight forward in Newton's Laws of Motion with the Galilean transformation, no longer works if the Lorentz transformation is applied (as will be discussed in more detail below).

The geometric character of the Lorentz transformation was first made explicitly by Herman Minkowski in 1908. Minkowski had been Einstein's teacher in Zurich.[291] By defining the fourth dimension of space-time as imaginary, $x^4 = ict$, the space-time invariant has the Euclidean appearance,

$$x^2 + y^2 + z^2 - (ct)^2 = (x^1)^2 + (x^2)^2 + (x^3)^2 + (x^4)^2.$$

The Lorentz transformation in this guise can then be seen as a rotation by an angle ψ of the primed frame about the origin of the unprimed frame with $\tan\psi = iv/c$ (and therefore, $\cos\psi = 1/\sqrt{1 - v^2/c^2}$, $\sin\psi = (iv/c)/\sqrt{1 - v^2/c^2}$, also see discussion of figure 8-1 (p. 231). The following form of the Lorentz transformation results,[292]

$$x'^1 = x^1 \cos \psi + x^4 \sin \psi.$$
$$x'^2 = x^2.$$
$$x'^3 = x^3.$$
$$x'^4 = -x^1 \sin \psi + x^4 \cos \psi.$$

In expressing a geometric view of Special Relativity, Minkowski went on to say, *"Henceforth space by itself and time by itself, are doomed to fade away into mere shadows, and only a kind of union of the two will preserve an independent reality."* Einstein later acknowledged his indebtedness to Minkowski's vision for facilitating the transition from Special to General Relativity.[293] cxv

In contrast to the loss of the Newtonian world view, Maxwell's equations of electromagnetism are consistent with the Lorentz transformation.[294] It is beyond the scope here to elaborate upon this result; however, the nature of space and time in the Lorentz transformation is crucial to the findings of Oersted and Faraday in which electrical phenomena involving the motion of charges results in the creation of magnetic fields and which is reciprocated in that the motion of magnetic sources creates electric fields. To this point, Einstein's revolutionary paper of 1905 is entitled, *"On the Electrodynamics of Moving Bodies."*[295] Einstein's results for the electromagnetic fields can be explicitly described by introducing tensor expressions for the electromagnetic fields in four-dimensional space-time.

---

cxv   The use of an imaginary time coordinate somewhat obscures the significant difference between the Euclidean three-dimensional invariant, which is always greater than zero, and the space-time invariant which can be less than zero, or in the case of light, equal to zero. If $x^4 = ct$, then the Lorentz transformation may be expressed in terms of hyperbolic functions highlighting this difference. Let $\tanh \alpha = \sinh \alpha / \cosh \alpha = v/c$ (see discussion of hyperbolic functions beginning with equation (4-13, p. 125), and note that $1 - \tanh^2 \alpha = 1/\cosh^2 \alpha$, and $\cosh^2 \alpha - \sinh^2 \alpha = 1$, then:

$$x'^1 = x^1 \cosh \alpha - x^4 \sinh \alpha.$$
$$x'^2 = x^2.$$
$$x'^3 = x^3.$$
$$x'^4 = -x^1 \sinh \alpha + x^4 \cosh \alpha.$$

Despite the deficiencies in the Newtonian view described by Einstein, the extraordinary success of Newton's Laws of Motion must be accounted for. With such success it is not surprising that the Lorentz transformations are approximately the same as the Galilean transformation for small relative velocities, that is $^v/_c \ll 1$. Our task is to find a formulation of the laws of motion in four-dimensional space-time consistent with Newton's Laws at low velocities.

11.1.2 Four vector momentum and $E = mc^2$

Once the Lorentz transformations are accepted as the relationship between coordinates of inertial frames moving at constant velocity relative to each other, the definitions for velocity and acceleration that are used in Newton's Laws of Motion are no longer valid as they do not transform properly.[cxvi] Assume for a moment that the motion of an object is only in the x direction. In the unprimed reference frame, as usual, the velocity of the object $u = \frac{dx}{dt}$. Similarly, when observed in another frame moving with relative velocity v to the unprimed frame, $u' = \frac{dx'}{dt'}$. From equations (11-1) and (11-4) the differentials dx' and dt' can be found in terms of those of the unprimed variables x and t. The result is,

$$\frac{dx'}{dt'} = u' = \frac{\frac{dx}{dt} - v}{1 - \frac{(dx/dt)v}{c^2}} = \frac{u - v}{1 - \frac{uv}{c^2}}.$$

For low velocities, v^2/c^2, $u^2/c^2 \ll 1$, the Galilean result is obtained: $u' = u - v$. Nevertheless, the velocity u does not transform as required by the Lorentz transformation. If $\mathcal{U}$ is a vector in four-dimensional space-time, it must transform as shown below.

[cxvi] As in the previous section, in the following presentations, it is assumed for simplicity that relative motion of reference frames is only in one direction, taken as the x axis.

Space-Time 4-Vector

$$\mathfrak{U}'^1 = (\mathfrak{U}^1 - (v/c)\mathfrak{U}^4)/\sqrt{1 - v^2/c^2}. \tag{11-9}$$

$$\mathfrak{U}'^2 = \mathfrak{U}^2. \tag{11-10}$$

$$\mathfrak{U}'_3 = \mathfrak{U}_3 \tag{11-11}$$

$$\mathfrak{U}'^4 = (\mathfrak{U}^4 - (v/c)\mathfrak{U}^1)/\sqrt{1 - v^2/c^2}. \tag{11-12}$$

Four vectors are part of a geometry with a metric tensor of four-dimensional space-time. Following Einstein's notation, the metric tensor is represented with Greek indices to remind the user that they span from 1 to 4. From the space-time invariant, $dS^2 = g_{\mu\nu}dx^\mu dx^\nu = g'_{\mu\nu}dx'^\mu dx'^\nu$, with $g_{\mu\nu} = g'_{\mu\nu}$ represented in its matrix form below (with $x^4 = ct$).

$$ds^2 = g_{\mu\nu}dx^\mu dx^\nu; \quad g_{\mu\nu} = \begin{pmatrix} 1 & 0 & 0 & 0 \\ 0 & 1 & 0 & 0 \\ 0 & 0 & 1 & 0 \\ 0 & 0 & 0 & -1 \end{pmatrix} \tag{11-13}$$

The inner product of the four-vectors forms an invariant,

$$\begin{aligned} \mathfrak{U}' \cdot \mathfrak{U}' &= g_{\mu\nu}\mathfrak{U}'^\mu \mathfrak{U}'^\nu, \\ &= (\mathfrak{U}'^1)^2 + (\mathfrak{U}'^2)^2 + (\mathfrak{U}'^3)^2 - (\mathfrak{U}'^4)^2, \\ &= (\mathfrak{U}^1)^2 + (\mathfrak{U}^2)^2 + (\mathfrak{U}^3)^2 - (\mathfrak{U}^4)^2, \\ &= \mathfrak{U} \cdot \mathfrak{U}. \end{aligned}$$

To obtain a four-vector of velocity we need to take the derivative of the space-time coordinates x^μ with respect to a time-like coordinate that does not change with the frame of reference. Such an invariant can be formed with the event interval, $\Delta S^2 = (\Delta x^1)^2 + (\Delta x^2)^2 + (\Delta x^3)^2 - (\Delta x^4)^2 = \Delta x^2 + \Delta y^2 + \Delta z^2 - (c\Delta t)^2$. Let $\Delta \tau$ be the time change of a stationary clock, that is, a clock in a frame (x', y', z, t') attached to the moving object. This is known as the proper time. In any other inertial frame (x, y, z, t):

$$(\Delta\tau)^2 = \Delta t'^2 = -\frac{\Delta S^2}{c^2} = (\Delta t)^2 - \frac{(\Delta x^2 + \Delta y^2 + \Delta z^2)}{c^2}.$$

We want to form the four-vector velocity $\mathfrak{U}= (dx/d\tau, dy/d\tau, dz/d\tau, d(ct)/d\tau) = (\mathfrak{U}^1, \mathfrak{U}^2, \mathfrak{U}^3, \mathfrak{U}^4)$, and to relate these four-vectors to our familiar velocities $\mathbf{u} = (dx/dt, dy/dt, dz/dt) = (u^1, u^2, u^3)$. Using the chain rule, $\mathfrak{U}^1 = \frac{dx}{dt}\frac{dt}{d\tau}$. Similarly, $\mathfrak{U}^2 = \frac{dy}{dt}\frac{dt}{d\tau}, \mathfrak{U}^3 = \frac{dz}{dt}\frac{dt}{d\tau}$, and $\mathfrak{U}^4 = \frac{d(ct)}{dt}\frac{dt}{d\tau}$. Now we need $dt/d\tau$.

$$(\Delta\tau/\Delta t)^2 = -\Delta S^2/(c\Delta t)^2 = 1 - (\Delta x^2 + \Delta y^2 + \Delta z^2)/(c\Delta t)^2 = 1 - \frac{u^2}{c^2}.$$

where u is the speed of the moving object since $\sqrt{\Delta x^2 + \Delta y^2 + \Delta z^2}$ is the distance moved in time Δt in a frame not attached to the object. Therefore, $(\Delta t/\Delta\tau) = 1/\sqrt{1-\frac{u^2}{c^2}}$. Taking the limit as $\Delta\tau \to 0, \frac{dt}{d\tau} = 1/\sqrt{1 - u^2/c^2}$.

$$\mathfrak{U} = \frac{d(x, y, z, ct)}{d\tau} = \frac{d(x^1, x^2, x^3, x^4)}{dt}\frac{dt}{d\tau},$$

$$= \frac{1}{\sqrt{1 - \frac{u^2}{c^2}}}\frac{d(x^1, x^2, x^3, x^4)}{dt}, \text{ or}$$

$$\mathfrak{U}^\mu = \frac{1}{\sqrt{1 - \frac{u^2}{c^2}}}\frac{dx^\mu}{dt}.$$

$$(11\text{-}14)$$

The three-dimensional velocity vector $\mathbf{u} = (u^1, u^2, u^3) = \frac{d(x^1, x^2, x^3)}{dt}$, and $\mathbf{u} \cdot \mathbf{u} = u^2$. The derivative. $dx^4/dt = d(ct)/dt = c$, so finally, the four-velocity may be written concisely as,

$$\mathfrak{U} = \frac{(\mathbf{u}, c)}{\sqrt{1 - \frac{u^2}{c^2}}}.$$

$$(11\text{-}15)$$

The inner product of the four-velocity is invariant since it is governed by the Lorentz transformation, that is,

$$\mathfrak{U} \cdot \mathfrak{U} = (u^1)^2 + (u^2)^2 + (u^3)^2 - (u^4)^2 = \frac{(u^2 - c^2)}{\sqrt{1 - \dfrac{u^2}{c^2}}^2} = -c^2. \qquad (11\text{-}16)$$

[117]

With the concept of a four-vector as a vector in four-dimensional space satisfying the Lorentz transformation, we are ready to see how to accommodate the Laws of Motion in four dimensions. Recall that one of the forms of Newton's Second Law for force is $F = \frac{dp}{dt}$, where the momentum is the product of the mass and the object's velocity, $p = mv$. We want to find the four-vector expression for this equation. First, let us find a four-vector for momentum. Taking advantage of our experience with velocity, a valid momentum four-vector appears to be,

$$p = m_0 \mathfrak{v} = m_0 \cdot (v, c)/\sqrt{1 - (v^2/c^2)}.$$

Here v is the velocity of the moving object in three-dimensions. Since the mass of an object is constant in the Newtonian world, multiplying mass times the four-vector of velocity $\mathfrak{v}$ should result in a valid four-vector. However, note that I have expressed the mass with a subscript 0. This is to indicate that it is the mass in a frame of reference where it is at rest. The reason that this is useful is suggested by looking at the implications of the Lorentz transformation equation (11-12, p. 283). In the object's rest frame, $p^4 = m_0 c$. In a (prime) frame moving with velocity v with respect to this rest frame, the momentum component $p'^4 = (m_0 c + 0)/\sqrt{1 - (v^2/c^2)} = m_0 c/\sqrt{1 - (v^2/c^2)}$. The four-vector of momentum can therefore be simply expressed as,

$$p' = m_0 \mathfrak{v} = (mv, mc), \text{ with} \qquad (11\text{-}17)$$

$$m = m_0/\sqrt{1 - (v^2/c^2)}. \qquad (11\text{-}18)$$

[cxvii] Einstein introduces a "light-time" l = ct in place of time in order that the constant c shall not enter into some formulas explicitly. As a result, the inner product $\mathfrak{U} \cdot \mathfrak{U} = -1$, and other formula developed here differ from Einstein's by a factor of c^2 (Einstein, e.g., pp.33–34,45,89).

In the equation for mass m, it can be seen that as v approaches c, mass increases without limit showing that the velocity of an object cannot exceed the speed of light. Another consequence of this equation is Einstein's discovery that mass represents energy and energy has mass. As Einstein showed, the increase in mass is associated with an increase in energy. At rest the energy $E = m_0c^2$.[296] This can be seen by expressing the right-hand side of equation (11-18) as a series of terms multiplied by increasing powers of v^2/c^2. Newton first stated the rule for doing this in what is known as the binomial theorem (shown below with $-1 < \Delta x < 1$).[297]

$$(x + \Delta x)^r = x^r + \frac{r}{1}x^{r-1}\Delta x + \frac{r(r-1)}{2\cdot1}x^{r-2}\Delta x^2 + \frac{r(r-1)(r-2)}{3\cdot2\cdot1}x^{r-3}\Delta x^3 + \cdots.$$

Equation (11-18) may be expressed as $m = m_0(1 - (v^2/c^2))^{-1/2}$. We can apply the binomial theorem with $x = 1$, $\Delta x = -v^2/c^2$, and $r = -1/2$

$$m = m_0\left(1 + (-1/2)(1)^{-\frac{3}{2}}\left(-\frac{v^2}{c^2}\right) + \frac{\left(-\frac{1}{2}\right)\left(-\frac{1}{2}-1\right)}{2}(1)^{-\frac{5}{2}}\left(-\frac{v^2}{c^2}\right)^2 + \cdots,\right.$$

$$m = m_0 + \frac{\frac{1}{2}m_0v^2}{c^2} + \cdots. \tag{11-19}$$

Recall that the term $\frac{1}{2}m_0v^2$ is the classical kinetic energy of a particle. So $1/c^2$ times the classical kinetric energy represents an increase in mass. Multiplying equation (11-19) by c^2, we identify the rest energy of the particle as m_0c^2. The remaining terms add to form the total energy, E. Thus we arrive at Einstein's famous equation, $E = mc^2$. It is important to remember that this extraordinary result comes essentially from the constancy of the velocity of light and the requirement that the physical laws be consistent with the Lorentz transformation between reference frames in relative uniform motion.

Taking into account the relation between energy and mass, the four-vector of momentum $\mathbf{p}$ may also be expressed as, $\mathbf{p} = (\mathbf{p}, E/c)$, and an invariant formed.

$\mathbf{p} \cdot \mathbf{p} = p^2 - E^2/c^2$. In the particles rest frame

$$= -m_0 c^2. \text{ Therefore,}$$

$$E^2 = p^2 c^2 + m_0^2 c^4.$$

A particle such as a photon which has no mass has energy $E = pc$.[cxviii]

At this point, the four-vector form replacing Newton's Second Law, $\mathbf{F} = d\mathbf{p}/dt$, should be apparent. The derivative of the four-vector of momentum with the proper time $d\mathbf{p}/d\tau$ is a four-vector and therefore the four-vector equation of force is,

$$\mathcal{F} = \frac{d\mathbf{p}}{d\tau} \qquad ,$$
$$= \frac{d(\mathbf{p}, mc)}{dt} \cdot \frac{dt}{d\tau},$$
$$\mathcal{F} = \frac{(d\mathbf{p}/dt, d(mc)/dt)}{\sqrt{1 - (v^2/c^2)}}.$$

At low velocities compared to that of light, mass is constant and Newton's Second Law is recovered. However, the component $d(mc)/dt$ represents the relativistic phenomenon of the dependence of mass on motion presenting another path to Einstein's famous relation between mass and energy, $E = mc^2$ (see Lawden, pp. 44-46).

[cxviii] The relationship of energy for a particlecxviii)" without mass provides an entrée to the other revolutionary discovery in physics of the twentieth century, Quantum Mechanics. In 1923, Louis de Broglie proposed that particles have a wavelength just like light, and that the wave length was inversely proportional to the particle's momentum, or $p = h/\lambda$ with h being Planck's constant. Earlier in the century, Max Planck had proposed that radiation was quantified in packets of energy $E = h\nu$ with ν being the frequency of radiation. Einstein subsequently showed that these quanta of energy gave light a particle character in his photoelectric effect for which he won the Nobel prize in 1921. De Broglie's proposal meant that everything had a dual wave - particle reality. For light $\nu = c/\lambda$, and $E = pc$ (Einstein) $= hc/\lambda$ (DeBroglie) $= h\nu$ (Planck), thus, a consistent relation between relativity and de Broglie's proposal is recovered (see Asimov, pp. 368-375, 582; Pais, pp. 378-382, 502-511)

Einstein's acceptance of the principle of the constancy of the velocity of light in *vacuo*, along with his demand that in reference frames moving uniformly relative to each other, the laws of nature be the same, led Einstein inescapably to conclusions that overthrew the Newtonian view of the universe. Special Relativity removed forever the confusions of absolute space and the ether; however with the disappearance of absolute space, there still is the problem of defining the systems where the laws of nature are valid. It appears a system is valid where the physical laws apply, and the laws apply in all systems in uniform relative motion to such a system, a triumph of circular reasoning. Einstein solved this quandary by developing in his General Theory of Relativity an approach in which all coordinate systems are valid and each observer can consistently account for the observations in all other coordinate systems. General Relativity builds on four-dimensional space-time of Special Relativity, but its broader scope required the extended geometric vision that followed from Riemann's insight of n-dimension geometry. The General Theory that Einstein announced in 1915 replaced Newton's universal gravitational force with a universe in which the geometry of space-time determined the motion of the stars and planets and in turn the geometry of space-time is in a balance with the distribution of mass in the universe. We now turn to what might be called the ultimate problem in geometry.

11.2   General Relativity: a universe driven by geometry

11.2.1   Einstein's insight: the Equivalence Principle

As discussed in the previous section, Special Relativity did away with absolute space, but it still restricted valid prescriptions of physical laws to privileged inertial systems moving with uniform velocity with respect to each other. Besides the problem of defining the inertial system, the question remained, why are these frames of reference privileged. In characteristic daring, Einstein sought a form of the physical laws that would be valid for reference frames in arbitrary motion. Ultimately this would lead Einstein to formulate his theory using the tensor methods of

Riemannian geometry. Such an approach allows transformations from one coordinate system to another assuring the invariance of the laws of physics, that is as Einstein explained, *"...the equations expressing the laws are co-variant with respect to arbitrary transformations.'*[298] Furthermore, Riemann's geometry developed for n-dimensional spaces would accommodate the four dimensional space-time created by Special Relativity.

In 1907, while considering how to extend Special Relativity to arbitrary reference frames and include gravitation, Einstein had an insight which he called in an unpublished manuscript (now in the Pierpont Morgan Library) *"...the happiest thought of my life..."*

"The gravitational field has only a relative existence in a way similar to the electric field generated by magnetoelectric induction. Because for an observer falling freely from the roof of a house, there exists - at least in his immediate surroundings - no gravitational field. Indeed, if the observer drops some bodies then these remain relative to him in a state of rest or of uniform motion, independent of their particular chemical or physical nature (in this consideration the air resistance is ignored.) The observer therefore has the right to interpret his state as at rest."[299] [cxix]

Noting that all experimental evidence shows that all bodies fall with the same acceleration, Einstein continued,

"...the observer lacks any means of perceiving himself as falling in a gravitational field. Rather he has the right to consider his state as one of rest and his environment as field-free relative to gravitation....and

---

[cxix] It would be interesting to know if Einstein was aware of the description of a free fall in a gravitational field given by Lewis Carrol in *Sylvie and Bruno*:

"Well, now, if I take this book, and hold it out at arm's length, of course I feel its weight. It is trying to fall, and I prevent it. And, if I let go, it falls to the floor. But, if we were all falling together, it couldn't be trying to fall any quicker, you know: for, if I let go, what more could it do than fall? And, as my hand is falling too – at the same rate – it would never leave it, for that would be to get ahead of it in the race. And it could never over take the falling floor!," in *The Complete Works of Lewis Carrol*, The Modern Library, Random House, New York, (p. 341).

[this] is a powerful argument for the fact that the relativity postulate has to be extended to coordinate systems which relative to each other are in non- uniform motion."

Speaking of this same thought in a lecture in Kyoto 1907, he said that, *"It impelled me toward a theory of gravitation."* [300]

These considerations eventually led in 1915 to Einstein's theory of General Relativity in which a gravitational field can be transformed locally (over the space of the differential invariant d S) to a free-falling inertial frame. [301] This result, known as the Equivalence Principle, is the broadest expression of those happy thoughts that Einstein had in 1907. At that time, he had not begun to use the tensor approach that he would need to more precisely express this principle and create a theory in which a valid physical description could be made for any reference frame. However, the core elements of the Equivalence Principle, can be expressed in terms of Einstein's original insight.[302]

Equivalence Principle: Locally, an accelerating frame of reference is completely physically equivalent to a frame at rest in a gravitational field, and a frame of reference falling freely in a gravitational field is equivalent to an inertial frame at rest.

Einstein notes, this principle is intimately connected to equality of inertial mass as employed in Newton's Second Law of motion and gravitational mass in his Universal Law of Gravity. As illustrated in Chapter 9, the equality is responsible for the observed fact that the acceleration imparted to a body by a gravitational field is independent of the nature of the body. The significance of the Equivalence Principle including the equality of inertial and gravitational mass has famously been illustrated in thought experiments involving various reference frames such as elevators in space.[303] Some of the details of these thought experiments and their implications for gravity and the geometry of space are discussed below.

First imagine a laboratory in deep space, reference frame K, with occupants who consider it to be at rest and inertial. They observe a rocket with uniform vertical acceleration a. The occupants of the rocket are blissfully unaware of their state of motion, but notice that they are "weighed down" at the base of their chamber in the rocket. Furthermore, they observe that all dropped objects take the same time to fall to the floor, falling with the same downward acceleration that they report as g = a. Again, unaware of their state of motion, they presume that they are at rest in a gravitational field. In the rocket, an experiment is performed with a laser pointer to see if the path of a light ray aimed horizontally across their chamber is affected by the gravitational field. Indeed, they see that the location where the light ray strikes the opposite wall is below the horizontal level. From this observation, they proclaim that light is deflected by a gravitational field! The observers at rest in reference frame K say, "not so fast. The light ray wasn't deflected downward, your reference frame accelerated upward." Who is right?

The correct answer, in the spirit of denying a privileged status to any reference frame, is that both reference frames give valid descriptions. However, by the Equivalence Principle, the accelerating frame is equivalent to one at rest with a gravitational field. Therefore, the conclusion that a light ray is deflected by gravity is correct. The experimental verification of this conclusion would be one of the key results that would gain acceptance for General Relativity. Within this thought experiment, it should also be clear that the ability to consistently describe the results from the two frames of reference depends on the equality of inertial and gravitational mass. This will be discussed in more detail in another thought experiment in a gravitational field.

The accelerating rocket chamber in the previous thought experiment can be used with the Equivalence Principle to predict another effect of gravity called the gravitational redshift. From the floor of the rocket's chamber, the laser pointer sends a light beam towards the ceiling of the chamber. As the light beam is from a laser, it has a specific wavelength λ_{floor}. At the ceiling, a rocket scientist measures the wavelength as $\lambda_{ceiling}$. The scientist finds that $\lambda_{ceiling} > \lambda_{floor}$ and proclaims that propagation of the laser light upward through the gravitational field has increased its wavelength. Once again, our deep space inertial friends cannot agree.

They point out that the wavelength appeared to increase because the measurement instrument on the ceiling of the rocket was accelerating upward. The experience in the rocket was therefore similar to the experience of hearing the pitch of a train whistle decrease as it moves away from you. The scientist on the ceiling measures a lower frequency v and since $\lambda = c/v$, a longer wavelength. Nevertheless, the Equivalence Principle says that the same effect will be seen in a gravitational field. The phenomenon is known as the gravitational redshift because in the visible spectrum, wavelengths increase towards the color red. Such a shift has been observed from the light of stars.

In the previous thought experiments, effects of a gravitational field were created through an accelerated reference frame. Now we will remove the gravitational field effects by switching to an accelerated frame. Imagine a reference frame K' in which a uniform gravitational field acts. A frame K'' formed by the legendary elevator in space is falling freely within the gravitational field. The occupant of the elevator is aware of the feeling of weightlessness. He assumes that he finds himself isolated in Newton's absolute space. Objects floating in space at rest move in straight lines if given a push $\mathbf{f}''$, and the light of his laser pointer traverses the room in a horizontal line. An observer from K' will be able to interpret the events in their gravitational frame by noting that all the occupants and objects at rest in K'' are actually moving downward at the acceleration of gravity $\mathbf{g}'$, and the light has been deflected. The validity of Newton's laws in K'' can be explained by taking into account the acceleration and forces relative to K'. (In this thought experiment, we assume the elevator is falling at a velocity much less than that of light.) From the perspective of K', an object is acted upon by a total force $\mathbf{F}' = \mathbf{f}'' + m\mathbf{g}', = m\mathbf{a}'$. The acceleration relative to K'' is $\mathbf{a}'' = \mathbf{a}' - \mathbf{g}'$. So we have the consistent result in both frames of reference, $\mathbf{f}'' = m\mathbf{a}'' = m(\mathbf{a}' - \mathbf{g}')$. Crucially, this interpretation requires the equality of inertial and gravitational mass.

The examples of the Equivalence Principle given above should make it clear that a theory of gravitation was required if Einstein was to meet his goal of a physics valid for all coordinate systems. Furthermore, the incompatibility of Euclidean geometry with such a theory is made apparent in the following thought experiment with a rotating frame of reference discussed by Einstein and by numerous other authors with variations. [304] In the version as adapted here, there is a space station in deep space in the form of a torus with spokes to its center. The space station spins around its central axis as shown in the next figure.

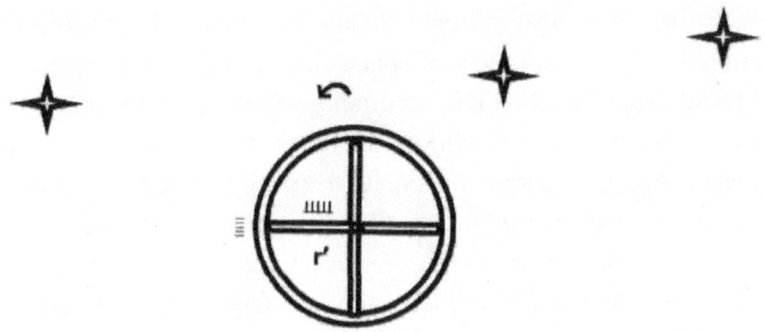

11-1 A rotating space station in deep space

An astronaut on the space station measures the radius of the space station as r′ and the circumference of the station as C′. A stationary observer in deep space outside the station agrees with the radius measurement (the spoke is moving perpendicular to the direction of motion as viewed by the stationary observer and does not undergo contraction). However, she sees that the measurements around the rim of the torus need to be corrected as the rods used for measurement on the space station are placed parallel to the direction of motion. Our deep-space observing astronaut computes the circumference in her frame as $C = 2\pi r'$ and compares it to the measurement C′. made around the rim of the space station in the rotating frame. She corrects for length contraction of the moving measuring rods finding that $C'\sqrt{1 - v^2/c^2} = 2\pi r'$ with v being the speed the rod passes by the observer due to rotation. Therefore, the astronaut on the space station makes the measurement $C' = 2\pi r'/\sqrt{1 - v^2/c^2}$. In other words, the astronaut on the space station finds the result $C'/2r' > \pi$. Because he considers himself to be stationary, he is adamant that the forces that he feels are due to a gravitational field; hence he observes that the space is not Euclidean in the presence of a gravitational field. The Equivalence Principle says that whether a frame of reference is accelerating or in a gravitational field, the laws of physics will be the same, so the astronaut on the space station is correct to conclude that Euclidean geometry is also not satisfied in frames of reference in a gravitational field.

Einstein saw the connection between the non-Euclidean geometry of space-time and gravitational fields as crucial. Furthermore, he recognized that the geometry of space-time could be characterized by its metrical properties in a Riemannian geometry in an analogous way to Gauss's theory of non-Euclidean surfaces. Therefore, Einstein chose to focus on the metric tensor of space-time and tensor methods in his approach to a theory of gravitation. With this decision, he needed to find a governing equation for the metric of space-time that would be consistent with the Equivalence Principle, reference frame independence, and Special Relativity in the case of reference frames in relative uniform motion. The search for a governing equation would take Einstein ten years beyond his publication of Special Relativity in 1905 due in part to only a gradual realization of the mathematical requirements of a new theory which included gravitation.[cxx] One of the most important guides in his search was the mathematical description of Newton's Law of Gravitation expressed in terms of the gravitational potential discussed in the next section. Einstein knew that his theory would have to match Newton's excellent results in the limit. The form of Newton's gravitational theory that guided Einstein will now be developed.

11.2.2 Newton's Law of Gravity revisited as an equation of gravitational potential

In Chapter 9, Newton's Laws of Motion and of Gravitation were introduced and in Chapter 10 the gravitational potential energy that is implicit in Newton's laws. The deficiencies of these laws have subsequently been discussed: the reliance on the concept of absolute space in the definition of inertial frames, the privileging of frames of

---

[cxx]   Pais (pp. 177-261) gives an excellent detailed account of the twists and turns of Einstein's path to his Field Equations and General Relativity.

reference that are in relative uniform motion, the incompatibility of the Galilean transformation with the constancy of the velocity of light, and the lack of recognition of the principle that inertial and gravitational mass are equivalent. However, the extraordinary success of Newton's Laws including the ability to predict such as things the existence of the previously unobserved planet Neptune[305] could not be dismissed by any theory meant to supplant Newton's. As yet another spectacular example among the countless, I remind you that the navigation of the Apollo spacecraft to the moon was based upon Newton's Laws. Some of the characteristics of Newton's Law of Gravitation that guided Einstein will be discussed below. Of particular importance is the equation for the gravitational potential derived from Newton's Law of Gravitation. I will begin the derivation with an analysis of the gravitation field for a single particle.

The Gravitational Field of a Single Particle

To start, I have reproduced Newton's Law Of Gravitation (equation (9-4, p. 264)) as it applies to a single particle of mass m attracted radially to a second particle of mass M with the distance between the two particles being r (measured from the mass M). Just below that, I have rewritten the equation in a form that will be useful later in dealing with more than two particles. Here the attractive force applies to a single particle of mass m_1, at location specified by the position vector $\mathbf{r}_1$, attracted radially to a second particle of mass m_2 at $\mathbf{r}_2$ with the distance between the two particles being $|\mathbf{r}_1 - \mathbf{r}_2| = |\mathbf{r}_2 - \mathbf{r}_1|$.

$$\mathbf{F}_{g(m)} = -\frac{GMm}{r^2}\mathbf{e}_r.$$

$$\mathbf{F}_{g(m_{1-2})} = \frac{Gm_1 m_2}{|\mathbf{r}_2 - \mathbf{r}_1|^3}(\mathbf{r}_2 - \mathbf{r}_1) \qquad (11\text{-}20)$$

Note that in equation (11-20) above, the radial direction is given by the vector $(\mathbf{r}_2 - \mathbf{r}_1)$ requiring the denominator to be $|\mathbf{r}_2 - \mathbf{r}_1|^3$, consistent with the inverse square law. The gravitational; attraction of the particle of mass m_2 by m_1 is given below.

$$F_{g(m_2-1)} = \frac{Gm_1m_2}{|r_1 - r_2|^3}(r_1 - r_2).$$

As any particle at position r will experience the same gravitational acceleration due to the equality of inertial and gravitational mass, it is useful to define the gravitational field $g(r)$ as the gravitational force per unit mass at location r. For example, the gravitational field due to a particle of mass M:

$$g(r) = F_g/m,$$

$$= -\frac{GM}{r^2}e_r.$$

An important consequence of the gravitational field's dependence on the inverse square of distance is the relation between the source of the field and the flux of the gravitational vector field through a closed surface. This is shown schematically in the following figure for the single mass M enclosed within a spherical surface S of radius r. The arrows depicted in the figure illustrate the magnitude and direction of the field at the enclosure surface. Of course, the actual field is continuous over the entire surface and everywhere perpendicular to it.

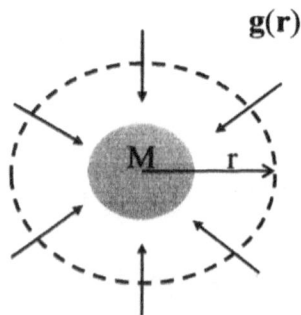

11-2 Flux of a gravitational field

The flux of the gravitational field through the spherical surface $S = 4\pi r^2$ is given by the integral below. The surface element $d\mathbf{S}$ is an outwardly oriented radial vector opposite to the direction of the gravitational field.

$$\iint \mathbf{g} \cdot d\mathbf{S} = -\iint \frac{GM}{r^2} \mathbf{e}_r \cdot d\mathbf{S},$$

$$= -\frac{GM}{r^2} \iint dS,$$

$$= -\frac{GM}{r^2} (4\pi r^2),$$

$$\iint \mathbf{g} \cdot d\mathbf{S} = -4\pi GM \tag{11-21}$$

The relationship between the flux of the gravitational field and its mass sources (M in the equation above) was among the most important guides for Einstein's choice of governing equations. However, the particular form that guided Einstein was as the partial differential equation of the gravitational potential. This will be derived, starting with the equation for single particle using equation (11-21).

From the definition of gravitational potential energy Φ in equations (10-2), p. 271) and (10-4, p. 272), the gravitational potential Φ_m due to a single particle of mass M is defined as the potential energy per unit mass.

$$\Phi_m = -\int_\infty^r \mathbf{F}_g \cdot d\mathbf{r}/m = -\int_\infty^r \mathbf{g}(r) \cdot d\mathbf{r} = -\int_\infty^r (-GM/r^2) dr = -GM/r]_\infty^r = -GM/r. \tag{11-22}$$

From equation (11-22),

$$d\Phi_m = -\mathbf{g} \cdot d\mathbf{r}; \text{ however, } d\Phi_m \text{ may also be expressed as,}$$

$d\Phi_m = \dfrac{\partial \Phi_m}{\partial x^i}\, dx^i$. Therefore,

$\mathbf{g}(\mathbf{r}) = -\dfrac{\partial \Phi_m}{\partial x^i}$, or in common vector notation ,

$\mathbf{g}(\mathbf{r}) = -\nabla \Phi_m$.

As noted in the discussion related to equations (8-17) and (8-18, pp. 235,236), $\frac{\partial \Phi_m}{\partial x^i}$ is a covariant vector known as the gradient. Equation (11-21) can now be written in terms of the gravitational potential, and the mass on the right-hand side replaced by an integral of density ρ over the volume V enclosed by the surface S.

$$\iint_S \nabla \Phi_m \cdot d\mathbf{S} = 4\pi GM = 4\pi \iiint_V \rho dV. \qquad (11\text{-}23)$$

The above equation can be simplified by making the left-hand side into an integral also over the volume V. This may be accomplished by introducing the concept of the divergence of a vector. For a vector $\mathbf{A}$, the divergence of $\mathbf{A}$, designated as $\nabla \cdot \mathbf{A}$ is defined as below.

$$\nabla \cdot \mathbf{A} = \lim_{\delta V \to 0} \frac{\iint_S \mathbf{A} \cdot d\mathbf{S}}{\delta V}.$$

From this definition, it should not be surprising that the following theorem, associated with Gauss, may be proved by integrating (summing in the limit) over the elements δV that make up V.[cxxi]

Divergence Theorem: If $\mathbf{A}$ is a vector defined over the volume V with surface S, then

$$\iint_S \mathbf{A} \cdot d\mathbf{S} = \iiint_V \nabla \cdot \mathbf{A} dV.$$

[cxxi] The Divergence Theorem applies over a wide range of conditions; however, constraints on the vector A (such as being continuous over V with continuous partial derivatives and a sufficiently "smooth" surface area S need to be stated for the theorem to be fully rigorous; however, this is beyond what can be introduced here.

To make the definition of the divergence concrete, I will calculate the divergence in Cartesian coordinates and from the resulting form generalize the result from our knowledge of tensors.

Let us use a cube of volume $\delta V = \Delta X \Delta Y \Delta Z$. The cube is shown in the figure below.

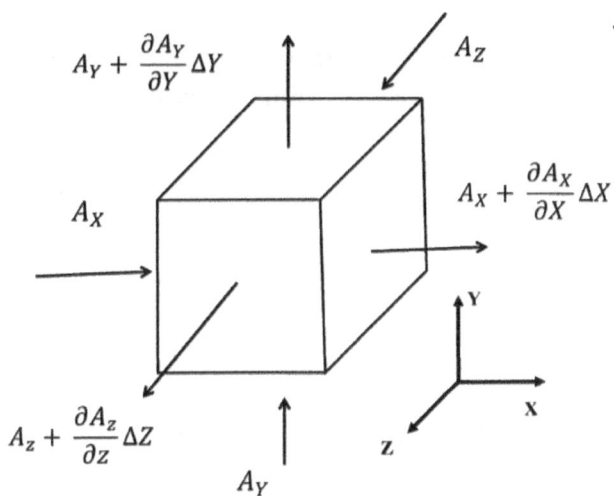

11-3 Divergence of the vector A

Note that the surface vector on all faces is outward. The flux in the X direction entering the YZ face of the cube, $\mathbf{A} \cdot d\mathbf{S}_{\text{YZ face}} = -A_X \Delta Y \Delta Z$. The flux leaving the other side then is $\left(A_X + \frac{\partial A_X}{\partial x} \Delta x\right) \Delta Y \Delta Z$ + higher order terms.[cxxii] The net flux out, neglecting higher order terms, is therefore, $\left(A_X + \frac{\partial A_X}{\partial x} \Delta X - A_X\right) \Delta Y \Delta Z = \frac{\partial A_X}{\partial x} \Delta X \Delta Y \Delta Z.$

Similarly the net flux from the other faces is $\left(\frac{\partial A_Y}{\partial Y} \Delta Y\right) \Delta X \Delta Z + \left(\frac{\partial A_Z}{\partial z} \Delta Z\right) \Delta X \Delta Y$. Applying the definition of the divergence,

---

[cxxii] A function of a single variable f(x) may be represented in many cases by a Taylor series about a point a as $f(x) = f(a) + \frac{df}{dx}\Big|_{x=a}(x-a) + \frac{1}{2 \cdot 1}\frac{d^2f}{dx^2}\Big|_{x=a}(x-a)^2 + \frac{1}{3 \cdot 2 \cdot 1}\frac{d^3f}{dx^3}\Big|_{x=a}(x-a)^3 + \cdots$. Using partial derivatives, powers series can be extended to multivariable functions. Here we are looking at changes in the single directions X, Y, and Z, so the partial derivatives takes the place of the derivative in the Taylor series of a single variable. Higher order terms are those such as ΔX^2 corresponding in the Taylor series to $(x-a)^2$.

$$\nabla \cdot \mathbf{A} = \lim_{\delta V \to 0} \frac{\iint_S \mathbf{A} \cdot d\mathbf{S}}{\delta V}.$$

$$= \lim_{\delta V \to 0} \frac{1}{\delta V} \left(\frac{\partial A_X}{\partial x} + \frac{\partial A_Y}{\partial Y} + \frac{\partial A_Z}{\partial Z} \right) \Delta X \Delta Y \Delta Z + (\text{higher order terms})/\delta V.$$

$$\nabla \cdot \mathbf{A} = \frac{\partial A_X}{\partial x} + \frac{\partial A_Y}{\partial Y} + \frac{\partial A_Z}{\partial Z}.$$

The divergence may be in expressed in tensor form for **A** as a contravariant vector A^i or covariant vector A_j using the covariant derivative (here expressed in Einstein's notation).

$$\nabla \cdot \mathbf{A} = A^i_{;i}$$
$$= \frac{\partial A^i}{\partial x^i} + \Gamma^i_{ik} A^k, \text{ and}$$
$$A^i = g^{ij} A_j.$$

The tensor forms may be used to express the divergence in any specific coordinate system;[306] however, here I primarily wish to show the compatibility of the divergence as an operator consistent with Einstein's program for a coordinate system independent formulation. The choice of the vector notation $\nabla \cdot (\ )$ for the divergence is helpful in the case of a rectangular Cartesian system as the divergence in these coordinate may be thought of as the inner product of the vector operator $\left(\frac{\partial}{\partial x}, \frac{\partial}{\partial Y}, \frac{\partial}{\partial Z} \right)$ with $\mathbf{A} = (A_X, A_Y, A_Z)$.

The tensor form of the divergence in terms of the covariant derivative makes it clear that the form of the divergence in other coordinate systems cannot be assumed to be the simple form shown above in Cartesian coordinates. However, the covariant derivative gives a result that is consistent with our derivation in a Cartesian system because in that coordinate system the Christoffel symbols Γ^i_{ik} are zero, leaving the tensor form to be $\frac{\partial A^i}{\partial x^i}$.

With this understanding of the divergence and $\mathbf{g(r)} = -\nabla \Phi_m$, we are now in a position to reinterpret equations (11-21) and (11-23).

$$\iint \mathbf{g} \cdot d\mathbf{S} = -4\pi GM, \tag{11-24}$$

$\iint_s \nabla \Phi_m \cdot d\mathbf{S} = 4\pi GM = 4\pi \iiint_V \rho dV.$ Using the Divergence Theorem,

$\iiint_V (\nabla \cdot \nabla \Phi_m) dV = 4\pi G \iiint_V \rho dV,$

$\iiint_V \{(\nabla \cdot \nabla \Phi_m) - 4\pi G\rho\} dV = 0.$ Thus,

$\nabla \cdot \nabla \Phi_m = 4\pi G\rho.$

The last equation is often written as $\nabla^2 \Phi m = 4\pi G\rho$ with the operator $\nabla^2(\ )$ known as the Laplacian. [cxxiii] Using the tensor definitions for the gradient $\nabla \Phi_m = \partial \Phi_m / \partial x^i$ as a covariant vector and the tensor expression for the divergence given above, the equation $\nabla^2 \Phi_m = 4\pi G\rho$ can be expressed in any coordinate system.[cxxiv] In a rectangular Cartesian coordinate system it reduces to,

$$\frac{\partial \Phi_m^2}{\partial x^2} + \frac{\partial \Phi_m^2}{\partial y^2} + \frac{\partial \Phi_m^2}{\partial z^2} = 4\pi G\rho,$$

or in Cartesian coordinates x^i,

$$\frac{\partial^2 \Phi_m}{\partial x^i \partial x^i} = 4\pi G\rho.$$

A partial differential equation has been derived for the gravitational potential outside a single particle. (This form of partial differential equation is often referred to as a Poisson equation after the mathematician Siméon-Denis Poisson (1781-1840).) The general equation for many particles has the same form and may be similarly derived because the effects of each additional particle are additive, a property known as

---

[cxxiii] Einstein (p. 82.) uses the notation $\Delta \phi = 4\pi K\rho$.

[cxxiv] As discussed in footnote cviii (p. 209), the expression given by the tensor forms are not necessarily the physical components. For example, in polar coordinates, the angular component may be $x^1 = \theta$. The Laplacian is usually expressed in the physical components, see Wrede, pp. 237-243.

superposition. (I also note that in deriving equation (11-24), a spherical surface surrounding the particle was used for simplicity, but any surface enclosing the particle gives the same result.) The derivation of the equation for the gravitational potential will now be outlined for the case of many particles following a similar approach.

The Gravitational Field of Many Particles

With the background of the gravitational field of a single particle, an extension to systems of many particles can now be outlined by analogy. Newton's Law of Gravitation applies to all material particles, so the net force on a particle at any location is the vector sum of all the forces from all material particles in the universe. (Because the gravitational force diminishes as the inverse square of the distance between two particles, the nearest particles are the most significant - for example, on the surface of the earth, the earth's gravity is dominant.) The net force on a particle of mass m at position $\mathbf{r}$ due to a system with particles of mass m_i at positions $\mathbf{r}_i$ is,

$$\mathbf{F}_{system} = \sum_i \mathbf{F}_i(\mathbf{r}) = \sum_i \frac{Gmm_i\,(\mathbf{r}_i - \mathbf{r})}{|\mathbf{r}_i - \mathbf{r}|^3}.$$

For comparison with the single particle result, see equation (11-20, p. 295). Because of the equivalence of inertial and gravitational mass, all particles at position $\mathbf{r}$ will experience the same acceleration due to gravity, known as the gravitational field.

$$\mathbf{g}(\mathbf{r}) = \frac{\mathbf{F}_{system}}{m} = \sum_i \frac{Gm_i\,(\mathbf{r}_i - \mathbf{r})}{|\mathbf{r}_i - \mathbf{r}|^3} = \sum_i \mathbf{g}_i.$$

The flux of the gravitation field through a surface S enclosing the particles may be calculated as in the case of a single particle. It can be proved that the result is independent of the shape of the enclosing surface.

$$\iint \mathbf{g} \cdot d\mathbf{S} = \iint \sum_i \mathbf{g}_i \cdot d\mathbf{S} = -4\pi G \sum_i m_i.$$

The sum of discrete masses m_i in the above equation may be replaced by an integral over the entire volume of space V associated with the surface S. The system of particles is described as a function of the mass density ρ (dm = ρdV) in space where $\rho = 0$ in the absence of matter. The sum over discrete mass particles is replaced by the integration over differential elements of mass, dm, and the right-hand side may then be also converted to a volume integral.

$$\iint \mathbf{g} \cdot d\mathbf{S} = -4\pi G \iiint \rho dV.$$

The gravitational potential $\mathfrak{G}$ is defined analogously to the potential Φ_m due to a single particle $(\mathbf{g(r)} = -\nabla\Phi_m$, see p.298)

$$\mathbf{g} = -\nabla\mathfrak{G}, \qquad (11\text{-}25)$$

$\iint -\nabla\mathfrak{G} \cdot d\mathbf{S} = -4\pi G \iiint \rho dV.$ By the Divergence Theorem,

$\iiint (\nabla \cdot \nabla\mathfrak{G})dV = 4\pi G \iiint \rho dV.$ Therefore,

$$\nabla^2\mathfrak{G} = 4\pi G\rho. \qquad (11\text{-}26)$$

Equation (11-26) was crucial in Einstein's search for the form of his equations for General Relativity. Just as the four-vectors of Special Relativity satisfied Newton's Laws of Motion in the limit of velocities much less than that of light, Einstein looked for tensor forms that would be compatible in a four-dimensional space-time in the limit of gravitational fields where Newton's Law of Gravity gave accurate results. We are now in a position to introduce Einstein's Field Equations of General Relativity.

11.2.3   Einstein's Field Equations

From the time of Einstein's insight in 1907 that led to the Equivalence Principle, it would take Einstein another eight years before he would finalize the set of equations that would define General Relativity. From his

Equivalence Principle, Einstein was already aware in 1907 of the phenomena in gravitational fields of the deflection of a beam of light and of a spectral redshift. He revisited these phenomena in 1911, with the difference that his efforts from that time were directed towards an approach in which the Equivalence Principle was to be derived from a new theory of gravitation within a generalized theory of relativity. However, he had not yet found a mathematical approach that would allow him to express his concepts in forms for arbitrary accelerating reference frames.[307] The mathematical approach that Einstein needed came eventually to him through an increasing appreciation for the close connection between gravity, accelerated reference frames, and the geometry of space. Reflecting in 1921 on his discovery of this connection, Einstein had this to say,

"The decisive step of the transition to generally covariant equations would certainly not have taken place [had it not been for the following consideration]. *Because of the Lorentz contraction in a reference frame that rotates relative to an inertial frame, the laws that govern rigid bodies do not correspond to the rules of Euclidean geometry, thus Euclidean geometry must be abandoned if noninertial frames are admitted on an equal footing."*[308]

Furthermore, Einstein, speaking a year later of the state of his thoughts in 1912, had these reflections,

If all [accelerated] *systems are equivalent, then Euclidean geometry cannot hold in all of them. To throw out geometry and keep* [physical] *laws is equivalent to describing laws without words. We must search for words before we can express thoughts. What must we search for at this point? This problem remained insoluble until 1912, when I suddenly realized Gauss's theory of surfaces holds the key for unlocking this mystery. I realized that Gauss's surface coordinates had a profound significance. However, I did not know at that time that Riemann had studied the foundations of geometry in an even more profound way."*[309]

A major breakthrough towards solving the problem that Einstein alluded to in the above quotation came about as a result of collaboration with the mathematician Marcel Grossmann (1878-1936). Grossmann had been a friend of Einstein when they were both students at the Swiss Federal Institute of Technology (ETH) in Zurich. Later he helped Einstein obtain his job in the Swiss patent office. Einstein dedicated his doctoral thesis to Grossmann.[310] As reported by Pais,[311] Einstein recollected that he told Grossman that he needed a geometry which allowed for the most general transformation that would leave the line element ds^2 invariant. Grossmann replied that he was looking for Riemannian geometry.

Even more specifically Einstein, recalls posing to Grossmann *"the problem of looking for generally covariant tensors whose components depend only on derivatives of the coefficients of the quadratic fundamental invariant $ds^2 = g_{ij}dx^i dx^j$."* [312] [cxxv] As we shall see, the key tensor which Grossmann drew to Einstein's attention was the Ricci tensor $R_{\mu\nu}$ formed from the Riemann curvature tensor $R^\lambda_{\mu\nu\kappa}$ (see equations (8-38, p. 251) and (8-41, p. 253).

Some of the motivation for the selection of the Ricci tensor and the eventual formulation of Einstein's Field Equations can be made apparent by describing the inspiration that came from the theories of Ernest Mach and from the equation for the gravitation potential discussed in the previous section. Ernest Mach (1838 - 1916) was at different times a professor of mathematics, physics, and philosophy. He greatly influenced Einstein through his writings rejecting Newton's concepts of absolute space and inertia. [313] [cxxvi] In Newtonian physics, only reference frames with uniform motion relative to absolute space are valid frames for Newton's laws. In essence, inertial mass is a property of absolute space. Thus, space exerted a physical effect, but is not itself affected. Mach sought to eliminate absolute space and absolute motion by rejecting preferred frames of reference. For Mach, only relative motion existed;

[cxxv] In regard to the specificity of this recollection, it is perhaps worth noting that Einstein had taken a course in differential geometry at the ETH which covered Gauss's theory of surfaces (Pais, p. 212).

[cxxvi] Mach is probably more well known to the public for his theoretical investigations of the effects of the motion of bodies moving with velocities in the range of the velocity of sound. For his research, the ratio of velocity of a body to that of sound is called the Mach number (Asimov, p. 165).

he could see *"no distinction between rotation and translation."* [314] Inertial mass, instead of being a property from absolute space, resulted from motion relative to all other masses in the universe. [315] In the paper with Grossmann in 1913, in which Einstein introduced tensor forms, Einstein expressed his indebtedness to Mach for inspiring some of his ideas, particularly *"Mach's bold idea that inertia originates in the interaction of [a given] mass point with all other [masses]."* [316] In 1930, Einstein wrote that *"...it is justified to consider Mach as the precursor of general relativity."* [317]

In the paper of Einstein and Grossmann in 1913, the distribution of mass in the universe that was central to Mach's concept was introduced into a second rank tensor, the energy-momentum tensor $T_{\mu\nu}$; [cxxvii] however, unlike Mach's vision, the role of space did not disappear. Einstein proposed that the energy-momentum tensor be balanced with a second rank tensor formed from differential operations on the space-time metric tensor $g_{\mu\nu}$. When the final form of the Field Equations was obtained in 1915, the equation of motion of a particle in a gravitational field was recognized as the geodesic equation of motion which depends through the Christoffel symbols on the metric tensor. Hence the distribution of mass, determines the metric of space-time which in turn determines the motion of particles. [318]

The final form of the Field Equations was guided to a great degree by the understanding of Einstein that in a weak gravitational field, Newton's Law of Gravitation, expressed in terms of the gravitational potential, should be recovered as a limit. The equation for the gravitational potential $\mathfrak{G}$, equation (11-26, p. 303) is reproduced below in vector notation and as it is formed in rectangular Cartesian coordinates.

$$\nabla^2\mathfrak{G} = \nabla\bullet\nabla\mathfrak{G} = \frac{\partial\mathfrak{G}^2}{\partial x^2} + \frac{\partial\mathfrak{G}^2}{\partial y^2} + \frac{\partial\mathfrak{G}^2}{\partial z^2} = 4\pi G\rho,$$

[cxxvii] That the distribution of energy is described by a second order symmetrical tensor follows from the energy tensor of the electromagnetic field (Einstein, pp. 49-50). The name energy-momentum tensor is used by Lawden. It is also designated as the energy tensor, by Einstein, and the stress-energy tensor by others.

As previously, ρ is the local mass density, and G, the gravitational constant in Newton's Law of Gravity. From Einstein's thought experiments such as that supplied by rotating coordinate systems and the Equivalence Principle, a connection can be inferred between sources of gravitational fields and the local geometry of space as described by the metric tensor. Einstein generalized the Newtonian gravitational potential as the metric tensor of space-time, $g_{\mu\nu}$. From the equation for the potential 𝔊, he reasoned that the the tensor balancing the second order energy-momentum tensor, $T_{\mu\nu}$ which includes the mass density ρ, must be a second rank tensor having derivatives of the metric tensor no higher than second order (see equation for 𝔊 above). Furthermore, he required that the divergence of the energy-momentum tensor be zero ($T_{\mu\nu;\nu} = 0$) as a conservation principle (discussed further below) restricting the form of space-time geometry. From these considerations, the Ricci tensor $R_{\mu\nu}$, and the curvature scalar $\mathcal{R}$ formed from it were selected to represent the space-time aspects of geometry in the Field Equations finalized in 1915.[319] These equations are given below followed by a discussion of their relationship to Newton's Laws of Motion and Law of Gravitation.

The Einstein Field Equations

The Field Equations may be specified in a number of equivalent forms.
[320] Below the Field Equations are expressed with covariant tensors.[cxxviii]

[cxxviii] By successively applying the contravariant tensors $g^{\alpha\mu}$ and $g^{\beta\nu}$ to raise indices (equation (8-22, p. 237)) in equation (11-27), the contravariant form may be determined.

$$R^{\alpha\beta} - \frac{1}{2}g^{\alpha\delta}\mathcal{R} = -\kappa T^{\alpha\beta}.$$

When Einstein introduced his Field Equations, he assumed that the conservation condition $T^{\mu\nu}_{;\nu} = 0$ was a constraint on the Riemannian geometry of space-time. However, unknown to Einstein, the divergence of the left-hand side of his Field Equations, $\left(R^{\mu\nu} - \frac{1}{2}g^{\mu\nu}\right)_{;\nu}$ had been proved to be zero by Aurel Voss in 1880, Ricci in 1889, and again independently by Luigi Bianchi in 1902. This result is frequently known as the Bianchi identities (Pais, pp.256, 275-276.). Thus the vanishing of the the divergence of the energy-momentum tensor in the Field Equations is automatically satisfied.

$$R_{\mu\nu} - \frac{1}{2} g_{\mu\nu} \mathcal{R} = -\kappa T_{\mu\nu}. \tag{11-27}$$

with κ being a constant related to Newton's constant G. I reproduce here (with Greek indices to remind you that they range from 1 to 4 in space-time) the definitions of the Riemann curvature tensor, Ricci tensor, Ricci scalar (equations (8-38, p. 251), (8-41), (8-42, p. 253)), and Christoffel symbols (equation (8-31, p. 244)) to make apparent the relationship of the metric tensor to the Field Equations.

$$R^{\alpha}_{\mu\nu\beta} = \Gamma^{\alpha}_{\rho\nu}\Gamma^{\rho}_{\mu\beta} - \Gamma^{\alpha}_{\rho\beta}\Gamma^{\rho}_{\mu\nu} + \frac{\partial \Gamma^{\alpha}_{\mu\beta}}{\partial x^{\nu}} - \frac{\partial \Gamma^{\alpha}_{\mu\nu}}{\partial x^{\beta}}.$$

$$R_{\mu\nu} = R^{\alpha}_{\mu\nu\alpha}. = \Gamma^{\alpha}_{\rho\nu}\Gamma^{\rho}_{\mu\alpha} - \Gamma^{\alpha}_{\rho\alpha}\Gamma^{\rho}_{\mu\nu} + \frac{\partial \Gamma^{\alpha}_{\mu\alpha}}{\partial x^{\nu}} - \frac{\partial \Gamma^{\alpha}_{\mu\nu}}{\partial x^{\alpha}}.$$

$$\mathcal{R} = g^{\mu\nu} R_{\mu\nu}.$$

$$\Gamma^{\alpha}_{\mu\nu} = \frac{1}{2} g^{\alpha\rho} \left(\frac{\partial g_{\nu\rho}}{\partial x^{\mu}} + \frac{\partial g_{\rho\mu}}{\partial x^{\nu}} - \frac{\partial g_{\mu\nu}}{\partial x^{\rho}} \right).$$

As the Christoffel symbols depend on the partial derivatives of the metric tensor, and the Ricci tensor and Riemann curvature tensor depend on the partial derivative of the Christoffel symbols, it should be clear that the left-hand side of the Field Equations (equation (11-27)) depends on the second partial derivative of the metric tensor.

A useful relation between the Ricci scalar $\mathcal{R}$ and the energy-momentum tensor $T_{\mu\nu}$ may be found by applying the contravariant metric tensor $g^{\mu\nu}$ to the Field Equations of (11-27). A scalar of the energy-momentum tensor is defined as $\mathcal{T} = g^{\mu\nu} T_{\mu\nu} = g_{\mu\nu} T^{\mu\nu}$. Also in four-dimensional space-time, $g^{\mu\nu} g_{\mu\nu} = 4$ (just as in three dimensions $\delta^{ij}\delta_{ij} = 3$).

$$g^{\mu\nu} R_{\mu\nu} - \frac{1}{2} g^{\mu\nu} g_{\mu\nu} \mathcal{R} = -\kappa g^{\mu\nu} T_{\mu\nu},$$

$$\mathcal{R} - \frac{4}{2} \mathcal{R} = -\kappa \mathcal{T},$$

$$\mathcal{R} = \kappa \mathcal{T}.$$

With the relation between the scalars of the Ricci and energy-momentum tensors, the dependence on the Ricci tensor, expressing the curvature of space-time, is explicitly shown to be a function of energy-momentum sources.

$$R_{\mu\nu} = \kappa \left(\frac{1}{2} g_{\mu\nu} \mathcal{T} - T_{\mu\nu} \right). \tag{11-28}$$

One more form was proposed by Einstein in which a term called the cosmological constant was added. This will be discussed in the next chapter, and its significance to the current understanding of the evolution of the universe. However, at this point, we will look at the relationship of the Field Equations to Newton's description of gravity in terms of the gravitational potential.

The Geodesic of a Particle in a Weak Gravitational Field

We begin our comparison of the Field Equations with Newton's laws with the description of the law of motion for a particle in a weak gravitational field. Einstein initially proposed as an independent postulate of his theory that particles would be governed by the geodesic equation characterized by the geometry defined by the Field Equations. Later he found that the geodesic equation of motion for a particle was a consequence of his Field Equations and did not need to be specified as an independent postulate.[321]

The geodesic equation is reproduced below from equation (8-37, p. 246) with Greek letters used for indices to account for the four-dimensional nature of space-time and its invariant S.

$$\frac{D(dx^{\alpha}/dS)}{dS} = \frac{d^2 x^{\alpha}}{dS^2} + \Gamma^{\alpha}_{\mu\nu} \frac{dx^{\mu}}{dS} \frac{dx^{\nu}}{dS} = 0.$$

Einstein notes that

"In order to learn whether the equations [(11-27)] are consistent with experience, we must above all else, find out whether they lead to the

*Newtonian theory as a first approximation. We already know that
Euclidean geometry and the law of the constancy of the velocity of
light are valid, to a certain approximation, in regions of great extent,
as in planetary systems.''* [322]

Therefore, Einstein introduced the approximation that the metric tensor
is equal to the metric of Special Relativity, designated here as $\eta_{\mu\nu}$, plus
a perturbation due to a weak gravitational field, $h_{\mu\nu}$.[323]

$$g_{\mu\nu} = \eta_{\mu\nu} + h_{\mu\nu}.$$

For convenience (and more conventionally) in the following derivations,
the metric tensor $g_{\mu\nu}$. is the negative of that defined in equations (11-13,
p. 283) and (11-16, p. 285): $\eta_{11} = \eta_{11} = \eta_{33} = -\eta_{44} = -1$, and $\eta_{\mu\nu} = 0$
for $\mu \neq \nu$.[cxxix] Thus, $dS^2 = \eta_{\mu\nu}dx^\mu dx^\nu = (cdt)^2 - (dx^i)^2$ using $i = 1$ to 3
for the spatial components. (Note that here, the invariant dS^2 takes on
the meaning of the proper time interval $d\tau^2$.) The perturbations $h_{\mu\nu}$ are
considered to be much less than $\eta_{\mu\nu}$, that is, $|h_{\mu\nu}| \ll |\eta_{\mu\nu}|$.

As we are interested in the Newtonian limit, the acceleration terms for
the component x^i in the geodesic equation are of particular interest. The
Christoffel symbols of interest are given below neglecting terms formed
by the products of the perturbations and their derivatives.

$$\Gamma^i_{\mu\nu} = \frac{1}{2}g^{i\rho}\left(\frac{\partial g_{\nu\rho}}{\partial x^\mu} + \frac{\partial g_{\rho\mu}}{\partial x^\nu} - \frac{\partial g_{\mu\nu}}{\partial x^\rho}\right),$$

$$= \frac{1}{2}\eta^{i\rho}\left(\frac{\partial h_{\nu\rho}}{\partial x^\mu} + \frac{\partial h_{\rho\mu}}{\partial x^\nu} - \frac{\partial h_{\mu\nu}}{\partial x^\rho}\right).$$

A further approximation in the Newtonian limit is that the velocities
$\frac{dx^i}{dt} \ll c$ ($i = 1$ to 3), or $dS \approx cdt$. In consequence, $\frac{dx^i}{dS} = \frac{dx^i}{dt}\frac{dt}{dS} = \frac{v^i}{c} \ll \frac{dx^4}{dS} \approx \frac{d(ct)}{d(ct)} = 1$
. In the geodesic equation reproduced above, among the terms given by
$\Gamma^\alpha_{\mu\nu}\frac{dx^\mu}{dS}\frac{dx^\nu}{dS}$, terms, with the Christoffel symbol multiplied by the spatial

---

[cxxix] In the weak gravitational field approximation, the contravariant metric tensor η^{11}
$= \eta^{22} = \eta^{33} = -\eta^{44} = -1$, and $\eta^{\mu\nu} = 0$ if $\mu \neq \nu$.

derivatives $\frac{dx^i}{dS}\frac{dx^j}{dS}$ are therefore neglected. The geodesic equation for the spatial component x^i is reduced to:

$$\frac{D(dx^i/dS)}{dS} = \frac{d^2x^i}{dS^2} + \Gamma^i_{44}\frac{dx^4}{dS}\frac{dx^4}{dS} = \frac{d^2x^i}{dS^2} + \Gamma^i_{44} = 0, \text{ and}$$

$$\Gamma^i_{44} \approx \frac{1}{2}\eta^{i\rho}\left(\frac{\partial h_{4\rho}}{\partial x^4} + \frac{\partial h_{\rho 4}}{\partial x^4} - \frac{\partial h_{44}}{\partial x^\rho}\right).$$

Assuming the gravitational field does not change with time, that is, $\partial/c\partial t = \partial/\partial x^4 = 0$, then

$$\Gamma^i_{44} = -\frac{1}{2}\eta^{i\rho}\left(\frac{\partial h_{44}}{\partial x^\rho}\right).$$

Finally, with the approximation $dS = cdt$, the equation for component i of the acceleration due to the gravitational field $\mathbf{g}$ may be written:

$$\frac{d^2x^i}{dt^2} = \eta^{i\rho}\left(\frac{c^2}{2}\frac{\partial h_{44}}{\partial x^\rho}\right),$$

$$\frac{d^2x^i}{dt^2} = -\frac{c^2}{2}\left(\frac{\partial h_{44}}{\partial x^1}, \frac{\partial h_{44}}{\partial x^2}, \frac{\partial h_{44}}{\partial x^3}\right),$$

or in vector notation,

$$\mathbf{a} = \mathbf{g} = -\nabla(c^2 h_{44}/2).$$

Recall that $\mathbf{g} = -\nabla\mathfrak{G}$, $\mathfrak{G}$ being the gravitational potential (equation (11-25, p. 303), therefore

$$\mathfrak{G} = c^2 h_{44}/2, \tag{11-29}$$

$$g_{44} = \eta_{44} + h_{44} = 1 + 2\mathfrak{G}/c^2. \tag{11-30}$$

This provides the connection between Newton's Law of Motion for a particle in a gravitational field with Newtonian gravitational potential $\mathfrak{G}$ and the space-time metric with its associated geodesic equation. The relationship supports Einstein's early intuition that space-time metric tensor would replace the gravitational potential. We now need to show that this interpretation is consistent with the Field Equations (11-27).

The Field Equations and Newton's Law of Gravitation

As we begin to investigate the relationship between the Field Equations and Newton's Law of Gravitation, this is perhaps a good time to remind you that the Field Equations, just like Newton's Laws of Motion and Law of Gravitation, cannot be derived. They are generalizations, inspired in part from empirical knowledge, but ultimately expressing more than that knowledge. The process of discovering the laws of science is inductive with reasoning going from specific facts to the general, but it is also a profoundly creative process. Einstein believed that *"the invention of scientific concepts and the building of theories upon them was one of the creative properties of the human mind."* As Pais points out in relation to Einstein's view, this was opposed to Mach's view that the laws of science were only an economical way of describing a large collection of facts.[324] To put Einstein's point more prosaically, a pile of bricks is not a house.

The fundamental laws of science are inductive truths, in contrast to the deductive truths of the propositions derived from postulates such as those of Euclid. The fundamental laws of science may be thought of as postulates that cannot be proved, but they are far from self-evident as Euclid's postulates were thought to be by the ancient Greeks. The acceptance of the laws of science depends entirely on their continuing agreement with experimental experience as that knowledge base grows. The perfect example of the superseding of one accepted truth by another is that of Newton's Law of Gravitation by General Relativity. We can never know, however, if a scientific law is a final truth or merely an approximation to a more general truth. As we shall see below, we can show how Newton's triumphant picture of the universe is embedded within the broader canvass of General Relativity.

We will start with the left-hand side of the Field Equations (11-27). As above, the metric tensor $g_{\mu\nu}$ is assumed to be that of the inertial reference frame of Special Relativity plus a small perturbation, $g_{\mu\nu} = \eta_{\mu\nu} + h_{\mu\nu}$. As the components of $\eta_{\mu\nu}$ are constants, and we are neglecting products of the perturbations of the second power and products of the perturbations and their derivatives, the Ricci tensor may be approximated as below.[cxxx]

$$R_{\mu\nu} = R^\alpha_{\mu\nu\alpha\cdot} = \Gamma^\alpha_{\rho\nu}\Gamma^\rho_{\mu\alpha} - \Gamma^\alpha_{\rho\alpha}\Gamma^\rho_{\mu\nu} + \frac{\partial\Gamma^\alpha_{\mu\alpha}}{\partial x^\nu} - \frac{\partial\Gamma^\alpha_{\mu\nu}}{\partial x^\alpha},$$

$$\approx \frac{\partial\Gamma^\alpha_{\mu\alpha}}{\partial x^\nu} - \frac{\partial\Gamma^\alpha_{\mu\nu}}{\partial x^\alpha},$$

$$\approx \frac{\eta^{\alpha\rho}}{2}\left\{\frac{\partial}{x^\nu}\left(\frac{\partial h_{\alpha\rho}}{\partial x^\mu} + \frac{\partial h_{\rho\mu}}{\partial x^\alpha} - \frac{\partial h_{\mu\alpha}}{\partial x^\rho}\right) - \frac{\partial}{x^\alpha}\left(\frac{\partial h_{\nu\rho}}{\partial x^\mu} + \frac{\partial h_{\rho\mu}}{\partial x^\nu} - \frac{\partial h_{\mu\nu}}{\partial x^\rho}\right)\right\}.$$

Focusing on $\mu = \nu = 4$, as in the above discussion of the geodesic path of a particle and assuming that the gravitational field does not change with time, then $\frac{\partial}{c\,\partial t} = \frac{\partial}{\partial x^\mu} = \frac{\partial}{\partial x^\nu} = 0$. Therefore,

$$R_{44} = \frac{\eta^{\alpha\rho}}{2}\left(-\frac{\partial}{x^\alpha}\left(-\frac{\partial h_{\mu\nu}}{\partial x^\rho}\right)\right),$$

$$= \frac{\eta^{\alpha\rho}}{2}\frac{\partial^2 h_{44}}{\partial x^\alpha\,\partial x^\rho},$$

$$= -\frac{1}{2}\left(\frac{\partial^2 h_{44}}{\partial x^1\,\partial x^1} + \frac{\partial^2 h_{44}}{\partial x^2\,\partial x^2} + \frac{\partial^2 h_{44}}{\partial x^3\,\partial x^3}\right),$$

$$= -\frac{1}{2}\nabla^2 h_{44}, \text{ and from equation } \textbf{(11-29)},$$

$$R_{44} = -\frac{1}{c^2}\nabla^2\mathfrak{G}. \tag{11-31}$$

Now it is time to look at the right-hand side of the Field Equations. Newton's Law of Gravitation, expressed in terms of the potential $\mathfrak{G}$ is

---

[cxxx] For convenience the definition of the Christoffel symbols (equation (8-31, p. 244) with appropriate Greek indices) is again repeated here: $\Gamma^\alpha_{\mu\nu} = \frac{1}{2}g^{\alpha\rho}\left(\frac{\partial g_{\nu\rho}}{\partial x^\mu} + \frac{\partial g_{\rho\mu}}{\partial x^\nu} - \frac{\partial g_{\mu\nu}}{\partial x^\rho}\right)$.

$\nabla^2 \mathfrak{G} = 4\pi G\rho$, equation (11-26, p. 303). The source of the gravitational field, on the right-hand side of the equation, is mass expressed here as the mass density ρ, the mass per unit volume. In the Field Equations, the metric tensor $g_{\mu\nu}$ takes the place of $\mathfrak{G}$, and the source term is the energy-momentum tensor $T_{\mu\nu}$. Energy and momentum must be included, since, as we have seen in Special Relativity, mass represents energy, energy has mass.

While the left-hand side of the Field Equations is well defined by the Ricci tensor, the right-hand side specified by the energy-momentum tensor represents specific physical situations and sources of energy, matter, and momentum. As we shall see in the next chapter, the modeling of this term gives rise to various descriptions of the evolution of the universe. Modeling must account for a universe in which the percentage of matter making up stars and planets is less than about 5% with the rest inferred as what is known as dark matter and dark energy.[325] While discussing some of the implications of new knowledge, we will follow Einstein's simplest model of the universe as a static dust cloud of matter at rest described by a distribution of density ρ.[326]

Einstein notes that the contribution of electromagnetic fields to the energy of matter is negligible. Furthermore, the energy due to deformation of matter and chemical energy is considered negligible. With these considerations Einstein approximates the energy-momentum tensor as the outer product of the four-vector velocity with the rest density ρ, [327]

$$T^{\mu\nu} = \rho \frac{dx^\mu}{ds} \frac{dx^\nu}{ds} = \rho \mathfrak{U}^\mu \mathfrak{U}^\nu. \tag{11-32}$$

The spatial components (i, j = 1 to 3) of T^{ij} are the flux of the ith component of momentum across the jth surface. The four-vector velocity is $\mathfrak{U}^\mu = \frac{1}{\sqrt{1-\frac{u^2}{c^2}}}(u, c)$, $\mathbf{u}$ being the velocities in three-dimensional space. Considering the velocity of the masses under consideration to be small compared to that of light ($ds \approx cdt$), the only component of the energy-momentum tensor is $T^{44} = \rho c^2$ (see footnote cxvii, p. 285). With this assumption and the previous assumption that the perturbation to the metric tensor is small, $T_{44} = \mathcal{T} = \rho c^2$.[328] Now, returning to The Field Equation as cast in (11-28, p. 309),

$$R_{\mu\nu} = \kappa\left(\frac{1}{2}g_{\mu\nu}\mathcal{T} - T_{\mu\nu}\right),$$

$$R_{44} = \kappa\left(\frac{\mathcal{T}}{2} - T_{44}\right), \text{ and using equation (11-31),}$$

$$-\frac{1}{c^2}\nabla^2\mathcal{G} = -\kappa\rho c^2/2, \text{ and}$$

$$\nabla^2\mathcal{G} = \kappa c^4 \rho/2.$$

Comparing the above result with equation (11-26, p. 303), $\kappa = 8\pi G/c^4$,[329] with G being Newton's gravitational constant. Thus, the Field Equations are able to account for all of the successful Newtonian predictions from his Law of Gravity.

At this point it is reasonable to ask what the relationship is between the Field Equations and the Equivalence Principle. The answer is that it is always possible to find a set of local coordinates in which the Christoffel symbols vanish. In other words to find a local coordinate system in which the spatial invariant ds^2 is expressed as in Special Relativity. This is the local inertial frame which acts as a free-falling infinitesimal laboratory.[330] Thus, the conclusions of the Equivalence Principle find expression in the Field Equations. The next step for Einstein was to use the Field Equation to predict new phenomena, including the gravitational redshift and the bending of light which are consequences of the Equivalence Principle.

12 Consequences of General Relativity

12.1 The first three predictions: the spectral redshift, Mercury's orbit, and the bending of light

A week before the finalization of the Field Equations on November 25, 1915, Einstein was able to predict the discrepancies of Mercury's orbit from Newton's laws and to predict the bending of light in a gravitational field. He made these calculations using approximate methods with essentially the equation $R_{\mu\nu} = 0$ which applies to empty space.[331] The prediction of the redshift had been calculated from the Equivalence Principle long before he was close to his final Field Equations.[332] Remarkably only two months after publication of the Field Equations, Einstein became aware of an exact solution for his equations by Karl Schwarzschild for a static, spherically symmetric, isotropic gravitational field of a single mass point surrounded by empty space ($R_{\mu\nu} = 0$).[333] Schwarzschild, having long been involved in astrophysics, had presented a paper in 1900 to the German Astronomical Society proposing that the geometry of the universe might be non-Euclidean and had been director of the astrophysical observatory in Potsdam. As one tragic event in the overwhelming tragedy that was World War I, Schwarzschild died in June 1916 two months after having contracted a disease as a German soldier at the Russian front.[334] Einstein had presented Schwarzschild's paper to the Prussian Academy in January 1916 and had commemorated his life in an address to the Prussian Academy in June, expressing his conviction that Schwarzschild's contributions would continue to play a stimulating role in science.[335]

The Schwarzschild solution for the proper time invariant $d\tau^2 = -d\mathcal{S}^2/c^2$ is,[336]

$$(cd\tau)^2 = \left(1 - \frac{2GM}{c^2r}\right)c^2dt^2 - \left(\frac{dr^2}{1 - \frac{2GM}{c^2r}} + r^2(\sin^2\phi d\theta^2 + d\phi^2)\right). \qquad (12\text{-}1)$$

The solution is formed by a proper time invariant which approaches that of Special Relativity with increasing coordinate r. Associating coordinates $(x^1, x^2 . x^3, x^4)$ with (r, θ, φ, ct),[cxxxi] then

$$g_{11} = (1 - 2GM/c^2r)^{-1}, g_{22} = -r^2\sin^2\phi, g_{33} = -r^2, g_{44} = 1 - 2GM/c^2r.$$

The solution, satisfying $R_{\mu\nu} = 0$, is consistent with Newton's Law of Gravitation for a single particle of mass M as discussed previously in that $g_{44} = 1 - 2GM/c^2r = 1 + 2\mathfrak{G}/c^2$ (equation (11-30, p. 311)). With these metric tensor components and using the geodesic equations and relationships implied by the proper time invariant, the spectral redshift, the orbit of Mercury, and the bending of light can be calculated. The predictions of these phenomena through the Field Equations and their experimental verification demonstrated the role of geodesics in the curved geometry of space-time. The calculations are discussed below.

The spectral redshift

Proper time τ is defined as the time *measured* by a standard clock stationary in its frame of reference. The proper time differential dτ is related to a *coordinate* time dt at fixed spatial coordinates through the Schwarzschild metric as shown below with $\mathfrak{G}$ being the gravitational potential.

$$d\tau = \left(1 - \frac{2GM}{c^2r}\right)^{1/2} dt,$$

---

[131] Here, as elsewhere, φ is the angle of colatitude and θ the azimuthal angle of spherical coordinates.

$$= \left(1 + \frac{2\mathfrak{G}}{c^2}\right)^{1/2} dt, \text{ at a given location r.}$$

$$\tau = \int_0^t \left(1 + \frac{2\mathfrak{G}}{c^2}\right)^{1/2} dt,$$

$$\tau = \left(1 + \frac{2\mathfrak{G}}{c^2}\right)^{1/2} t.$$

(12-2)

Note that the coordinate time, distinguished from the proper time, increases and approaches equality with the proper time as the coordinate radius r increases from the mass point. The proper time is that which would be given by standard clock such as an atomic clock.[337]

As a thought experiment,[338] suppose light at a given wavelength is emitted in a fixed gravitational field from location 1 towards location 2, and time t is the time between the successive waves of light emitted as measured at location 1. The coordinate time at location 1 is given by equation (12-2) with $\mathfrak{G} = \mathfrak{G}_1$. Now if T is the difference in coordinate time between the emission of light at location 1 and its reception at location 2, then a wave emitted at t_0 will arrive at location 2 at $t_0 +$ T as measured a location 2. The next wave will be emitted at location 1 at $t_0 + t$ and arrive at location 2 at $t_0 + t +$ T. In other words, the time between waves at location 2 will also be t. From equation (12-2), standard clocks will measure time differently at different locations in a gravitational field. Recall that this also follows from the Equivalence Principle (see pp. 291-292). Let τ_1 be the proper time between waves at location 1 and τ_2 be the proper time interval at location 2, then

$$\frac{\tau_1}{\tau_2} = \frac{\left(1 + \frac{2\mathfrak{G}_1}{c^2}\right)^{1/2} t}{\left(1 + \frac{2\mathfrak{G}_2}{c^2}\right)^{1/2} t}$$

If $\mathfrak{G}_1 < \mathfrak{G}_2$, then $\tau_1 < \tau_2$. For example, location 1 might be the surface of a star and location 2 at some distance beyond in the star's increasing potential field such as our observatories. As the wavelength λ of light is proportional to the time period of the waves, then the wavelength λ_2 is shifted to longer wavelengths relative to that measured at location 1, that is the wavelengths are redshifted.[cxxxii]

In the laboratory, Robert Pound and Glen Rebka in 1959 observed the change in frequency of a gamma ray beamed downward 24 m. Now, the gravitational time effect is routinely and necessarily accounted for in Global Positioning System satellites which orbit about 20,100 km above the earth.[339] More related to our subject, the gravitational redshift has been observed in the light emitted from stars.[340]

Mercury's orbit

In 1859, Urbain Jean Joseph Verrier submitted a letter to the Academy of Sciences in Paris in which he reported that the point in the planet Mercury's orbit which is the closest to the sun, its perihelion, advanced in the plane of Mercury's orbit by 38 arcseconds per century. This effect, despite being extremely small, was very significant because it could not be accounted for by Newton's laws. As Verrier noted, the advance was due to *"some as yet unknown action on which no light had been thrown...a grave difficulty, worthy of attention by astronomers."* [341] In the intervening years before Einstein was able to calculate this effect, numerous unsatisfactory approaches were made to explain this phenomenon.

Abraham Pais, the physicist and biographer of Einstein, believes that, *"This discovery* [the calculation of the advance of Mercury's perihelion] *was ...by far the strongest emotional experience in Einstein's scientific life, perhaps in all his life. Nature had spoken to him."* Of the discovery based upon approximate solution methods, Einstein said, *"For a few days, I was beside myself with joyous excitement."* [342]

[cxxxii] Wavelength λ and frequency ν are related as $\lambda = c/\nu$, therefore the increase in wavelength is accompanied by an decrease in the frequency of the wave.

The calculation of the advancement of Mercury's perihelion falls out of Schwarzschild's solution directly. The orbit of Mercury is predicted from the geodesic equations for the coordinate t and azimuthal angle θ. It can be shown from the solution that Mercury's orbit stays within a plane, that is, the colatitude ϕ is constant. [343] Here the results are shown taking the orbit to be in the plane of the equator, $\phi = \pi/2$. The geodesic equations for coordinates θ and t as functions of the invariant path parameter τ are given below with the constant $m = GM/c^2$. The geodesic equations can be calculated with the same procedures used in to calculate the geodesics on a sphere (p. 246; see also, Lawden, p. 147; Wrede, p. 377). The Schwarzschild metric components are given below from (12-1) and from which the geodesic equations follow.

$$(\phi = \frac{\pi}{2}, m = \frac{GM}{c^2}, g_{11} = (1 - 2m/r)^{-1}, g_{22} = -r^2, g_{33} = -r^2, g_{44} = 1 - 2m/r.)$$

cxxxiii

$$\frac{d^2\theta}{d\tau^2} + \frac{2}{r}\frac{dr}{d\tau}\frac{d\theta}{d\tau} = 0, \tag{12-3}$$

$$\frac{d^2t}{d\tau^2} + \frac{2m}{(1 - 2m/r)r^2}\frac{dt}{d\tau}\frac{d\theta}{d\tau} = 0. \tag{12-4}$$

These equations can be put into a form in which a first integration may easily be performed. Equations (12-3) and (12-4) become, respectively,

$$\frac{d}{d\tau}\left(r^2\frac{d\theta}{d\tau}\right) = 0.$$

$$\frac{d}{d\tau}\left(\frac{(r - 2m)}{r}\right)\frac{dt}{d\tau} = 0.$$

---

cxxxiii The geodesic equations are calculated from: $\frac{d^2x^i}{d\tau^2} + \Gamma^i_{jk}\frac{dx^j}{d\tau}\frac{dx^k}{d\tau} = 0$ with, in the case of Relativity, i, j, k, going from 1 to 4. As an example, for $i = 2 = \theta$, j = 1 = r, $\frac{d^2x^i}{d\tau^2} + \Gamma^i_{jk}\frac{dx^j}{d\tau}\frac{dx^k}{d\tau} = \frac{d^2\theta}{d\tau^2} + \Gamma^\theta_{r\theta}\frac{dr}{d\tau}\frac{d\theta}{d\tau} + \Gamma^\theta_{\theta r}\frac{d\theta}{d\tau}\frac{dr}{d\tau} = 0.$ Using the Schwarzschild metric, $\Gamma^\theta_{r\theta} = \Gamma^\theta_{\theta r} = \frac{1}{2}(-\frac{1}{r^2})(\frac{\partial(-r^2)}{\partial r}) = \frac{1}{r}$, and $\frac{d^2\theta}{d\tau^2} + \frac{2}{r}\frac{dr}{d\tau}\frac{d\theta}{d\tau} = 0.$

With h and k being constants of integration,

$$\frac{d\theta}{d\tau} = \frac{h}{r^2}, \text{and} \qquad (12\text{-}5)$$

$$\frac{dt}{d\tau} = \frac{kr}{r - 2m}. \qquad (12\text{-}6)$$

Now the Schwarzschild equation (12-1, p. 317) may be recast in terms of the derivatives $d/d\tau$ and be simplified by taking $\phi = \pi/2$.

$$c^2 = \left(1 - \frac{2GM}{c^2 r}\right) c^2 \left(\frac{dt}{d\tau}\right)^2 - \left(\frac{1}{1 - \frac{2GM}{c^2 r}} \left(\frac{dr}{d\tau}\right)^2 + r^2 \left(\frac{d\theta}{d\tau}\right)^2\right). \qquad (12\text{-}7)$$

Using $m = GM/c^2$ and Newton's notation, $\dot{r} = dr/d\tau$, $\dot{\theta} = d\theta/d\tau$, $\dot{t} = dt/d\tau$ and rearranging,

$$\frac{r}{r - 2m} \dot{r}^2 + r^2 \dot{\theta}^2 - \frac{c^2}{r}(r - 2m)\dot{t}^2 = -c^2. \qquad (12\text{-}8)$$

Substituting equations (12-5) and (12-6) into equation (12-8) and rearranging,

$$\dot{r}^2 + \frac{h^2}{r^3}(r - 2m) = c^2(k^2 - 1) + \frac{2mc^2}{r}.$$

By taking the derivative $d/d\tau$ of the above result, it can be compared with the radial acceleration dr^2/dt^2 calculated from Newton's laws. Taking the derivative and rearranging, the above equation becomes,

$$\ddot{r} - \frac{h^2}{r^3} = -\frac{mc^2}{r^2} - \frac{3mh^2}{r^4}. \qquad (12\text{-}9)$$

The relationship between the radial coordinate r and azimuthal coordinate θ may similarly be made explicit by using the chain rule with equation (12-5) to form the second derivative $d^2r/d\theta^2$.[cxxxiv]

$$\frac{d^2r}{d\theta^2} - \frac{2}{r}\left(\frac{dr}{d\theta}\right)^2 - r = -\frac{mc^2r^2}{h^2} - 3m. \tag{12-10}$$

Now we want to compare this to the result obtained from Newton's laws $\mathbf{F} = -(GMm/r^2)\,\mathbf{e}_r = \mathbf{ma} = m\frac{d^2\mathbf{r}}{dt^2}$. The vector equation may be expressed as the radial and azimuthal acceleration using the methods employed to calculate the acceleration of a hypothetical planet in a circular orbit about its sun (pp. 262-263).[cxxxv] The radial force is the force of gravity and there is no azimuthal force. As the velocity of Mercury around the sun is much less than the velocity of light, dt = dτ. Therefore,

$$\frac{F}{m_{Mercury}} = -\frac{GM}{r^2} = -\frac{mc^2}{r^2} = \ddot{r} - r\dot{\theta}^2, \text{and} \tag{12-11}$$

$$0 = r\ddot{\theta} + 2\dot{r}\dot{\theta}. \tag{12-12}$$

Equation (12-12) for the azimuthal acceleration may be expressed as equation (12-5) using the same integration constant h and the approximation t = τ. Note that, as seen below, these assumptions are justified by the comparisons between the Newtonian and Relativistic results. Substituting the expression $\dot{\theta} = h/r^2$ into the equation for the radial force, equation (12-11) results in the following expression,

---

[cxxxiv] $\frac{dr}{d\tau} = \frac{d\theta}{d\tau}\cdot\frac{dr}{d\theta} = \frac{h}{r^2}\cdot\frac{dr}{d\theta};\ \frac{d^2r}{d\tau^2} = \frac{d\theta}{d\tau}\cdot\frac{d}{d\theta}\left(\frac{dr}{d\tau}\right) = \frac{h}{r^2}\frac{d}{d\theta}\left(\frac{h}{r^2}\cdot\frac{dr}{d\theta}\right) = \left(\frac{h}{r^2}\right)^2\frac{d^2r}{d\theta^2} - 2\frac{h^2}{r^5}\left(\frac{dr}{d\theta}\right)^2.$

[cxxxv] Let the position vector $\mathbf{r} = r(\cos\theta, \sin\theta)$, then with Newton's notation,
$\frac{d\mathbf{r}}{dt} = \dot{r}(\cos\theta, \sin\theta) + r(-\sin\theta\,\dot{\theta}, \cos\theta\,\dot{\theta}) = \dot{r}\mathbf{e}_r + r\dot{\theta}\mathbf{e}_\theta.$

$\frac{d^2\mathbf{r}}{dt^2} = \ddot{r}(\cos\theta, \sin\theta) + \dot{r}(-\sin\theta\,\dot{\theta}, \cos\theta\,\dot{\theta}) + \dot{r}\dot{\theta}(-\sin\theta, \cos\theta)$

$\qquad + r\ddot{\theta}(-\sin\theta, \cos\theta) + r\dot{\theta}(-\cos\theta\,\dot{\theta}, -\sin\theta\,\dot{\theta}),$

$\frac{d^2\mathbf{r}}{dt^2} = (\ddot{r} - r\dot{\theta}^2)\mathbf{e}_r + (r\ddot{\theta} + 2\dot{r}\dot{\theta})\mathbf{e}_\theta.$

$$\ddot{r} - \frac{h^2}{r^3} = -\frac{mc^2}{r^2}. \tag{12-13}$$

Similarly, as previously, an expression can be developed for $\frac{d^2r}{d\theta^2}$,

$$\frac{d^2r}{d\theta^2} - \frac{2}{r}\left(\frac{dr}{d\theta}\right)^2 - r = -\frac{mc^2r^2}{h^2}. \tag{12-14}$$

Comparing equations (12-9) and (12-10) with their Newtonian counterparts (12-13) and (12-14), there is an addition term on the right-hand side which accounts for the advance of Mercury's perihelion. The ratio of this term to the gravitational force terms is quite small; however, its impact is cumulative accounting for the small change of 43 arcseconds per century. Lawden provides the details of the analysis to find the functional form of Mercury's orbit and its advancement.[344]

Accounting for the discrepancy of Mercury's orbit from the predictions of Newton's laws was an extraordinary success for General Relativity as it demonstrated motion along a geodesic in the curved space-time of $R_{\mu\nu} = 0$.. It was particular notable that the theory was able to explain an anomaly known decades before the formulation of the Field Equations and which was not part of its development. The next important success also focused on the effect of space-time on a pathway, but this time it was that of light. The verification in this case would be delayed by World War I, but when it was accomplished, Einstein became world famous. The calculation of the bending of light from the Field Equations and its verification will be discussed next.

The deflection of light through curved space-time

The discovery from Special Relativity that energy is associated with mass leads to the thought that light rays, having energy, should be affected by a gravitational field. This thought is further advanced by the Equivalence Principle. In a thought experiment following the introduction of the Equivalence Principle (p. 290), a comparison is made between the observations of those in a free-falling laboratory with those outside, presuming that they are stationary in a gravitational field observing the free-falling laboratory. The conclusion from the Equivalence Principle is that a light ray is deflected in a gravitational field. The deflection can be calculated using the Schwarzschild solution. As an example, the path of a celestial light ray that just grazes the sun and is observed on earth may be calculated. The calculation and the story of its verification are given below.

The method of calculating the path of a light ray passing near an object of mass M in the plane $\Phi = \frac{\pi}{2}$ is similar to that used in the calculations of Mercury's orbit; however, because the path is for a light ray, the invariant $d\tau^2 = 0$. Instead of taking derivatives with respect to τ, we take derivatives with respect to a path parameter σ. With these changes equation (12-7, p. 321) becomes:

$$\left(1 - \frac{2GM}{c^2 r}\right) c^2 \left(\frac{dt}{d\sigma}\right)^2 - \left(\frac{1}{1 - \frac{2GM}{c^2 r}}\left(\frac{dr}{d\sigma}\right)^2 + r^2 \left(\frac{d\theta}{d\sigma}\right)^2\right) = 0.$$

Reinterpreting, derivatives in terms of the path parameter σ, equations (12-5) and (12-6) remain the same. Substituting these into the above equation, results in the following equation for a light ray analogous to that following the substitutions into equation (12-8),

$$\dot{r}^2 + \frac{h}{r^3}(r - 2m) = c^2 k^2. \tag{12-15}$$

Our goal is to determine the impact of the gravitational field of the sun with mass M and radius R on a light ray that otherwise would propagate in a straight line. Let the path of the light ray in the absence of the sun be $r \sin \theta = R$, or equivalently in the x-y plane $y = R$. When the star is just about to be eclipsed (occulted) by the sun, the light ray can be considered to just graze the sun. In 1919 the scientist Arthur Eddington made exactly this observation to measure the deflection of the light ray compared to its position at a time when the sun was not in this position. With this description, it is helpful to recast equation (12-15) with the angle θ as the independent variable as previously in the case of Mercury's orbit. To that end, using the chain rule, $dr/d\sigma = (d\theta/d\sigma) \cdot (dr/d\theta) = (h/r^2) \cdot (dr/d\theta)$.

Substituting this relation into equation (12-15),

$$\left(\frac{dr}{d\theta}\right)^2 + r(r - 2m) = \left(\frac{ck}{h}\right)^2 r^4.$$

From this point, a common transformation, $r = 1/u$, is used to facilitate determination of the angle that the light ray is deflected (see Lawden, p. 148, Wrede, p. 378). Substituting the variable u and $dr/d\theta = -(1/u^2) \cdot (du/d\theta)$, we arrive at,

$$\left(\frac{du}{d\theta}\right)^2 + u^2 - 2mu^3 = (ck/h)^2.$$

And finally, taking the derivative with respect to θ,

$$\frac{d^2u}{d\theta^2} + u = 3mu^2 \,.^{136} \qquad\qquad (12\text{-}16)$$

Note that the equation for the undeflected light ray, $u = \frac{1}{r} = \frac{\sin\theta}{R}$ is a solution to $\frac{d^2u}{d\theta^2} + u = 0$. The right-hand side of equation (12-16) is the term causing deflection. Because the deflection is small, the right-hand side may be approximated as $3mu^2 = 3m\left(\frac{\sin\theta}{R}\right)^2$ and we seek a solution for the deflection u_δ with $u \approx \frac{\sin\theta}{R} + u_\delta$.

$$\frac{d u_\delta{}^2}{d\theta^2} + u_\delta = 3m\left(\frac{\sin\theta}{R}\right)^2.$$

[cxxxvi] If the transformation, $r = 1/u$, used for the light ray had been followed in the analysis of Mercury's orbit, the equation would be found to be $\frac{d^2u}{d\theta^2} + u = \frac{mc^2}{h^2} + 3mu^2$ (Lawden, p. 148).

The solution is $u_\delta = \frac{m}{R^2}(2 - \sin^2\theta)$,[cxxxvii] and

$$u = \frac{1}{r} = \frac{\sin\theta}{R} + \frac{m}{R^2}(2 - \sin^2\theta).$$

As $r \to \infty, u \to 0, \sin\theta \to \theta_{\delta+}, \sin^2\theta \to 0$, and $0 = \frac{\sin\theta_{\delta+}}{R} + \frac{2m}{R^2}$. . Therefore,

$$\theta_{\delta+} = -\frac{2m}{R}.$$

Similarly, as $r \to -\infty, u \to 0, \theta \to \pi - \theta_{\delta-}$, and $\theta_{\delta-} = -\frac{2m}{R}$. Adding the deflections, or simply by symmetry, the magnitude of the total deflection $\theta_\delta = \frac{4m}{R}$.

Aware that the Equivalence Principle predicted the phenomenon of the bending of light by gravity, Einstein looked for experimental verification. In 1912 an Argentinian eclipse expedition to Brazil had on its agenda the measurement of the light deflection, but it was rained out. A German expedition was on its way to make observations in the Crimea in August 1914 which became impossible with the outbreak of World War I. Finally in 1919 two expeditions were initiated, one to Principe Island off the coast of Spanish Guinea, led by Arthur Eddington, and the other to Sobral, Brazil led by Andrew Crommelin. The expeditions targeted May 29 as providing an excellent star configuration. Of his prospects, Eddington said,

"The present eclipse expedition may for the first time demonstrate the weight of light; or they may confirm Einstein's weird theory of non-Euclidean space; or they may lead to a result of yet more far-reaching consequences - no deflection."

[cxxxvii]The solution for u_δ may be found by assuming a solution of the form $u_\delta = A + B \sin\theta$, substituting this into the equation for u_δ, and collecting terms with $\sin^2\theta$ while recalling that $\sin^2\theta + \cos^2\theta = 1$. Then $A = -2B = \frac{2m}{R^2}$.

On November 6, 1919, it was announced that the measurements were consistent with Einstein's theory. The measurements indicated a deflection of 1.61" ± 0.30" (arcseconds) from the two expeditions.[345]

With $G = 6.673 \cdot 10^{-11} m^3 kg^{-1} s^{-2}$, $M = 1.991 \cdot 10^{30} kg$, $c = 2.998 \cdot 10^8 m/s$, and $R = 6.95 \cdot 10^8 m$,[346] the deflection is predicted to be,

$$\theta_\delta = 4m/R = 8.507 \times 10^{-6} \text{ radians} = 1.75".$$

At a joint meeting of the Royal Society and the Royal Astronomical Society when the results of the measurements were announced, the chair of the meeting, Joseph Thomson had this to say,

> *"This is the most important result obtained in connection with the theory of gravitation since Newton's day, and it is fitting that it should be announced at a meeting of the Society so closely connected with him...The result* [is] *one of the highest achievements of human thought."* [347]

12.2   Black holes and the expanding universe

The first predictions of Einstein' General Relativity dealt with effects that are extremely small: the perihelion advance of Mercury's orbit by 43" (arcseconds) per century and the bending of a light ray from a star by 1.75". The effect of the spectral redshift due to gravity was so small that it was only definitively observed in the laboratory and from stars, with improvements in measurements, after Einstein's death in 1955. Although these effects are small, the implications for our understanding of the universe are enormous. Solutions to Einstein's Field Equations with various models for the energy-momentum tensor $T_{\mu\nu}$ provided the vocabulary for very different descriptions of the universe with the results dependent on actual values of the parameters of the models. A common feature of many models is a universe, beginning with what became known as the Big Bang, expanding forever; while the models also show the possibility of a universe expanding from birth, then contracting to its end to form an eternal cycle of birth and death. The possible universes, as we shall see, surprised even Einstein. Looking backward in time, General Relativity predicts a universe so small at its creation that a complete physical description must take into account the other seminal discovery of the first half of the twentieth century, Quantum Mechanics and the particle physics which followed from a new formulation encompassing Special Relativity, Quantum Field Theory. The search for a theory to encompass both General Relativity and particle physics which arose from Quantum Mechanics continues. The search has led to a multitude of speculative descriptions of the universe including multiverses in which our universe is just one of many (see Barrow for descriptions as strange as any science fiction). Where will this end, no one knows. In this section I will outline the results that led to the theory of the Big Bang and some variants. But first I will start with a phenomenon of the universe that certainly is not small in its impact and was forseen as a mathematical singularity in Schwarzschild's solution.

Black holes

The Schwarzschild solution for the region outside a spherically symmetric body of mass M and radius R has a metric, equation (12-1, p. 317) in which the coefficients of dt^2 and the reciprocal of the coefficient of dr^2 are equal to zero if $r = 2m$. In this case, the coefficient of dr^2 becomes infinite. The radius at which this occurs is called the Schwarzschild radius, r_S.

$$r_S = 2m = 2GM/c^2$$

The Schwarzschild radius for the sun is 2.96 km. If the body has a radius R greater than r_S, as is the case with the sun, then analyses for $r > R$, are valid without further discussion. When $R < r_S$, then it is not immediately clear how to analyze the space for $r > R$. Initially, it was thought that the Field Equations were no longer valid across the Schwarzschild radius. However, the singularity at $r = r_S$ is not intrinsic, that is, it can be removed by a coordinate transformation.[348] Nevertheless, r_S has very important physical significance. Coordinate transformations discussed by Lawden[349] lead to the following conclusions. For $r < r_s$, all bodies must move towards the center $r = 0$. Bodies may move towards the center across the spherical boundary formed by $r = r_s$, but then they cannot escape. This is also true for the paths of photons, hence the term black hole.[cxxxviii] While the Schwarzschild singularity is not intrinsic, the singularity at $r = 0$ cannot be removed. However, the validity of the Field Equations as r approaches zero without accounting for quantum effects makes the validity of the theory questionable in any case.

Concerns about the Schwarzschild radius ceased to be academic in the 1930s due to a number of papers on the regulation of the stability of stars by the balance of gravitational and nuclear forces. These considerations provided conditions for which a star's radius R could become less than the Schwarzschild radius r_s, resulting in a black hole. In a series of papers by J. Robert Oppenheimer (famous for his direction of scientists in the development of the atomic bomb in World War II) and co-workers, the limit on the size of stars that would remain stable was set (called the Oppenheimer - Volkoff limit). In 1939, further analysis in a paper by Oppenheimer and Hartland Snyder concluded that

> *"When all thermonuclear sources of energy are exhausted, a sufficiently heavy star will collapse;* [a contraction follows which] *will continue indefinitely."* [350]

The condition for collapse to occur leading to $R < r_s$ is that a star's mass be about three times greater than that of the sun. A star with a mass between 1.4 and 3 times that of the sun will stabilize as a neutron star, stabilized by the outward pressure of neutrons. Those stars of a mass less than 1.4 suns will become what is known as white dwarfs, stabilized against gravity by electrons.[351] The condition of a

---

[cxxxviii]   The idea that light could be trapped by a massively dense body of radius R was conceived of as long ago as 1763 when John Mitchell, using Newton's laws, expressed the condition under which the velocity of light c would be the escape velocity, $v_{escape}^2 = c^2 = 2GM/R$ (Muller, p. 48; also see development of equation (10-5, p. 273). Despite the extraordinary differences in the physical assumptions of Newton's laws and General Relativity, this is the same relationship as that defining the Schwarzschild radius.

collapsed star resulting from a thermonuclear supernova explosion was given the name black hole by the physicist John Wheeler in 1967. [352] In 1972, Cygnus X-1 was identified as a black hole. [353] However, due to differences in the reference frame of the earth observer compared to the local frame of the black hole, the time to fully complete the process of becoming a black hole will take an infinite time despite being 99.999 % of the way there.[354] While electrons and neutrons may stabilize some stars, if the mass of a dying star is greater than 3 suns, nothing can stop it from eventually forming a black hole. [355]

Any casual browsing of news from the Internet will find black holes to be the subject of new and fascinating discoveries. However, it is time to move on to the bigger picture painted by General Relativity, the evolution of the universe. To do this we will first look more closely at key observations of the universe that guided the development of cosmological models in General Relativity.

Discovering an unexpected universe

The scientific view of the universe began with Copernicus's displacement of the earth as the center of the universe, Galileo's astronomical observations through a telescope, and Kepler's planetary laws based on the observations of Tyco Brahe. Newton's laws then gave order to the motion of the planets on the universe's stage of absolute space and absolute time. For over two hundred years, the universe was seen as our galaxy extended without limit with some speculative contrarian views, including those of Immanuel Kant, that there were many other galaxies.[356] In Lord Kelvin's nineteenth century model of the universe, we remained somewhat special with the earth being near the center of our universe.[357] However, by 1921, high powered observations of smudges of light formerly thought to be gaseous nebulas in our galaxy were generally accepted by scientists to be distant galaxies with their own distinct stars.[358] Edwin Hubble in 1924 announced results clearly identifying the Andromeda nebula as an extragalactic object.[359]

In 1917, Einstein used his Field Equations to model the universe as finite, curved, and static. However, another major change in our picture of universe came about through Hubble's observations, reported in 1929, of nebular redshifts. Hubble inferred from his observations that the nebulas were receding with a velocity proportional to their distance from the earth, an observation that became known as Hubble's Law.[cxxxix] It became increasingly accepted that the universe was expanding, not static.[360] Such a possibility, as shall be discussed, had already been foreseen in mathematical papers in 1922 and 1924 by Alexander Friedmann (1888 - 1925)[361] and independently, with physical interpretations by the physicist and Roman Catholic priest Georges Lemaître in 1927.[362] Hubble had this to say of his discovery, *"The outstanding feature...is the possibility that numerical data may be introduced into discussions of the general curvature of space."* [363]

A consequence of the picture of an expanding universe was the implication that the universe had begun its evolution at some finite time in the past. While accepting the experimental evidence of an expanding universe, some did not find a universe with a beginning or one that was constantly changing to be aesthetically pleasing. In 1948, the physicist Fred Hoyle championed a steady-state theory of the universe. The theory assumed that on average the universe was the same at all times and all places. The expansion of the universe, which would change properties such as average mass density, was theorized to be counteracted by the continual creation of matter. The proponents argued that the theory was no more fantastic than a theory in which all of the matter in the universe was created at once in a in a "big bang," as Hoyle derisively referred to it. The name stuck. Despite that in hindsight Hoyle's theory seems unlikely, the debate continued for several decades with opponents noting spatial distributions of stellar objects that would contradict the steady-state theory.[364]

The nail in the coffin of the steady state theory was the discovery by Arno Penzias and Robert Wilson in 1965 of a cosmic background

---

[cxxxix] The redshifts here refer to changes in the frequency of emitted light due to the motion of the nebulas (the Doppler effect), not the much smaller gravitational redshift.

radiation of 3.5° ± 0.1 K[cxl] determined while trying to reduce noise in satellite communications. In 1948, Gamow and his students Ralph Alpher (1921 - 2007) and Robert Hermann (1914 - 1947) had shown from consideration of nuclear physics that a residual background heat radiation of 5° K was expected to be observed from events near in time to the beginning of the universe. The predictions of Gamow and his coworkers were ignored and forgotten, but they were in essence reproduced by others who explained the significance of the background "noise" to Penzias and Wilson. Penzias and Wilson received the Nobel Prize - Gamow and his coworkers did not.[365]

Further measurements of the background radiation provided significant guidance to modeling the evolution of the universe. In 1967, David Wilkinson and Bruce Partridge made measurements that showed temperature variations were less than 0.1 % implying a universe that had expanded in an extremely isotropic manner.[366] Cosmologists regarded the question of how the universe of galaxies evolved in such a smooth expansion as key to any understanding of the universe and to efforts to model the evolving universe from the Field Equations.

A potential answer of how the universe expanded isotropically was offered by Alan Guth in 1980. He proposed mechanisms from particle physics that would temporarily accelerate the expansion rate of the universe early in its history. The temporary increase in the expansion rate, which he termed inflation, would result in irregularities being smoothed out. These effects could be modeled with the Field Equations essentially as gravitational repulsion. Just as important, the inflation, by amplifying small quantum fluctuations, created the seeds of irregularities that would result in galaxies; however, without an end to inflation, there would be no stars and galaxies. The mechanism for inflation, arising from elementary particles randomly moving from their local energy minimum to lower energy levels could create many distinct regions of inflation with the unsettling possibility of the creation of multiple expanding universes. Comfortable in our own universe,

---

[cxl]   The degree Kelvin (K) is the absolute unit of temperature defined by Lord Kelvin. The lowest possible temperature is 0° K ≈ −273° C.

we would have no way of knowing the others with communication limited by the velocity of light.[367] Less spectacularly, but of great importance, the inflationary theory predicts patterns in the spectra of the background radiation. To date the predictions, seem to be in agreement with available satellite data.[368] Inflationary models have unleashed an assortment of new visions of the universe; however, they all have an expanding universe as a common feature first discovered by Alexander Friedmann using the Field Equations.

The Friedmann Universes

In 1917, Einstein applied his Field Equations to describing the universe. To do this, he modelled the mass distribution as one of constant density, ignoring the exact details of the objects of the universe. He proposed a metric for a universe that was homogeneous and isotropic, that is, an observer at any location would see what any other observer elsewhere would see, and the direction of sight would not make a difference. These two assumptions came to be known as the Cosmological Principle.[cxli] Another assumption of his model was that the universe was static - Hubble's discovery of an expanding universe had yet to come. The result was a finite universe in curved space, but unbounded. [369] However, to achieve this he had to add an additional term to the Field Equations called the Cosmological Constant. With empirical knowledge that the universe is expanding, and that this could be accommodated without the constant, Einstein abandoned the additional term and called it, *"the biggest blunder of my life"* He never used the Cosmological Constant again.[370] The constant, however, would be used by others to model such effects as inflation (accelerated expansion). With this background, we will now look at the models of Friedmann that illustrate many aspects of the universes that arise from the Field Equations.

[cxli] If you add the condition that the universe is also the same at all times, then you have the Perfect Cosmological Principle. This would accommodate Hoyle's steady state theory as it would allow for an infinite universe with no beginning in which expansion effects were counteracted by the continuous creation of matter.

Alexander Friedmann (1888 - 1825) was a mathematician and meteorologist whose life, like Schwarzschild's, was tragically cut short as he died of typhus fever at the age of thirty-seven. While a professor of mathematics and physics at Perm State University in Russia, he survived the surrounding armed struggles of Trotsky's Red Army against the "Whites." He moved on in 1920 to the Geophysical Observatory in St. Petersburg where, in addition to his theoretical mathematical studies, he participated in high-altitude balloon flights investigating the effects of high altitude on the human body. In 1925, he held the record for the highest ascent (7,400 m). While at the Geophysical Observatory, he learned of Einstein's theory of General Relativity, along with Einstein's model of the universe. Friedmann investigated the Field Equations looking to see what kinds of solutions would be allowed, keeping Einstein's assumptions of a homogeneous and isotropic universe.[371] The resulting metric was adopted and analyzed by a number of researchers and is sometimes referred to as the Friedmann-Lemaître-Robertson-Walker (FLRW) model.[372] The form of the metric is given below with spatial coordinates σ, ϕ. and θ[cxlii]

$$ds^2 = S(t)^2 \left\{ \frac{d\sigma^2}{1 - k\sigma^2} + \sigma^2(d\phi^2 + \sin^2\phi d\theta^2) \right\} - c^2 dt^2. \qquad (12\text{-}17)$$

The FLRW metric is derived from the Cosmological Principle with the following considerations.[373] The metric applies to observers from any location in the universe allowing observers from any galaxy to take the origin of the coordinate system as their own location. Similarly, as all observers will have the same large-scale view of the universe, the frames of all observers are moving with the common galactic motion. Therefore, the clocks of all observers read the same proper time, consistent with the form of the metric. From considerations of isotropy, the FLRW metric is required to be of the form $ds^2 = g_{ij}dx^i dx^j - c^2 dt^2$ with the metric g_{ij} being a spatial metric over the three coordinates (σ, ϕ, θ). The variable S(t) scales the spatial coordinates and provides the variable accounting

---

[cxlii] Here, as throughout the text, the usage of spherical coordinates ϕ and θ is interchanged with that of Lawden. The angle ϕ is the colatitude and θ is the azimuthal angle, consistent with plane polar coordinate usage.

for the expansion of space in the universe. Finally the Cosmological Principle requires that the universe be one of constant curvature (being the same for all observers). This is represented in the FLRW metric by the curvature constant k which may assume the values of 0, 1, or -1 corresponding to universes which, respectively, are flat or have either positive or negative curvature.

That the FLRW metric is one of constant curvature may be made more apparent by making the transformation $\sigma = r/\left(1 + \frac{1}{4}kr^2\right)$. Substituting the transformation into equation (12-17), the following form results,[cxliii]

$$ds^2 = S(t)^2 \left\{ \frac{dr^2 + r^2(d\phi^2 + \sin^2\phi d\theta^2)}{\left(1 + \frac{1}{4}kr^2\right)^2} \right\} - c^2t^2. \qquad (12\text{-}18)$$

The form of the FLRW metric in brackets shown above may remind you of the Poincaré disk model of the hyperbolic plane, equation (7-10, p. 213) reproduced below.

$$ds^2 = \frac{dr^2 + r^2d\theta^2}{\left(1 - \frac{r^2}{4}\right)}.$$

[cxliii] The square of the differential $d\sigma$ from the derivative $d\sigma/dr$ is given below.

$$\frac{d\sigma}{dr} = \frac{1}{1 + \frac{1}{4}kr^2} - \frac{\frac{1}{4}rk2r}{\left(1 + \frac{1}{4}kr^2\right)^2},$$

$$= \frac{1 - \frac{1}{4}kr^2}{\left(1 + \frac{1}{4}kr^2\right)^2},$$

$$d\sigma = \left(\frac{1 - \frac{1}{4}kr^2}{\left(1 + \frac{1}{4}kr^2\right)^2} \right) dr.$$

Therefore, the first term in brackets in equation (12-17) transforms as:

$$\frac{d\sigma^2}{1 - k\sigma^2} = \frac{dr^2}{\left(1 + \frac{1}{4}kr^2\right)^2}$$

Recall that the surface of the hyperbolic plane, modelled as above, has a constant curvature of -1. Taking $k = -1$ and $\phi = \pi/2$ in the FLRW metric, the disk model may be seen as a restriction of the interior of the three-dimensional FLRW metric to a planar section. Furthermore the FLRW is itself a restriction to three dimensions of the metric of an n-dimensional space of constant curvature α given by Riemann in his *Foundations of Geometry*.[374]

$$ds^2 = \frac{\sum_{i=1}^{n}\left(dx^i\right)^2}{\left(1 + \frac{\alpha}{4}\sum_{i=1}^{n}(x^i)^2\right)^2}.$$

With these considerations, the identification of k in the FLRW metric with curvature should be clear.

The scale factor S(t) is closely connected to Hubble's Law which should not be surprising as it is a coefficient of the spatial differentials in the FLRW metric. Hubble's Law says that at a fixed time, the velocity $\dot{R}_g$ at which a galaxy is receding from an observer is proportional to its distance from the observer R_g. The constant of proportionality $H(t)$, in keeping with the Cosmological Principle, is the same for all observers and galaxies at a fixed time, but, despite its name, may vary with time.

$$\dot{R}_g = H(t)R_g.$$

The radial distance of a galaxy from an observer at a fixed time may be calculated using the FLRW metric, equation (12-17).

$$R_g = S(t)\int_0^{\sigma_g} \frac{d\sigma}{\sqrt{1 - k\sigma^2}} = C_{gk}S(t).$$

In the above result, C_{gk} is a constant that depends on the curvature k used in the metric. However, for each galaxy and curvature, $\dot{R}_g = C_{gk}\dot{S}$, and therefore,

$$\dot{R}_g = H(t)R_g = C_{gk}\dot{S}, \text{ and } \quad \dot{R}_g/R_g = \dot{S}/S. \tag{12-19}$$

The above relation is very important in interpreting results using the FLRW metric in the Field Equations. The reciprocal of the Hubble Constant has the unit of time and is estimated to be about 18 billion years. Lawden noted in 1982 that the age of the universe may be modelled as $\frac{2}{3}$H corresponding to an age of about 12 billion years. Current measurements and models estimate the age as between 12 and 14 billion years.[cxliv]

With this discussion as background, the metric tensor components $g_{\mu\nu}$ associated with the contravariant coordinates $(\sigma, \phi, \theta, t))$ can be listed simply by inspecting equation (12-17):

$$g_{11} = S^2/(1 - k\sigma^2), g_{22} = (S\sigma)^2, g_{33} = (S\,\sigma\sin\phi)^2, g_{44} = -c^2.$$

All the other components of $g_{\mu\nu}$ are zero. The contravariant metric tensor components $g^{\mu\nu}$ for $\mu = \nu$ are the reciprocal of $g_{\mu\nu}$ with the other components equal to zero (see equation (8-21, p. 236). The Field Equations are reproduced below from equation (11-27, p. 308) where the equations were introduced, along with a discussion of the relationship of the Ricci tensor $R_{\mu\nu}$ and its scalar $\mathcal{R}$ to the metric tensor. As previously established (p. 315), $\kappa = 8\pi G/c^4$.

$$R_{\mu\nu} - \frac{1}{2}g_{\mu\nu}\mathcal{R} = -\kappa T_{\mu\nu}.$$

The determination of the components of the $R_{\mu\nu}$ and its invariant $\mathcal{R}$ from the metric tensor and the Christoffel symbols is indeed tedious. First the Christoffel symbols are calculated from the metric tensor as in equation (8-31). The procedure is similar to that used in obtaining the Christoffel symbols for a spherical surface shown in Section 8.3 (p. 247). Then, $R_{\mu\nu}$ and $\mathcal{R}$ are calculated from, respectively, equations (8-41) and (8-42). However, the required results from the FLRW metric may be obtained from the more general results given by Lawden.[375] These are listed below with dotted variables being time derivatives.

[cxliv] https://map.gsfc.nasa.gov/universe/uni.age.html.

$$R_{11} = R_{\sigma\sigma} = -P/1 - k\sigma^2\,; \ R_{22} = R_{\phi\phi} - P\sigma^2; R_{33} = R_{\theta\theta} = -P\sigma^2\sin^2\phi;$$
$$R_{44} = R_{tt} = 3\,\ddot{S}/S \text{ with } P = 2k + (S\ddot{S} + 2\dot{S}^2)/c^2. \tag{12-20}$$

$$\mathcal{R} = -6(S\ddot{S} + \dot{S}^2 + kc^2)/(cS)^2. \tag{12-21}$$

Now we need to specify the right-hand side of the Field Equations. We start with the contravariant energy-momentum tensor $T^{\mu\nu}$. The simplest model of this tensor was given with equation (11-32, p. 314). In this model, the universe is described as consisting of a uniform distribution of dust with a rest mass density ρ.

$$T^{\mu\nu} = \rho\mathcal{U}^\mu\mathcal{U}^\nu.$$

With the assumption that the FLRW reference frame is moving with the common galactic motion, the spatial components of the four-velocity $\mathcal{U}^i = \frac{dx^i}{d\tau} = 0$ ($i = 1,2,3$), $\mathcal{U}^4 = 1$, and $T^{44} = \rho$.

Since we are using the covariant form of the Field Equations, the covariant form of the energy-momentum tensor $T_{\mu\nu}$ is needed. This can be accomplished using the index lowering property of the metric tensor, equation (8-22, p. 237), for example $V_i = g_{ij}V^i$. In the case of the energy-momentum tensor, two indices must be lowered. From the FLRW metric, $g_{44} = -c^2$, therefore $T_{44} = g_{44}g_{44}T^{44} = \rho c^4$.

The Field Equations therefore take the following form,

$$R_{11} - \frac{1}{2}g_{11}\mathcal{R} = R_{22} - \frac{1}{2}g_{22}\mathcal{R} = R_{33} - \frac{1}{2}g_{33}\mathcal{R} = 0, \text{ and}$$

$$R_{44} - \frac{1}{2}g_{44}\mathcal{R} = -\kappa c^4\rho.$$

Substituting the Ricci tensor components, equation (12-20), and the invariant $\mathcal{R}$, equation (12-21), into the Field Equations, only two separate conditions are found. The equations with R_{11}, R_{22}, or R_{33}, all simplify to:

$$2S\ddot{S} + \dot{S}^2 + kc^2 = 0. \tag{12-22}$$

The equation for R_{44} simplifies to:

$$3(\dot{S}^2 + kc^2) = \kappa c^4 \rho S^2. \tag{12-23}$$

Without solving the differential equations, a number of interesting characteristics can be found which have counterparts in other Friedmann universes using different models of the energy-momentum tensor. Using Hubble's Law, $H = \dot{S}/S$ (equation (12-19)), equation (12-23) may be solved for the curvature k.

$$k = \frac{\kappa c^2 S^2}{3}\left(\rho - 3\frac{H^2}{\kappa c^4}\right).$$

The above equation defines a critical density $\rho_c = 3\frac{H^2}{\kappa c^4}$ that corresponds to a universe in which space from a large-scale perspective is Euclidean ($k = 0$). If the density is greater or less than the critical density then the curvature is, respectively, positive or negative. Friedmann found, as we shall see below, that a universe with positive curvature is finite and closed, expanding to a maximum size. Otherwise, the universe is infinite, expanding forever.[376] Another important result can be obtained from the differential equations by multiplying equation (12-23) by S and taking the derivative as shown below.

$$\frac{d}{dt}\left[S\left(3(\dot{S}^2 + \kappa c^2)\right)\right] = \frac{d}{dt}[S(\kappa c^4 \rho S^2)],$$

$$(\dot{S}\dot{S}^2 + 2S\dot{S}\ddot{S} + \kappa c^2\dot{S}) = \frac{d}{dt}(\kappa c^4 \rho S^3/3),$$

$$\dot{S}(2S\ddot{S} + \dot{S}^2 + \kappa c^2) = \frac{d}{dt}(\kappa c^4 \rho S^3/3).$$

From equation (12-22),

$$\frac{d}{dt}(\kappa c^4 \rho S^3/3) = 0. \text{ Therefore,}$$

$$\kappa c^4 \rho S^3/3 = \text{constant} = C.$$

It is not surprising that the density is inversely proportional to S^3. As the space of the universe expands, the density decreases (assuming no continuing mass creation as in Hoyle's steady-state theory). Going backwards in time, the density increases enormously as S approaches zero. This result is a direct consequence of the Field Equations and the energy-momentum tensor model, not just an additional assumption. Putting this additional information into equation (12-23) allows a simple interpretation to be made of the behavior of the universe at times near the universe's creation and also after significant expansion has occurred. The interpretation depends upon the assumed curvature k. Substituting the constant C, equation (12-23) may be written,

$$\dot{S}^2 = \frac{C}{S} - kc^2. \tag{12-24}$$

If the large-scale geometry of the universe is Euclidean (k = 0) and S at time t = 0 is zero, then the solution the for all times is,

$$S = \left(\frac{9}{4}C\right)^{1/3} t^{2/3}. \tag{12-25}$$

The space of the universe expands forever; however, the rate of expansion $\dot{S}$ is decreasing, approaching zero at infinite time. Early on when S is small, the expansion rate for universes of positive or negative curvature will be the same as the Euclidean case with $S \propto t^{2/3}$ (S increasing proportionally with time as $t^{2/3}$). This is because the term C/S on the right-hand side of equation (12-24) will be dominant. If the curvature is negative, the expansion of space will also continue forever, but $S \propto t$ as a limit for long times.

In contrast to the behavior at long times for universes with zero or negative curvature, if the space curvature is positive (k = 1), the expansion does not continue indefinitely. Following Einstein,[377] equation (12-24) may be rewritten as,

$$\dot{S}^2 = \frac{c^2(S_0 - S)}{S}, \text{ with } S_0 = C/c^2.$$

The expansion factor S reaches a maximum when $S=S_0$ and $\dot{S}=0$. A detailed solution shows that the expansion factor as a function of time follows the equation of a cycloid. The expansion factor then decreases until $S = 0$. [378] [cxlv] In this universe, the Big Bang is followed by the Big Crunch. Friedmann noted that the Field Equations do not forbid the universe from then expanding again to begin an endless cycle of expansion and contraction.[379]

Friedmann was the first to show that the Field Equations contain the possibility of an expanding universe. More detailed models of the energy-momentum tensor would be made; however, these models would similarly show relationships between the universe's mass density, the large-scale geometry of space, and the experimental observations of an expanding universe. The specific values in these relationships, when known, will reveal much of the history and destiny of the universe.

Some of the additional physics that have been modelled in the energy-momentum tensor are radiation and pressure,[cxlvi] first modelled by Georges Lemaître in 1927.[380] Lemaitre's studies showed that Einstein's static universe was unstable, and his model predicted and expanding universe. This was notable as he was unaware of Friedman's work, and prior to Hubble's publication of his redshift observations and conclusions, Lemaître used his expanding universe model to explain reported previous redshift observations that had not been understood.

[cxlv] Note also that for k =1, in addition to $\dot{S}(S_0) = 0$, in equation (12-22), $\ddot{S}(S_0) < 0$ consistent with S_0 being a maximum.

[cxlvi] Pressure is included when modelling the mass-energy distribution as a perfect fluid, in contrast to the dust model. The pressure term is analogous to that of a liquid or gaseous fluid, but as Einstein (p. 113) noted, *"This must not, however, be confused with a hydrodynamical pressure, as it serves only for the energetic presentation of the dynamical relations inside matter."*

Lemaître also included the Cosmological Constant which Einstein eventually rejected. Its inclusion resulted in universes in which the expansion accelerated. This would become important in the 1980s when an accelerating expansion became an explanation for the isotropy of the universe and when acceleration was observed in 1998, as previously discussed.

The inclusion of additional physics by Lemaître resulted in numerous possibilities for modelling the universe. In 1932, Einstein and the astronomer Willem de Sitter published a two-page paper singling out the simplest of the possible models: no pressure, no Cosmological Constant, and zero curvature. The result is a universe expanding forever from a beginning (i.e., a finite past) with an expansion factor given by equation (12-25). This would be considered to be the best description of the universe for decades.[381] [cxlvii] In the 1980s, a major change in modelling approaches came with the acceptance of the need to model an inflationary period in the early stages of the universe to account for its observed isotropy. A role for the quantum effects of particles to account for the acceleration in expansion was proposed by Guth as discussed previously. These effects could be brought into the Field Equations through Einstein's rejected Cosmological Constant that could be used to counteract gravitational effects.[382] The use of the Cosmological Constant received another boost in 1998 when improved measurements of distances to supernovas with resulting improvements in the determination of Hubble's Constant revealed a universe undergoing accelerated expansion, not just in the early stages of the universe, but a few billion years ago.[383] The Cosmological Constant Λ is shown below in the Field Equations as originally proposed by Einstein.

$$R_{\mu\nu} - \frac{1}{2}g_{\mu\nu}\mathcal{R} - g_{\mu\nu}\Lambda = -\kappa T_{\mu\nu}, \text{ or}$$

$$R_{\mu\nu} - \frac{1}{2}g_{\mu\nu}\mathcal{R} = -\kappa\left(T_{\mu\nu} - \frac{\Lambda}{\kappa}g_{\mu\nu}\right).$$

[cxlvii] De Sitter (1873 - 1934) published a paper in 1917 in which he retained the Cosmological Constant, but assumed the mass density of the universe was zero, resulting in an expanding universe (Pais, pp. 287, 537; Barrow, pp. 57 -60.).

The effect of a positive Λ is to counteract the gravitational terms from $T_{\mu\nu}$.[cxlviii] But what are the physical sources for Λ?

In placing Λ on the right-hand side of the Field Equations, the Cosmological Constant takes on the role of an energy term. This is consistent with Goth's analysis of the early universe in which quantum energy transitions of sub-atomic particles randomly occur from one local minimum energy level to another, releasing energy. These transitions, occurring in "empty" space, lend themselves to the name "vacuum energy" which some physicists have associated with Λ.[384] More generally with the increased appreciation for the accelerating expansion of the universe, Λ is associated with the name "dark energy" which, as its name suggests, is a mysterious energy of unknown origin causing the acceleration.

Another observation that underscores how far we have to go to understand the universe is that physicists estimate that the observable matter in the universe only constitutes about 5 % of the total, based on inferences from galactic motion and accelerating expansion. In order for galaxies to maintain their structures at their observed rotational speeds, there must be additional unobserved matter to supply the necessary gravitational force. This unobserved material has been named dark matter. It is estimated to constitute another 27 %. Finally, the mysterious dark energy makes up the remaining 68%. It is thought to be associated with the vacuum in space, to be uniformly distributed in space and time, and to be responsible for the accelerated expansion of the universe.[cxlix]

What do we know for certain? Just as General Relativity had to account for the extraordinary successes of Newton's Laws of Motion

---

[cxlviii] The addition of the term $g_{\mu\nu}\Lambda$ does not invalidate $T_{\mu\nu}$ as a divergence free tensor ($T_{\mu\nu;\nu} = 0$) as it can be proved that the metric tensor is divergence free ($g_{\mu\nu;\nu} = 0$); Wrede, p. 338, Lawden, p. 110.

[cxlix] The percentage estimates are from the CERN website, \\https://home.cern/science/physics/dark-matter (accessed January 30, 2020). We can get a feel for the stability of these estimates from the 2011 discussion by Barrow (p. 285.) in which the estimates for normal matter, dark matter, and dark energy were given as, respectively, 5 %, 23 %, and 72 %.

and Gravitation, future theories will have to account for the successes of General Relativity's geometric description of the dynamics of the universe, the motion of celestial objects and light along geodesic paths, and its incorporation of Special Relativity in a more general theory. The experimental observations that have supported General Relativity may become more accurate, but the qualitative implications will not change - we live in an expanding universe.

Will the universe expand forever? Did our universe begin in a single event, or are we one of many creations in a multiverse that has no beginning? Or if we live in a universe of positive curvature, are we just one universe among many in an eternal cycle of birth and death?

Someday we may be able to answer the questions of the origins and future of the universe. Remarkably, they will be the endpoint of a story that goes back thousands of years. The ancient Greeks discovered a geometry that arose from just five postulates that were thought to be self-evident. The geometry came into being from the five postulates through another of their discoveries, deductive reasoning. The resulting geometric system was thought to be absolute and to encompass all the spatial relations of the universe, and for over two thousand years, it was accepted as such. Yet the desire to make the geometric system more perfect then led to the shock that the truths were not absolute, but only consistent results from a set of assumptions. Lobachevsky and Bolyai added their hyperbolic system, and then Riemann expanded geometry to include an infinite variety of n-dimensional spaces with their forms defined by their metric of length. Who could have foreseen that the new abstract understanding of geometry and the tensor calculus developed to describe it would be used to return to the search for an absolute truth: the geometric nature of space-time that defines the past and future of our universe.

Epilogue

Philosophic Thoughts

We have now completed a journey covering, well over two thousand years, if we see the starting point as Thales' and Pythagoras' glimmerings of truths in geometric forms and numbers. But we could just as well see the start as beginning with civilizations such as that of the Egyptians and of the Babylonians who recognized crude forms of these truths simply as practical matters gained through experience. In either case, these beginnings appear to have arisen from an innate disposition of the minds of ancient ancestors to express number and form. As I noted at the beginning of our journey, such behavior has been discovered in cave paintings and expressions of counting in notches on wolf bones. These artifacts, remarkably surviving tens of thousands of years, make clear our connection with the primordial past from the beauty that we find in these cave paintings and from our continuing expansion of the mathematics of relationships. Ultimately all goes back to the cauldron of the creation of the universe, the source of time, space and all matter. The evolution of our understanding of form and number has led us to the conclusion that this creation began some 14 billion years ago. It is interesting to reflect that in regard to the process that led to the stars, planets, and life, including self-aware, cognitive humans, our search to understand the universe is in some sense the universe reflecting and discovering its own self. Speaking of the relationship of the mind to nature, Arthur Eddington, whose expedition provided the earliest support for General Relativity, had this to say,

"We have found that where science has progressed the farthest, the mind has regained from nature that which the mind has put into nature. We have found a strange footprint on the shores of the unknown. We have devised profound theories, one after another to account for its origins. At last we have succeeded in reconstructing the creature that made the footprint, And Lo! It is our own." [385]

The investigation of the origins of the universe is necessarily intimately connected with our ability to discover that which is true and to understand the nature of truth which can be expressed in many forms. The development of mathematical truth is illustrated by the change that occurred from the mathematics of early civilizations, such as the Egyptians, to that of the Greeks who searched for truths about general geometric relations that could be proved. The Egyptians knew that stretched cords of lengths 3, 4, and 5 would form a right triangle, but they could not prove the general conditions that would establish this as true. The Greeks' method of proving such general relations through an unbroken line of logic going back to apparently unassailable concepts, established the meaning of mathematical truth. Their concept of truth, like their understanding of geometry, lasted over two thousand years. When a non-Euclidean geometry was discovered by Lobachevsky and Bolyai, mathematics lost its special place as a provider of absolute truths that could be proved. From that time, the truths were no longer absolute, but if you accepted the foundational premises, the postulates, then the propositions that followed were true. Another way of saying this is that all proved propositions were expected to be consistent in that no proposition could be proved that turned out to contradict another proposition. Then in 1931, the mathematician Kurt Gödel dropped a bombshell at least as great as the one dropped by the discovery of a non-Euclidean geometry. Gödel proved that for mathematical systems, at least as complicated as that encompassing the arithmetic of the natural numbers, consistency could not be proved. Furthermore, he proved that there are truths of such a system that cannot be proved from the postulates and propositions within the system.[cl]

So what is left of mathematical truth? Most importantly, concerns about consistency in accepted lines of mathematics have not occurred in practice. However, it means that the truthfulness of a particular mathematical system must ultimately be based on experience with its use, thereby bringing it closer to the concept of scientific truth.

Scientific truth provides a contrast with mathematical truth in that the truths of science must be anchored in the observed facts of our universe. That a system is consistent with a set of postulates is not sufficient for scientists. Scientific theories must be in agreement with scientific measurements that are objective in the sense that they were obtained through accepted scientific methodologies and confirmed by others. This is in contrast to personal truths such as those of aesthetics, religious faith, and moral judgments. These truths are subjective. They are arrived at through complex introspection. There is no objective way to settle disagreements among individuals concerning these truths. For example, when speaking of ethical teachings such as those of Jesus, the philosopher, mathematician, and social activist, Bertrand Russell (1872 - 1970) noted,

> "Such ethical innovations obviously imply some standard other than majority opinion, but the standard, whatever it is, is not objective fact, as in a scientific question. This problem is a difficult one, and I do not profess to be able to solve it." [386]

The scientific method of discovering truth is known as inductive reasoning, the formation of generalizations from specific facts. Inductive reasoning cannot prove that the generalizations follow from the specific facts in the same manner propositions are proved from postulates;

[cl] The details of Gödel 's proofs are well beyond the scope of this book; however, an excellent outline of the proof for the non-specialist is given by Ernest Nagel and James R. Newman, *Gödel's Proof*, New York University Press, 2001.

however, facts have the power to invalidate a theory. If the theory does not agree substantially with experimental observations, then it is invalid. Another possibility is that a theory is an approximation, limited to certain conditions, but valid within those conditions. The classic example presented in this book is that of Newton's laws being superseded by General Relativity. Newton's laws still accurately predict the orbits of the planets (except for the small deviation of Mercury's orbit), but General Relativity is able to account for such phenomena and others not predicted from Newton's Laws. Newton's Laws are not wrong, but of limited scope. Of as much or greater importance than the improved accuracy of some predictions is General Relativity's impact on our understanding of the foundations of space and time and its role in the dynamics of the universe. Although, we cannot prove a scientific truth such as General Relativity, we can have greater and greater confidence in the theory as more supporting evidence is observed. Nevertheless, it seems probable that someday General Relativity may also be superseded by including Quantum effects. Of the inductive method, Russell had this to say,

"...induction is an independent logical principle, incapable of being inferred either from experience or from other logical principles, and without this principle, science is impossible." [387]

Despite the difference between the mathematical and scientific approaches to truth, much of science, particularly physics, is expressed in the language of mathematics. If we view the fundamental laws of physics as postulates, such as General Relativity, the Laws of Thermodynamics, Quantum Mechanics (and the theories that followed from it), and Maxwell's Electromagnetic Equations, then mathematical methods may be used to deduce the implications of these fundamental laws. A good example of the role of mathematics in physics is the prediction from Maxwell's Equations of electromagnetic waves which was verified experimentally.

The reduction of phenomena being investigated to measurements, number, and mathematical theory is vital to the progress of scientific theory and its objectivity, allowing one scientist to confirm the results of another. Although experimental observations are crucial to the method of induction, the leap to a theory to comprehend the observations requires inspiration as creative and mysterious as any other expression of human understanding. To this point, Einstein had this to say,

"Our experience hitherto justifies us in believing that nature is the realization of the simplest conceivable mathematical ideas. I am convinced that we can discover by purely mathematical construction the concepts and laws connecting them with each other, which furnish the key to the understanding of natural phenomena. Experience may suggest the appropriate mathematical concepts, but they most certainly cannot be deduced from it. Experience remains, of course, the sole criterion of the utility of mathematical construction. But the creative principle resides in mathematics. In a certain sense, therefore, I hold it true that pure thought can grasp reality as the ancients dreamed." [388]

Because of the necessary connection between measurements and theory, it would seem that our understanding of the variables of theories such as General Relativity should be straightforward. However just as one jarring example, the Lorentz transformation, in which the coordinate of time is included in the spatial transformations and is no longer the same in all frames, tells us that we have left the familiar world of space *and* time, to enter the four-dimensional world of space-time. We may attempt to visualize space-time, and it is often illustrated graphically by suppressing one or two of the spatial coordinate, but to quote the physicist Stephen Hawking (1942 - 2018), *"It is impossible to imagine a four-dimensional space."* [cli] We can make analogies to provide insight, but I agree with Hawking that it is beyond our perception. Space-time is akin to the undefined terms of geometry with accepted measurements and relations like the Lorentz transformation expressing its meaning in mathematical terms. Other examples that were discussed at length in the last chapter are the curved space of the Friedmann universes and its expansion. Again we can use analogies such as the curvature of a two-

---

[cli]   Stephen Hawking, *A Brief History of Time*, Bantam Books, New York, 1996 (p. 24).

dimensional surface and the expansion of the volume contained by a closed surface. However, such analogies invite very confusing questions. For example, if the universe is expanding, what is it expanding into? The physicists Charles Misner, Kip Thorne, and John Wheeler respond to the question of where the additional space for the expansion comes from by saying,

> *"Rather than look for an answer, one had better reexamine the question...There is no such thing as a flowmeter to tell `how fast space is streaming past.' The very idea that `space flows' is mistaken...To try to pinpoint where those cubic kilometers of space get born is a mistaken idea, because it is a meaningless idea."*[clii]

Another approach to an answer might be to admit that, similar to Hawking, it is impossible to imagine the four-dimensional universe given our modes of perception. The mathematical equations are the closest we can come to "seeing" the universe. What we can say objectively is that the metric defining space is changing with time. The truthfulness of the theory is judged by its connections to measurements and its consistency with accepted theories. Our understanding of the physical universe at its deepest level appears to be only expressible in the abstract language of mathematics. This idea is especially evident in the world of Quantum Mechanics where particles behave sometimes as a wave and sometimes as a particle (wave-particle duality) directly challenging the notion that we may create pictures of reality. Along these lines, Bertrand Russell defined matter, *"as what satisfies the equations of physics."* [389]

One might ask if it may someday be possible for mathematical expression to become just another mode of perception. Speaking of the mathematics that governs the complex random motions of turbulent fluids, the physicist Richard Feynman (1918 - 1988) had this to say of our future ability to see phenomena in the equations of mathematics.

[clii]   Quoted from Misner, Thorne, and Wheeler, *Gravitation,* Princeton University Press, 2017 (p. 739). This is a classic text for the specialist. Although crammed with much detail and vivid explanations, it is not for the faint of heart.

"The next great era of awakening of human intellect may well produce a method of understanding the qualitative content of equations."[cliii]

Mathematics, like our images of the everyday world, occurs in the mind.[cliv] Science tells us, however, that the images of the natural world are initiated by profoundly complex realities far beyond familiar images. As a species, the human mind evolved in reaction to the external world, and children's minds develop as they encounter the world, fashioning for themselves the sense of space and time, and cause and effect that brings comprehension to the familiar world. Yet within the mind are the capabilities of discovering and expressing complex patterns of the universe that are completely foreign to our everyday experiences of the external world. Of this miraculous ability, I can only say that it seems to express an inherent connection between the mind and the structure of the universe. Of the modes of expression of the mind, mathematics has been unsurpassed in its ability to reveal the universe beyond the surfaces of familiar perception. We may have to accept, and indeed embrace, the idea that some realities may only be perceived through mathematics.

[cliii] Richard Feynman, Robert Leighton, Matthew Sands, *The Feynman Lectures on Physics*, Addison-Wesley Publishing Company, Reading, Massachusetts, 1977 (p. 41-12).

[cliv] Alfred North Whitehead, summarizing John Locke's view of sensations such as color and scent states that, "But the mind in apprehending also experiences sensations which, properly speaking, are qualities of the mind alone. These sensations are projected by the mind so as to clothe appropriate bodies in external nature. Thus, the bodies are perceived as with qualities which in reality do not belong to them, qualities which in fact are purely the offspring of the mind." (*Science and the Modern World*, The Free Press, Simon and Schuster, New York, 1997 (pp. 55-56) Whitehead collaborated with Bertrand Russell in an unsuccessful attempt to define mathematics entirely from logic.

Appendix A

Proof that √2 is irrational

The proof by contradiction cited by Aristotle is surprisingly simple.[clv] Let us assume that $\sqrt{2}$ is rational. Then $\sqrt{2}$ must be equal to the ratio of two natural numbers (1,2,3…) p and q with p/q = $\sqrt{2}$. We can assume that p and q have no factors in common since if they did the common factors would cancel.

If,

$$p/q = \sqrt{2}, \text{then}$$
$$(p/q)^2 = \left(\sqrt{2}\right)^2,$$
$$p^2/q^2 = 2,$$
$$p^2 = 2q^2.$$

Therefore, p^2 being a multiple of 2, it must be an even number.

If p^2 is an even number, then p must also be an even number.

Let p = 2n, for some natural number, n, then

$$p^2 = (2n)^2 = 2q^2, \text{ or}$$
$$q^2 = 2n^2.$$

[clv]   The proof of the irrationality of $\sqrt{2}$ has been cited in the past as Proposition 117 of Book X in Euclid's Elements (see Ball, p. 60); however, it is now thought to be a later addition although undoubtedly discovered much earlier by the Pythagoreans, Heath, Vol. 3, Book X, p. 2.}

As in the case of p, therefore, q is an even number. This contradicts the assumption that p and q have no factors in common; hence, $\sqrt{2}$ cannot be equal to the ratio of two natural numbers.

Numbers that cannot be represented as the ratio of two natural numbers (or more generally, integers with the inclusion of negative numbers) are called irrational numbers. Similar arguments can be made to prove that numbers such as the square root of any prime number or any number that is not a square of another number are irrational. In these cases, a contradiction will result due to the unique prime factorization of composites (numbers that are a product of prime numbers). Since there are an infinite number of primes, there must be an infinite number of irrational numbers. As discussed in Chapter 4, in fact there are "more" irrational numbers (stemming from irrational numbers such as π called transcendental numbers), than rational numbers, even though there are also an infinite number of rational numbers!

Appendix B

Euclid's Book I Propositions: 1 to 28 ^{clvi}

IE1 On a given finite straight line to construct an equilateral triangle.

IE2. To place at a given point (as an extremity) a straight line equal to a given straight line.

IE3. Given two unequal straight lines, to cut off from the greater a straight line equal to the less.

IE4. If two triangles have the two sides equal to the two sides respectively, and have angles contained by the equal straight lines equal, they will also have the base equal to the base, the triangle will be equal to the triangle, and the remaining angles will be equal to the remaining angles respectively, namely those which equal sides subtend.

IE5. In isosceles triangles the angles at the base are equal to one another, and if the equal straight lines be produced further, the angles under the base will be equal to one another.

IE6. If in a triangle two angles be equal to one another, the sides which subtend the equal angles will also be equal to one another.

IE7. Given two straight lines constructed on a straight line (from its extremities) and meeting in a point, there cannot be constructed on the same straight line (from its extremities) and on the same side of it, two

---

clvi   Propositions of Euclid as translated by Heath (Vol. 1).

other straight lines meeting in another point and equal to the former two respectively, namely each to that which has the same extremity with it.

IE8. If two triangles have the two sides equal to the two sides respectively, and have also the base equal to the base, they will also have the angles equal which are contained by the equal straight lines.

IE9. To bisect a given rectilinear angle.

IE10. To bisect a given straight line.

IE11. To draw a straight line at right angles to a given straight line from a given point on it.

IE12. To a given infinite straight line, from a given point which is not on it, to draw a perpendicular straight line.

IE13. If a straight line set up on a straight line make angles, it will make either two right angles or angles equal to two right angles.

IE14. If with any straight line, and at a point on it, two straight lines not lying on the same side make the adjacent angles equal to two right angles, the two straight lines will be in a straight line with one another.

IE15. If two straight lines cut one another, they make the vertical angles equal to one another.

IE16. In any triangle, if one of the sides be produced, the exterior angle is greater than either of the interior and opposite angles.

IE17. In any triangle two angles taken together in any manner are less than two right angles.

IE18. In any triangle the greater side subtends the greater angle.

IE19. In any triangle the greater angle is subtended by the greater side.

IE20. In any triangle two sides taken together in any manner are greater than the remaining one.

IE21. If on one of the sides of a triangle, from its extremities there be constructed two straight lines meeting within the triangle, the straight lines so constructed will be less than the remaining two sides of the triangle, but will contain a greater angle.

IE22. Out of three straight lines, which are equal to three given straight lines, to construct a triangle; thus it is necessary that two of the straight lines taken together in any manner should be greater than the remaining one.

IE23. On a given straight line and at a point on it to construct a rectilineal angle equal to a given rectilinear angle.

IE24. If two triangles have the two sides equal to the two sides respectively, but have the one of the angles contained by the equal straight lines greater than the other, they will also have the base greater than the base.

IE25. If two triangles have the two sides equal to the two sides respectively, but have the base greater than the base, they will also have the one of the angles contained by the equal straight lines greater than the other.

IE26. If two triangles have the two angles equal to the two angles respectively, and one side equal to one side, namely, either the side adjoining the equal angles, or that subtending one of the equal angles, they will also have the remaining sides equal to the remaining sides and the remaining angle to the remaining angle.

IE27. If a straight line falling on two straight lines make the alternate angles equal to one another, the straight lines will be parallel to one another.

IE28. If a straight line falling on two straight lines make the exterior angle equal to the interior and opposite angle on the same side, or the interior angles on the same side equal to two right angles, the straight lines will be parallel to one another.

Appendix C

An Alternative Approach to Covariant Derivatives and the Christoffel Symbols

A vector or more generally a tensor is not transformed into a tensor by partial differentiation. This is because the base vectors of general coordinate systems are a function of position. This problem is resolved in Section 8.3 by determining the form that the derivative of a vector must take in order to transform as a tensor and, in the case of a rectangular Cartesian coordinate system, simplify to the usual partial derivative. The significance of the resulting covariant derivative and Christoffel symbols is clarified in terms of the parallel displacement of vectors. Another approach to developing covariant derivatives is to look at the partial derivatives of contravariant vectors expressed with their covariant base vectors, i.e., $V^k \mathbf{e}_k$.

$$\frac{\partial \mathbf{V}}{\partial x^q} = \frac{\partial\left(V^k \mathbf{e}_k\right)}{\partial x^q} = \frac{\partial V^k}{\partial x^q}\mathbf{e}_k + V^k \frac{\partial(\mathbf{e}_k)}{\partial x^q}. \tag{C.1}$$

If the base vectors are, for example, rectangular Cartesian vectors, then as throughout previous chapters, the partial derivative of the vector is simply $\frac{\partial V^k}{\partial x^q}$. More generally, the base vectors are a function of position. For example, this is illustrated in the following figure in which the orthogonal bases at different positions on a spherical surface are shown.

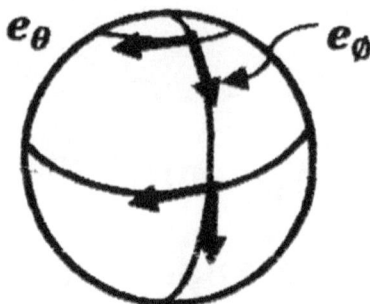

e_θ e_ϕ

C- 1 Orthogonal bases on a spherical surface

Following the calculations of r_θ and r_ϕ for the surface of a sphere in equations **(6-3)**, in rectangular Cartesian coordinates the position vector on a the surface of a unit sphere is $\mathbf{r} = (\sin\phi\cos\theta, \sin\phi\sin\theta, \cos\phi)$, and $e_\theta = \frac{\partial \mathbf{r}}{\partial\theta} = (-\sin\phi\sin\theta, \sin\phi\cos\theta, 0)$, $e_\phi = \frac{\partial \mathbf{r}}{\partial\phi} = (\cos\phi\cos\theta, \cos\phi\sin\theta, -\sin\phi)$. As the base vectors are a function of position, vectors with the same components at different positions are different vectors. This is another example of the need to modify the definition of a derivative to preserve the covariant and contravariant transformations from one coordinate system to another.

The Christoffel symbol can be defined in terms of the derivative of the base vectors,

$$\frac{\partial \mathbf{e}_k}{\partial x^q} = \Gamma^s_{qk}\mathbf{e}_s, \text{ then equation (C. 1) becomes}$$

$$\frac{\partial(V^k\mathbf{e}_k)}{\partial x^q} = \frac{\partial V^k}{\partial x^q}\mathbf{e}_k + V^k\Gamma^s_{qk}\mathbf{e}_s.$$

Switching dummy indices k and s in the second term on the right-hand side,

$$\frac{\partial(V^k\mathbf{e}_k)}{\partial x^q} = \frac{\partial V^k}{\partial x^q}\mathbf{e}_k + V^s\Gamma^k_{qs}\mathbf{e}_k = \left(\frac{\partial V^k}{\partial x^q} + V^s\Gamma^k_{qs}\right)\mathbf{e}_k, \text{ or}$$

the covariant derivative $V^k_{;q} = \frac{\partial V^k}{\partial x^q} + \Gamma^k_{qs}V^s.$

360

By defining the Christoffel symbol in terms of the partial derivative of the base vector, we have recovered the equation for the covariant derivative, equation (8-27, p. 241) developed by seeking to define a derivative that would transform as a tensor. Both approaches provide insight into the meaning and use of the covariant derivative.

Bibliography

Aczel, A. D., *The Mystery of the Aleph; Mathematics, the Kabbalah, and the Search for Infinity*, Washington Square Press, Pocket Books, New York, 2000.

Aleksandrov, A. D., Kolmogorov, A. N., Lavrent'ev, M. A., (eds.) *Mathematics, Its Content Methods and Meaning,* Three Volumes Bound as One, Dover Publications, Mineola, New York, 1969.

Asimov, I., *The History of Physics*, Walker Publishing Co., Inc., New York, 1984.

Ball, W. W., Rouse, A., *A Short History of Mathematics,* Dover Publications, Inc., New York, 1960.

Barrow, J. D., *The Book of Universes*, W. W. Norton & Co., New York, 2011.

Bill, R. G., *Images of Mathematics Viewed through Number, Algebra, and Geometry*, Xlibris, 2014.

Bonola, R., *Non-Euclidean Geometry*, with translations by G. B. Halsted of: *The Science of Absolute Space*, by John Bolyai and *The Theory of Parallels,* by N. Lobachevski, Dover Publications, Inc., New York, 1955

Brumbaugh, R. S., *The Philosophers of Greece*, State University of New York Press, Albany, 1981.

Bunt, L.N.H., Jones, Phillip S., and Bedient, Jack D., *The Historical Roots of Elementary Mathematics*, Dover Publications, New York,1988.

Courant, R. and Robbins, H., *What is Mathematics?* 2nd edition, revised by I. Stewart, Oxford University Press, 1995.

de La Croix, Horst and Tansey, Richard G., *Gardner's Art Through the Ages*, Fifth Edition, Harcourt, Brace and World, Inc, New York, 1970.

Drake, S., *Discoveries and Opinions of Galileo*, translated by Stillman Drake, Anchor Books, Random House, New York, 1957.

Einstein, A., *The Meaning of Relativity*, Institute for Advanced Study, Princeton University Press, Princeton, New Jersey, 1956.

Einstein, A. and Infeld, L., *Evolution of Physics* with a foreword by W. Isaacson, Touchstone, A Division of Simon & Schuster, Inc., New York, 2007.

Eves, H., *Foundations and Fundamental Concepts of Mathematics*, Dover Publications, Inc., Mineola, New York, 1990.

Gauss, C. F., *General Investigations of Curved Surfaces* including *Abstract, and New General Investigations of Curved Surface*s edited with introduction and notes by Pesic, P., Dover Publications, Inc., Mineola, N.Y., 2005.

Gamow, G., *The Great Physicists from Galileo to Einstein*, Dover Publications, Inc., New York, 1988.

Hamilton, A. G., *Numbers, sets, and axioms: the apparatus of mathematics*, Cambridge University Press, Cambridge, 1982.

Heath, T. L., *The Thirteen Books of Euclid's Elements*, Vol. 1 (Books I and II), 2nd edition, unabridged, Dover Publications, Inc., New York, 1956.

Heath, T. L., *The Thirteen Books of Euclid's Elements*, Vol. 2 (Books III - IX), 2nd edition, unabridged, Dover Publications, Inc., New York, 1956.

Heath, T. L., *The Thirteen Books of Euclid's Elements,* Vol. 3 (Books X - XIII), 2nd edition, unabridged, Dover Publications, Inc., New York, 1956.

Kline, M., *Mathematics, The Loss of Certainty,* Oxford University Press, Oxford, 1980.

Labarre, A. E., Jr., *Intermediate Mathematical Analysis,* Dover Publications, Inc., Mineola, New York, 2008.

Landau, Edmund, *Foundations of Analysis,* Chelsea Publishing Co., New York, New York, 1941.

Larson, R., Boswell, L., and Stiff, L., *Geometry,* McDougal Littell, a division of Houghton Mifflin Co., Evanston, IL, 2004.

Lawden, D. F., *Introduction to Tensor Calculus, Relativity and Cosmology,* Dover Publications, Inc., Mineola, New York, 2002.

McCleary, J., *Geometry From a Differentiable Viewpoint,* Cambridge University Press, Cambridge, U. K., 1997.

Merzbach, U. C. and Boyer, C. B., *A History of Mathematics,* John Wiley and Sons, Inc, Hoboken, N. J., 2011.

Meserve, B. E., *Fundamental Concepts of Algebra,* Dover Publications, Inc., New York, 1981.

Muller, R. A., *Now, the Physics of Time,* W. W. Norton & Company, Inc., New York, 2016.

Newton, I., *The Principia,* translated by Cohen, I. B. and Whitman, A., preceded by *A Guide to Newton's Principia,* Cohen, I. B., University of California Press, Berkeley, 1999.

Pais, A., *Subtle is the Lord, The Science and Life of Albert Einstein,* Oxford University Press, New York, 1982.

Russell, B., *A History of Western Philosophy*, Simon and Schuster, New York, 1945.

Saccheri G., *Euclides Vindicatus*, edited and translated by G. B. Halsted, Chelsea Publishing Co., New York, N. Y., 1986.

Stahl, S., *Geometry From Euclid To Knots*, Dover Publications, Inc., Mineola, New York, 2010.

Struik, D. J., *Lectures on Classical Differential Geometry*, Dover Publications, Inc., New York, 1961.

Todhunter, I., *Spherical Trigonometry for the Use of College and Schools with Numerous Examples,* www.gutenberg.org/ebooks/ 1977 Cached Nov. 12, 2006.

Wolfe, H. E., *Introduction to Non-Euclidean Geometry*, Dover Publications, Inc., Mineola, New York, 2012.

Wrede, R. C., *Introduction to Vector and Tensor Analysis*, Dover Publications, Inc., New York, 1972.

Notes

Prologue

1. de La Croix, H. and Tansey, R. G., pp. 14-22.
2. Bunt, Jones, and Bedient, pp. 2-3.
3. Ibid., pp. 33-37; Ball, pp. 5-7.
4. Bunt, Jones, and Bedient, p. 33; Ball, p. 7.
5. Bunt, Jones, and Bedient, pp. 58-62.
6. Ibid., pp. 6-7.
7. I bid., pp. 8-32.}
8. Ibid., pp. 43-46; Merzach and Boyer, pp. 23-25.
9. Bunt, Jones, and Bedient, pp. 49-54; Merzach and Boyer, pp. 28-30.
10. Ibid., pp. 30-31.

1.   Lessons From School: Euclid's Legacy

11. Ball, p. 14-17;Bunt, Jones and Bedient, p. 69.
12. Merzbach and Boyer, p. 43.
13. Bunt, Jones, and Bedient, pp.69-70.
14. Ibid., p. 70.
15. Eves, p. 11.
16. Brumbaugh, pp. 12-17.
17. Ibid., p. 11.
18. Ibid., pp. 78-92.
19. Ibid., p. 14.
20. Ball, p. 19.
21. Ibid., p. 22.
22. Heath, pp. 351-352.
23. Ball, p. 21; Eves, p. 13.
24. Merzbach and Boyer, p. 58.
25. Bunt, Jones, and Bedient, p. 142.
26. Ibid., p. 72.
27. Ibid., p. 82-83.

28. Ibid., pp. 79-80.
29. Merbach and Boyer, pp. 308-322.
30. Ibid., pp. 66-68.

1.1 Euclid's self-evident truths

31. Heath, Books I and II, Vol 1; Books III – IX, Vol. 2; Books X – XIII, Vol. 3.
32. Heath, Vol.1, pp. 117-124.;Eves, pp. 29-31.
33. Heath, Vol. 1, p. 118.
34. Ibid., p.119.
35. Larson, Boswell, and Stiff, Chapter 1, p. 10.
36. Stahl, Appendix E, p. 421.
37. Ibid., pp. 421-424;Eves, pp.82-86.
38. Heath, Vol. 1, p. 119.
39. Ibid., p. 121.
40. Ibid., pp.154-155.
41. Ibid., p. 223.
42. Ibid., p. 195.
43. Ibid., p. 200.

1.2 Consequences of the first four truths

44. Eves. pp. 16, 165.
45. Ibid., pp. 253, 255.
46. Stahl, p. 46.
47. Heath, Vol. 1 (Book I), pp. 241 – 243.
48. Ibid., pp. 234-240, 242; Eves, p. 39.
49. Stahl, pp. 415-424.
50. Ball, p. 16.
51. Heath, Vol. 2, pp. 61-63.

1.3 A not so self-evident truth: The Parallel Postulate and its consequences

52. Ibid., pp. 315-316.}
53. Ibid., pp. 323-328, 332-333. 338-339, 347-349.
54. Ibid., Vol.2, p. 188.
55. Ibid., pp. 194-195, 200-202.
56. Heath, Vol. 1, (Book II), pp. 403 – 409.
57. Merzbach and Boyer, pp. 146 – 147.

2.1 The search for simpler truths

58. Ball, pp. 96-99.
59. Eves, p. 53.
60. Heath, Vol. I, pp. 204-220.
61. Merzbach and Boyer, pp. 220-221.
62. Heath, p. 220; Eves, pp. 52-53; McCleary, pp. 32-33.

2.2 Saccheri vindicates Euclid – and misses a breakthrough

63. Eves, pp. 61-62.
64. Ibid., pp. 39, 64-65; Merzbach and Boyer, pp. 496-498.

2.2.2 The Saccheri Hypotheses of the Acute (HRA), Obtuse (HOA), and the Right Angle (HRA)

65. Saccheri, pp. 21-27.
66. Ibid. pp. 61-65;McCleary, pp. 28-29.
67. Merzbach and Boyer, pp. 420 – 421.
68. Saccheri, p. 173.

2.3.1 Visions of the HAA on a sphere of imaginary radius

69. Ibid. pp. 59 – 61.
70. Eves, pp. 58-60.
71. Heath, Vol. 1, pp. 212-213.
72. Ibid., pp. 212.

2.3.2 Relating the HOA to a real sphere

73. Ibid., pp. 216 – 217; Eves, p. 60; McCleary, pp.15-16.
74. Heath, Vol. 1, pp. 216-217.
75. Eves, pp. 63–65; Merzbach and Boyer, pp. 496-497; see Riemann's Lecture *Foundations of Geometry* in McCleary pp. 269-278.
76. Ibid., pp. 4-5.
77. Merzbach and Boyer, pp. 494-496.

3.1 Gauss's insight

78. Merzbach and Boyer, pp. 466-474.

79. Wolfe, pp. 48-49.
80. Bonola, pp. 65-66.
81. Wolfe, p. 49.
82. Bonola, pp. 66-67.
83. Ibid., pp. 77-83.
84. Wolfe, pp. 46-47.
85. Ibid., p. 48.
86. Bonola, p. 64.
87. Eves. p. 61.
88. Wolfe, pp. 48-53.
89. Ibid., pp 53-56.
90. Ibid., p. 63.
91. Bonola, Appendix: Lobachevsky, N., *Geometrical Researches on the Theory of Parallels*, translated and edited by G. B. Halsted (first published in 1891), p.13.
92. Ibid., pp. 15-17.
93. Ibid., pp. 18-19.
94. Ibid., pp. 17-18.

3.3   Modeling the hyperbolic plane – a first look

95. Eves, pp. 58, 65-67.
96. Merzbach and Boyer, pp. 500-501:McCleary, p. 236.
97. Eves, pp. 67, 88-92.

4.1   Preclude to a revolution – symbolic algebra

98. Heath, Vol1., p. 375.
99. Ball, pp. 62-77; Merzbach and Boyer, pp. 109-126.
100. Ibid., pp. 127-141; Ball, pp. 77-83.
101. Merzbach and Boyer, pp. 160 – 164; Ball, pp. 104-106.
102. Ball, pp. 106-109.
103. Merzbach and Boyer, pp. 186 – 202.
104. Ibid., pp. 203-212.
105. Ball, pp. 155-158.
106. Ibid., pp. 221-222.
107. Ibid. p. 164.
108. Ibid., pp. 217–221.
109. Ibid pp. 221–225; Merzbach and Boyer, pp. 255 -259, 260.
110. Ibid., p. 259.
111. Ibid., p. 228; Merzbach and Boyer, pp. 260-262.

[112.] Ball., p. 228; Merzbach and Boyer, pp. 282-286.

[113.] Ibid., pp 249-250.

[114.] Ibid., p. 259.

[115.] Ibid., pp. 251-253; Ball, pp. 208-212.

[116.] Merzbach and Boyer, p. 254; Ball, p. 215.

[117.] Merzbach and Boyer, pp. 262-263; Ball, pp. 214-215.

[118.] Ibid., pp. 237-238, 241-242.

[119.] Ibid., pp. 238-239.

[120.] Merzbach and Boyer, p. 274.

[121.] Ball, p. 229.

[122.] Merzbach and Boyer, p. 273.; Ball, pp. 229-230.

[123.] Ibid., p. 232.

[124.] Merzbach and Boyer, pp. 274-277; Ball, pp. 231-233.

[125.] Merzbach and Boyer, p. 274.

[126.] Ibid., p. 276.; Ball, 242.

4.2 Euclid finds his place on the Cartesian plane – analytic geometry

[127.] Ibid., pp. 268–278.

[128.] Merzbach and Boyer, p. 317.

[129.] Ibid., pp. 320 - 322.

[130.] Ibid., pp. 146 – 147.

[131.] Ibid., p. 409.

[132.] Heath, Vol. 1, p. 269.

4.3.1 The postulates

[133.] Merzbach and Boyer, p. 317.

[134.] Labarre, Jr., pp. 33, 239-241; Eves, pp. 179-183.

[135.] Ball, p. 276.

[136.] Labarre, Jr., p. 240.

4.3.2 Rational numbers as decimals – a starting place for the real numbers

[137.] Merzbach and Boyer, pp. 452 – 460.

4.3.3 Filling in the holes in Euclid's line with the infinity of real numbers

138. Ibid., pp. 531-532.
139. Ibid., pp. 450, 545-546; Eves, 290 - 292.
140. Merzbach and Boyer, pp. 520 - 521.
141. Ibid., pp. 536 -537; Eves, pp. 196-197.
142. Ibid., pp. 197 -199; Hamilton, pp. 26 - 49.
143. Merzbach and Boyer, p. 89.

4.3.4 Counting to infinity

144. Aczel, pp. 55-56.
145. Eves. pp. 227-228.
146. Ibid., pp. 226-227.
147. Merzbach and Boyer, pp. 545-546.
148. Ibid., p. 546.
149. Eves, pp. 228-229.

4.4 The real joins the imaginary – the complex plane

150. Kline, p. 117.
151. Merzbach and Boyer, pp. 454-455, 473; Kline, p. 89.
152. Courant and Robbins, pp. 269-271; Merzbach and Boyer, p. 465.
153. Meserve, p. 139-140; Courant and Robbins, pp. 101-103.
154. Merzbach and Boyer, pp. 421 – 422.
155. Ibid., pp. 421-422.

5.1 Overview

156. Ibid., pp. 358-372; Ball, pp. 319-352.
157. Merzbach and Boyer, pp. 382-389; Ball, pp. 353-366.
158. Ibid., 254-257.
159. Merzbach and Boyer, p. 366.

5.2.1 Changes along the path: differential calculus

160. Labarre, Jr., p. 115.
161. Merzbach and Boyer, pp. 378-379.

5.2.2 Summing up along the path: integral calculus

162. Ibid., pp. 322-324.
163. Ibid., pp. 356-358.
164. Ibid., pp.115-120.
165. Ibid., pp. 303-306.
166. Ibid., pp. 324-325.
167. Courant and Robbins, p. 398.
168. Merzbach and Boyer, pp. 326, 358.
169. Ibid.,,p. 409.
170. Ibid., pp. 115-116.
171. Courant and Robbins, pp. 404-406, 464-465; Labarre,pp. 145-148.
172. Ibid., pp. 160-164.
173. Struik, p. 111.

5.2.3 Numbers with direction: vectors

174. Ball, pp. 48 – 49, 245 – 247; Wrede, R. C., p. 1
175. Merzbach and Boyer, pp. 300 – 301.
176. Newton, p. 417.
177. Wrede, pp. 7-11.

6 Gauss reveals curvature as the heart of geometry

178. Gauss, *Introduction to the Dover Edition* by P. Pesic, pp. vi-viii.

6.1 Curves on the Euclidean Plane

179. Wrede., pp. 5-8.

6.2 Surfaces in space

180. Gauss, p. 20.
181. Struik, p. 59.
182. Ibid., p. 63.

6.3 Gaussian curvature and the demystification of Non-Euclidean geometry

183. Ball, p. 426.

184. Struik, pp. 73-82.
185. Gauss, p. 15.
186. Ibid., p. 29.
187. Ibid., p. 48.
188. Struik, p. 83.
189. Ibid., pp. 106-113.
190. Ibid., p. 113.
191. Gauss. p. 45.
192. Ibid., p. 46.
193. Gauss, p. 95; Struik, p. 157.
194. Struik, pp. 108, 156.
195. Ibid., pp. 153-159

7.1 The Hypothesis of the Obtuse Angle (HOA) and the sphere revisited

196. Eves, p. 57.
197. Merzbach and Boyer, p. 420.
198. McCleary, pp.4-5.
199. Merzbach and Boyer, pp. 496-498; Eves, pp. 39, 63-64; Riemann's lecture is given in McCleary, pp. 269-278.
200. Wolfe, p. 177.
201. Heath, Vol. 1, pp. 195-196.
202. Eves, pp. 64-66; Wolfe, pp., 173-174.
203. Ibid., pp. 174-178; Eves, pp. 64-65.
204. Wolfe, pp.178-179.
205. Todhunter, p. 24.
206. Bonola, pp. 77-83.

7.2.1 The sphere of imaginary radius

207. Ibid., pp. 89-90.
208. Aleksandrov, A. D. in Aleksandrov, A.D., Kolmogorov, A.N., Lavrent'ev, M.A. (eds.),Vol. 3, Chapter XVII, pp. 108– 110.
209. Bonola, p. 89; see also in Bonola, Appendix, *Theory of Parallels*, Lobachevsky, p. 41.

7.2.2 Poincaré's half-plane and disk models

210. Wolfe, pp.62, 203; Struik, p.152; Eves, pp. 66-67; McCleary, pp. 217-224, 300.
211. Ibid., pp. 236, 304.

[212] Ibid., p. 228.
[213] Ibid., pp. 226, 227.

7.2.3 The pseudosphere – the model that changed mathematics

[214] Bonola, pp. 112-113; Wolfe, pp. 53, 56.
[215] Bonola, pp. 94-95.
[216] Ibid., pp. 122-125.
[217] Struik, p. 152.
[218] Ibid., p. 152.
[219] Merzbach and Boyer, p. 391.
[220] McCleary. p. 158; Bonola, pp. 137; Aleksandrov, Vol. III, p. 115.
[221] Ibid., p. 115.
[222] Merbach and Boyer, pp. 496-498; McCleary, pp. 269-278.

8.1   The answer to "what is a straight line?" - geodesics

[223] Heath, Vol. 1, pp. 159-165.
[224] Ibid., p. 154, with commentary from pp. 195 - 196.
[225] Ibid., pp. 286-288.
[226] Gauss, pp. 29-30.
[227] Merzbach and Boyer, pp. 396, 431-432, 435.
[228] Sobolev, S. L. and Ladyzenskaja in Alelsandrov, Kolmogorov, and Lavrent'ev, Vol. 2, pp. 124-128.
[229] McCleary, p. 163; Struik, p. 133.
[230] Ibid., p. 131; McCleary, p. 160.
[231] Struik, pp. 140-142.
[232] Krylov, V. I. in Alelsandrov, Kolmogorov, and Lavrent'ev, Vol. 2, pp. 130-131.
[233] Wrede, p. 6.
[234] McCleary, pp. 160-161; Wrede, pp. 344-345; Struik, pp. 131-132.

8.2   The foundation of geometry: the metric tensor

[235] Wrede, pp. 324-325.
[236] Ibid.,, p. 326.
[237] Ibid., p. 228.
[238] Ibid., p. 228.
[239] Ibid., p. 229.
[240] Ibid., pp. 230-232.
[241] Ibid., pp. 331-335.

8.3 Tensors and a universal geodesic equation for straight kines and the orbits of planets

[242.] Einstein, p. 72.
[243.] Wrede, pp. 6, 331.
[244.] Ibid., pp. 68-74.
[245.] Einstein, pp. 69-70.
[246.] Ibid., p. 70.
[247.] Wrede, p. 333-334; Lawden, pp. 108-110; Einstein, p. 71.
[248.] Lawden, pp. 98-100.
[249.] Lawden, pp. 95-98, 100-102.
[250.] Ibid., p. 101; Einstein, pp. 73-74.
[251.] Struik, pp. 131-134; Wrede, pp. 348-350.
[252.] Struik, pp. 131-134; Wrede, pp. 344-346.

8.4 Leaving Gauss' surface – the curvature of space

[253.] Einstein, pp. 75-77.
[254.] McCleary, p. 257.
[255.] Wrede, p. 359.
[256.] McCleary, pp. 258-260, Wrede, p. 366-367.
[257.] Einstein, p. 77; Lawden, pp. 105, 114.
[258.] Ibid., p. 77.
[259.] Wrede, pp. 366-367.

9.1 On the shoulders of giants

[260.] Merzbach and Boyer, p. 358.
[261.] Kline, pp. 35 – 41. Newton, pp. 789, 800, 805 with preceding background: Cohen, *Guide to Newton's Principia*, pp. 21, 67, 123, 128.
[262.] Newton, p. 424; Ball, pp. 248–249.
[263.] Newton, pp. 416-417.
[264.] Ibid., p. 404.
[265.] Ibid,, p. 424.
[266.] Ibid., p. 405.
[267.] Ibid., p. 446.
[268.] Ibid., p. 468.
[269.] Ibid., p. 800.
[270.] Ibid., pp. 806.
[271.] Ibid., pp. 810, 811.
[272.] Ibid., p. 943.

[273] Ibid., pp. 794-796.

9.2 The stage of Newton's universe: space and time

[274] Ibid., p. 408.
[275] Einstein, p. 28.
[276] Newton, pp. 408-409.
[277] Ibid., p. 819.
[278] Einstein, p. 25.

10 Beyond Newton: Conservation of Energy

[279] Newton, pp. 528-532, with the preceding clarifying introduction by Cohen, pp. 120-122.
[280] Einstein, pp. 81-82, Wrede, pp. 373-374.
[281] Einstein and Infeld, pp. 42-43, 48.
[282] Ibid., pp. 47-50.

11.1.1 The Lorentz transformation

[283] Kline, p. 67; see Ball for details of Laplace's life and work, pp. 412-421.
[284] Wrede, p. 177; Ball, p. 450.
[285] Ibid.,, pp. 450-451; Wrede, p. 178.
[286] Einstein and Infeld, pp. 142-153; Asimov, pp. 476-479.
[287] Ibid,, pp. 119-122, 171-176; Einstein, pp. 26-27; Wrede, p. 179.
[288] Einstein and Infeld, p. 177.
[289] Wrede, p. 178 -179.
[290] Einstein, pp. 28-29.
[291] Pais, pp. 151-152.
[292] Einstein, pp. 31-34; Lawden, pp. 9-11.
[293] Pais, p. 152.
[294] Einstein, pp. 40-43.
[295] Wrede, p. 176.

11.1.2 Four vector momentum and $E = mc^2$

[296] Einstein, pp. 46-47.
[297] Ball, pp. 327-328.

11.2.1 Einstein's insight: the Equivalence Principle

298. Einstein, p. 61.
299. Pais, p. 178.
300. Ibid., p. 179.
301. Ibid., p. 219.
302. Einstein, pp. 57-58.
303. Einstein and Infeld, pp. 214-222.
304. Einstein, pp. 61-63; Einstein and Infeld, pp. 226-230; Gamow, pp. 201-203; Lawden pp. 130-131.

11.2.2 Newton's Law of Gravity revisited as an equation of gravitational potential

305. Wrede, p. 171.
306. Lawden, pp. 112-113.

1.1.3 Einstein's Field Equations

307. Pais, pp. 180-181, 194-200.
308. Ibid., p. 213.
309. Ibid., pp. 211-212.
310. Ibid., p. 208.
311. Ibid., p. 213.
312. Ibid., p. 212.
313. Ibid., pp. 282-285.
314. Ibid., p. 282.
315. Einstein, p. 56.
316. Pais, pp. 216, 285.
317. Ibid., p. 283.
318. Ibid., pp. 220-221.
319. Einstein, pp. 83-84; Pais, p. 256; Lawden, p. 137.
320. Ibid., p. 137.
321. Pais, pp. 220, 243, 290; Lawden, p. 142.
322. Einstein, pp. 85-86.
323. Ibid., pp. 79-81, 86-90.
324. Pais, p. 283.
325. Barrow, pp. 284-290.
326. Einstein, p.117-125.
327. Ibid., p. 88.
328. Ibid., pp. 87-88, 118.

[329.] Lawden, p. 140.

[330.] Pais, p. 219.

12.1   The first three predictions: the spectral redshift, Mercury's orbit, and the bending of light

[331.] Ibid., pp. 253-256.

[332.] Ibid., pp. 196-198.

[333.] Ibid., p. 255.

[334.] Barrow, 44-46.

[335.] Pais, p. 255.

[336.] Lawden, p.146-147.

[337.] Ibid., p. 152.

[338.] Ibid., pp. 153-154.

[339.] Muller, p. 66.

[340.] Lawden, p. 154.

[341.] Pais, pp. 253-254.

[342.] Ibid., pp. 253.

[343.] Lawden, p. 148.

[344.] Ibid., pp. 147-150.

[345.] Pais, pp. 303-305.

[346.] Lawden, pp. xiii, 152.

[347.] Pais., p. 305.

12.1   Black holes and the expanding universe

[348.] Ibid., p. 289, fn.

[349.] Lawden, pp. 155-159.

[350.] Pais, p. 269.

[351.] Barrow, pp. 280-281.

[352.] Pais, p. 269.

[353.] Ibid., p. 270; Muller, pp. 49 -50.

[354.] Ibid., pp. 86-87.

[355.] Barrow, p. 281.

[356.] Ibid., pp. 26-31, 31-32.

[357.] Ibid., pp. 34-35.

[358.] Ibid., p. 28.

[359.] Pais, p. 268.

[360.] Ibid, p. 268.

[361.] Ibid., p. 288; Barrow, pp. 61-63.

[362.] Ibid., pp. 65-72.

363. Pais, p. 268.
364. Barrow, pp. 125-127.
365. Ibid., pp. 140-144; Muller, pp. 14-142.
366. Barrow, pp. 160-161.
367. Ibid., pp. 198-204.
368. Ibid., pp. 204-208.
369. Ibid., pp. 54-57; Pais, pp. 285-286.
370. Barrow, p. 57; Pais, p. 288.
371. Barrow, pp. 60-64.
372. Muller, p. 130.
373. Lawden, pp. 180-181.
374. McCleary, p. 275.
375. Lawden, p. 186.
376. Barrow, pp. 62-63; Lawden, pp. 179-180.
377. Einstein, pp. 121- 123.
378. Lawden, pp. 189-191.
379. Barrow, pp. 62-63.
380. Ibid., pp. 65-69.
381. Ibid., pp. 74-76.
382. Ibid., pp. 199-201.
383. Ibid., pp 279-283.
384. Ibid., pp. 200-201.

Epilogue: Philosophic Thoughts

385. Kline, p. 341.
386. Russell, p. 118.
387. Ibid., p. 118.
388. Kline, pp. 346-347.
389. Russell, p. 658.

Index

A

analytic geometry: algebraic description of a circle, 94; algebraic description of Euclidean geometry, 93-4; Cartesian coordinate system, 91-2; dependence on real number system, 81; equation of a straight line, 91-2 157; recognition by Descartes and Fermat of equations as curves, 90-1

ancient Greek number system, 14

Apollonius of Perga: properties of conics, 84

Archimedes: aproximating areas and volumes of geometric figures, 140; estimate of pi, 84

Argand, Robert: 120

Aristotle: axiomatic system, 15, 18; common notions and principles, 19; logic, 22-3; points and lines, 15; understanding of infinity, 112

axiomatic method, 14, 85, 214, 218

B

Babylonian number system, 3-4, 102

Barrow, Isaac: method of tangents, 140

Beltrami, Eugenio. *See* Beltrami pseudosphere model; makes connection between surfaces of negative Gaussian curvature and the HAA, 215; models of hyperbolic geometry, 77

Beltrami pseudosphere model, 204; equation of tractrix, 215-7; formed by revolution of tractrix about an axis, 215; new understanding of axiomatic systems, 218; proved Parallel Postulate is an independent unproveable postulate, 218; refutes Kant's notion of Euclidean geometry as an a priori truth, 218; differential length, First Form, and Gaussian curvature, 216; tractrix curve defined, 215

Berkeley, Bishop George, 130

Bolyai, Johann, 46, 66. *See* hyperbolic geometry; letter to father announcing new discovery, 69; uncertainty whether Parallel Postulate could be proved, 214; ungenerous response by Gauss, 70; work of Lobachevsky unknown until 1848, 214

C

calculus. *See* chain rule, *See* differentiation, *See* Fundamental Theorem of Calculus, *See* integration; differentiation; *See* partial derivative; velocity as an example of, 127-8; importance to mathematics and science, 126; integral as anti-derivative, 128; table of derivatives and integrals, 134

Cantor, Georg: 112-3; comparison of real, rational, and natural numbers, 114-7; comparison of sizes of infinite sets, 112; transfinite numbers, 118-9

Cardano, Girolamo: analysis of cubic equations, 86

Cauchy, Augustin-Louis: definition of limit, 105; major contributions to mathematics, 105

Cauchy sequence: definition, 109

chain rule, 135, 137, 139, 164, 234-5, 240, 284, 322, 325

Christoffel symbols. *See* covariant derivative, *See* parallel displacement (transport); alternative derivation using derivatives of base vectors, 359-61; expressed in terms of the metric tensor, 244; for spherical surface, 252; introduction of, 226

complex numbers: complex plane, 120; definition as ordered pairs and operations with, 120-3; observed by Cardano, 86

Cosmological Constant. *See* Einstein, Albert; additional term in Field Equations, 333, 342-3; modelling acceleration in expansion of the universe, 342; use by Lemaître, 341-2

covariant derivative: absolute derivative defined in terms of, 246; defined to transform as a tensor, 241; defined with Christoffel symbols, 241; different approaches to derivation, 242-4, 359-61; Einstein notation, 241; mathematicians responsible for development of, 241-2; of tensors with covariant and contravariant indices, 244; partial derivatives of tensors do not transform as tensors, 240

curvature. *See* Gaussian curvature, *See* integral (total) curvature, *See* normal curvature, *See* spherical surfaces ; Euler's Theorem, 181; geodesics and the curvature vectors, 175-9

curves on a plane, 162-6. *See* surfaces; Cartesian position and tangent vector of circle with polar coordinates, 164-5; curvature vector, 165-6; tangent vector, 162-5

D

Dedekind, Richard: one-to-one correspondence of real numbers and the number line, 109-10

Democritus: atomic theory, 11

differentiation. *See* chain rule; calculate slope of curve, 131; criticism of by Bishop Berkeley, 130; definition, 129-30; importance of limit concept to, 130; linear operator, 135; notation, 130; of sums and products of functions, 135; slope of a parabolic curve, 131-3; as tangent (slope) to unit circle, 136-7

Descartes, René, 13, 90-1, 98, 110, 120, 257. *See* analytic geometry; tradition of x as the unknown, 101

differential: of length, 152

E

Eddington, Arthur, 324, 326; comment on the relation of the mind and nature, 345-6

Egyptian mathematics, 3

Einstein, Albert. *See* Cosmological Constant, *See* Special Relativity, *See* General Relativity; collaboration with Marcel Grossmann to adapt Riemannian geometry for General Relativity, 304-5; indebtedness to Minkowski's geometric vision of space-time, 281; invention of scientific concepts and creativity, 312, 349; need to abandon Euclidean geometry for General Relativity, 304; publication of theory of Special Relativity, 276, 281; publishes final form of General Relativity (1915), 290, 306, 316; role of the propagation of light in the definition of time, 280

elliptic geometry, 214, 227; consistency of the HOA with spherical geometry, 194-5; lines are boundless, 189; modifications of Euclid's postulates, 188-9; spherical geometry as a model, 192; sum of interior angles of a triangle is greater than 180 degrees, 188

energy: calculation of escape velocity from the earth, 273; conservation of mechanical energy, 271-2; First Law of Thermodynanics, 273-4; gravitational potential energy and General Relativity, 306-7, 311, 313-5; kinetic energy from Newton's Second Law, 270-1; potential energy from Newton's Universal Law of Gravitation, 271-2; relation to mass in Relativity, 285-6

equations as curves: recognition by Descartes and Fermat, 90-1

Equivalence Principle, 294, 303, 315, 318, 323-4, 326; Einstein- "the happiest thought of my life", 289-90; deflection of light by gravitational field, 290-1; equivalence of inertial and gravitational mass, 290; gravitational frame replaced by accelerated frame, 292; gravitational redshift, 291, 317-9; non-Euclidean nature of space in a gravitational field, 293; statement of, 290

Euclid's *Elements*: areas of rectilinear figures, 38; assumptions not expressed in proofs, 31-2; Common Notions, 19; difficiencies, 25-6; Definitions, 15-7; Postulate E5, the Parallel Postulate, 31; Postulates, 19; Propositions IE1 to IE28, 355-8; similar triangles, 38-9; types of proofs, 23-4

Euclid's Propositions using the Parallel Postulate: IE29, 31-3:; IE32, 34; IE34, 35; IE35,35; IE37, 36; IE41,35; IE46, 35; IE47, The Pythagorean Theorem, 36-8

Euclid's Propositions without the Parallel Postulate: IE1 24-5; IE4, 26; IE8, 30; IE9, 30; IE10, 27, 30; IE11, 30; IE12, 30; IE15, 26, 30; IE16, 26-8, 52, 58; IE17, 59; IE20, 30; IE26, 30; IE27, 28- 9; IE28, 29-30

Euler, Leonard:, 86, 100; defines e, 118; notation for imaginary numbers and relation to trigonometric functions, 120

Euler-Lagrange equation, 220-6, 239, 246, 248; applied to spherical surface, 225-6; Euclidean straight line, 221; minimization of an integral, 220; multivariable form, 224-5

F

Fermat, Pierre, 13, 126, 140-1. *See* analytic geometry; insights of, 91

Field Equations, 341-3. *See* Cosmological Constant; covariant tensor form, 307-8; energy-momentum tensor, 307, 338; form of Field Equations and Newtonian gravitational potential, 306-7; Newtonian gravitational potential and space-time metric, 311-2; paper of Einstein and Grossman (1913)with energy-momentum tensor balancing the metric of space-time, 306; Ricci scalar and energy momentum tensor, 308-9; Ricci tensor and energy-momentum tensor, 309; Ricci tensor components and Newtonian gravitational potential, 313-35

Friedmann, Alexander, 331, 334

Friedmann universes of FLRW model, 333-6; comparison with surface of hyperbolic geometry and the Poincaré disk model, 335-6; covariant and contravariant metric components, 337; covariant Ricci tensor components, 338; critical mass density to determine universe's curvature, 339; differential equations for the scale factor S(t) from Field Equations, 336-41; energy-momentum tensor, 338; expansion and contraction of space for positively curved universe, 340-1; expansion of space for flat and negatively curved universes continues forever, 340; Friedmann (1922) and Lemaître (1927) models of an expanding universe, 331; mass density inversely proportional to scale factor cubed, 339-40, 268; as a metric of constant curvature k, 334-6; scale factor dependence on time for flat universe and for early times in curved universes, 340; metric, 334; relation to assumptions of the Cosmological Principle, 334-5; scale factor S(t) and Hubble's Law, 336

Fundamental Theorem of Algebra, 122

Fundamental Theorem of Calculus: formal statement of inverse relation of differentiation and integration, 146; recognition by Newton and Leibniz, 141

G

Galileo Galilei, 112, 258; kinematics of a body accelerated under the force of gravity, 258; vector description of projectile motion, 153

Gauss, Karl Friedrich, 64, 215; attempt to prove the Parallel Postulate, 64-5; defines Gaussian curvature in terms of auxiliary sphere, 183; did not publish insights on a new geometry, 65-7, 227; First Form, 171; major contributions to mathematics, 64; on Euler's Theorem, 181; surface curvature, 161; *Theorema egregium*, 182

Gaussian curvature, 161, 197-9, 203, 205, 210, 252-3. *See* integral (total) curvature; defined geometrically in terms of auxiliary sphere, 183-4; defined in terms of principal curvature, 181-2; expressed entirely in terms of the First Form, 182

General Relativity. *See* Equivalence Principle, *See* Field Equations; geodesic equation in weak gravitational field and Newton's Laws of motion, 309-15; influence of Ernest Mach on Einstein's concepts, 305-6; motivation, 288-90, 305-06;

General Relativity predictions and verification: black holes, 327-30; deflection of light, 323-7; Mercury's orbit, 319-23; response to verification of light deflection by Eddington expedition, 327; Schwarzschild solution to Field Equations, 316-7; spectral redshift, 317-9; verification of deflection of light by Eddington expedition, 326

General Relativity-models of the universe. *See* universe, *See* Cosmological Constant, *See* Friedmann-Lemaître-Robertson-Walker (FLRW) model; Cosmological Principle, 334-5; Einstein and de Sitter (1932), 342; Einstein's – finite, curved, and static universe (1917), 331; Friedmann searches for types of solutions compatible with the Field Equations leading to the FLRW model, 334

geodesics: application of governing differential equation on spherical surface, 246-8; definitions, 179, 219, 223-4; differential equation with Christoffel symbols, 226, 239, 246; equation of a great circle, 221-22; Euclidean straight lines, 221; great circle as shortest path determined by Euler-Lagrange equations, 222-3; infinite number of great circles connect two poles, 219; name originated with Liouville and Jacobi, 223-4; used by Gauss to define triangles on general surfaces (shortest lines), 219

geometric algebra, 83-84

Gödel: theorem on the inability to prove the consistency of mathematical systems, 346-7

H

Heath, Thomas L., 10, 12, 20, 24-5, 28, 189, 191, 353, 355

Hilbert, David, 18; axiomatic system, 18; difficiencies in Euclid's geometric system, 25; Hilbert Hotel, 115

Hipparchus of Nicaea: trigonometric tables, 40

Hoyle, Fred: in opposition, names the creation of an expanding universe as the Big Bang, 331; Steady State Theory of the Universe(1948), 331

Hubble's Law, 331, 336

hyperbolic functions, 124-5

hyperbolic geometry: angle of parallelism, 71, 200-01; derivation of functional relation for angle of parallelism, 201-3; discovery by Lobachevsky and Bolyai, 69-70; functional form of angle of parallelism discovered by Lobachevsky and Bolyai, 203; Hyperbolic Postulate as 16th theorem of Lobachevsky, 70-1; Lobachevsky Angle Theorem, 72-3; Lobachevsky Right Angle Theorem, 73-4; proof that angle of parallelism decreases with increased distance of point from a line, 200-01; sum of triangle interior angles less than two right angles, 73-4

hyperbolic geometry models, 77-9. *See* hyperbolic trigonometry, *See* Poincaré Half-plane and disk models, *See* Beltrami pseudosphere model; Beltrami (Cayley-Klein-Beltrami) disk model, 204; Klein's disk model, 79; Taurinus' model of sphere with imaginary radius, 198-9, 201-3

hyperbolic trigonometry: example of hyperbolic equilateral triangle, 200; hyperbolic trigonometry using Taurinus' sphere with an imaginary radius, 199

I

Indian communities' developments in mathematics: decimal system, 85

infinite geometric series: 106

Infinite series: pi, sin x, cos x, e, 123-4

infinite sets, 101-2, 111-2, 118-19; Cantor's method of determining equivalence, 112-3; David Hilbert's Hilbert Hotel with an infinity of rooms, 115; equivalence of integers and rational numbers as countably infinite, 114-5; equivalence of natural numbers and integers as, 113-4; Galileo's concerns about the infinity of numbers, 112

integral (total) curvature, 185-6, 250; Gauss Triangle Theorem, 186; Gauss-Bonnet Theorem, 186; implications of Gauss Triangle Theorem for geometry, 186-7; related to sum of interior angles of a triangle, 182

integration: definite Riemann integral, 145; example using limit process, area under parabola, 144; example using Riemann integral, area of trapezoid, 145-46; impact of countability of rational numbers, 149-50; integral using trigonometric substitution, 147-8; length of curves, 150-1; method to calculate area, 141-4; numerical integration of length of curves, 152; numerical integration to estimate pi, 148-9, 152; precursors : Archimedes, Cavalier, Fermat, and Barrow, 140; properties of integrals, 145

irrational numbers, 106-11. 149-50. *See* square root of 2; e as base of natural logarithms, 118; Liouville numbers, 118; proof that pi is transcendental, 118; roots of quadratic equations rejected by Diophantus, 84-5; transcendental numbers as uncountably infinite, 117-8

K

Kant, Immanuel, 330; a priori nature of Euclidean geometry, 67-8, 227; a priori view of time, 68

Kepler, Johann: elliptical orbits of planets, 258, 261

Klein, Felix: disk model of hyperbolic geometry, 77; 204 naming of hyperbolic geometry, 70

L

Lambert, Johann, 42, 43; additivity of angle defect, 56; area of spherical triangle and the HOA, 56-7, 188; area of triangle under HOA, 62; area of triangles under HOA and HAA, 55-6; equality of area of similar triangles under HAA and HOA, 54-5; HAA and a sphere of imaginary radius, 57; hypotheses of acute (HAA), obtuse (HOA), and right angle (HRA), 53; quadrilateral of , 43, 53

Legendre: Adrien Marie, 42, 53, 57-9, 74, 188

Leibniz, Gottfried Wilhelm, 135, 140-1, 258, 259; independent discovery of the calculus, 126; notation for derivative, 128

Lemaître, Georges, 331; independently predicts expanding universe, 341

limit: definition, 105, 108

Lobachevsky, Nikolai I., 198-9, 202. *See* hyperbolic geometry; impossibility of determining whether the universe followed his new geometry or Euclid's, 214-5; steps to a new geometry, 70

logic, 346, 348; excluded middle and law of contradiction, 22; implication, 22; proof by contradiction, 23; syllogisms, 22-3

M

Mach, Ernest, 305-6, 312

mathematical induction, 142-3

metric tensor, 227-230, 237, 243-4, 247, 251-3, 283, 294, 306-8, 310, 337-8, 343; covariant and contravariant forms, 234; Euclidean metric in spherical coordinates, 236-7; form in Cartesian coordinates, Kronecker delta, 229; relation to base vectors, 234, 237; relation to First Form and definition, 228-9; spherical surface, 229;

N

natural logarithm, 146-7

Newton, Isaac, 100, 126. *See* Newton's Laws of Motion, *See* Universal Law of Gravitation; absolute space, 267; absolute time, 266; beneficiary of Copernicus, Kepler, and Galileo, 127, 257, 261, 264; binomial theorem, 286; discovery of the calculus and use in laws of motion and gravitation, 126, 257-62; Einstein summarizes Newton's views on space and time, 268; inductive nature of physics, 265; Second Law of Motion and Universal Gravity , 127, 258-9, 261-2; vector sum as diagonal of parallelogram, 153-4; view of Euclidean geometry, 227

Newtonian gravitational field: relation of flux to mass, 295-7, 302-3

Newtonian gravitational potential, 297-8, 300-1, 307, 311-2; importance of partial differential (Poisson) equation to formulation of Einstein Field Equations, 303; partial differential equation (Poisson), 301-3

Newton's Laws of Motion, 253, 261, 268, 270, 282, 294-5 inertial systems and absolute space and time, 268-9; predicts Galileo's results for falling bodies, 257, 259-60; Second Law expressed in terms of time derivative of momentum, 259; three laws as specified by Newton, 258-9; vectorial nature of laws, 259

non-Euclidean geometry, 28, 31, 39, 58, 65-76, 161, 174, 182, 198-9, 217-8, 227, 292-4, 346. *See* Riemannian geometry; comparisons of impact of hypotheses (HAA, HOA, and HRA), 75-6; relation of Gaussian curvature to geometry of circles on curved surfaces, 197-8

normal curvature, 187; from First Form, 177-9

number systens. *See* complex numbers; 3-4, 26, 93, 98-101, 122; rational numbers as decimals, 102-6; real numbers as decimals, 106-9; real numbers as Dedekind cuts, 109-10

P

Pacioli, Luca, 88

parallel displacement (transport). *See* parallel vector displacement (transport) around a closed curve; covariant derivative of contravariant vector, 242-4; definition of Christoffel symbols in terms of, 242-3; differential equation for vector along path, 245; equation for parallel vectors in terms of the absolute derivative, 245-6

Parallel Postulate, 14-5, 26, 28, 30-1, 33, 35, 38-9, 41-5, 53-4, 60, 64-5, 69-70, 189, 214, 218; investigation by mathematicians, 2nd to 19th century, 42; investigations by Alhazen and Kayyam introducing novel quadrilaterals, 43; investigation with Lambert and Saccheri quadrilaterals, 43; postulates equivalent to, 43; shown to be independent, unproveable assumption, 218; use by Euclid of, 31-41

parallel vector displacement (transport) around a closed curve, 249-51; calculation of change in vector's initial direction related to Riemann curvature tensor, 250-1; on a sphere, 249-50; relation to integral curvature on a spherical surface, 250

partial derivative, 139, 171, 180, 232, 233-6, 240-7, 298-300, 308, 359-61; chain rule, 138; definition, 138; forming total derivative and differential, 138

physics prior to Special Relativity: discoveries of electromagnetic phenomena challenge Newtonian viewpoint, 275-6; electromagnetic waves explained by the hypothesis of an ether, 276; Laplace's view of a deterministic universe, 275; First Law of Thermodynamics, 273-4

Plato: ideal forms, 21-2

Playfair's Postulate: alternate to the Parallel Postulate, 42 ; compared with the Hyperbolic Postulate, 69; proof from the Parallel Postulate, 44-5;

Poincaré disk model, 78-9; differential length, First Form, and Gaussian curvature, 213

Poincaré Half-plane model, 77-8, 204-12;: calculation of angles of parallelism, 211-2; calculation of distance, 205-8; differential length ds, First Form, and Gaussian curvature, 205; geodesics, 77, 205; numerical calculation of distance, 208; parallels and triangles, 77-8; triangle formed from geodesics, 208-10

position vector, 160, 162-5, 167, 184, 233, 251, 262

principal curvature, 181-2, 185

Proclus, 10, 15; comment on Ptolemy's attempt to prove the Parallel Postulate, 42; on Hippocrates of Chios, 12; self-evident principles, 18

Pythagoras: proofs, 11-2; relation of length of vibrating string to tune, 12-3

Pythagoreans: importance of whole numbers, 13

Pythagorean Theorem, 12, 35-8; dependence on the Parallel Postulate, 35; irrational numbers, 13; square root of 2, 13, 106-9

R

rational numbers: as decimals, 102-6; as limit of sequence of partial sums, 103-6; denseness, 114 ; ordered field, 100, 110

real numbers, 25-6, 122; algebraic numbers as countably infinite, 117; as uncountably infinite, 115-7; defined as Cauchy equences, 109; defined as Dedekind cuts, 109-10; exponentiation, 110-11; inclusion of natural numbers, integers, and irrational numbers, 99, 101; least upper bound property, 100, 108, 115; one-to-one correspondence with continuous number line, 90-1, 99, 109-10; Postulate of Continuity, 100, 102, 108-10; postulates, 99-100 postulates unknown until 19th century, 20, 98-9 reality and perception: Bertrand Russell's definition of matter, 350; conceiving of an expanding universe, 349-50; mathematics as a mode of perception, 350-1; space-time, 349; Stephen Hawking: "impossible to imagine a four-dimensional space", 349; wave-particle duality, 350

Ricci tensor and scalar, 305, 308-9, 313-4. *See* Riemann curvature tensor; selected as geometric components of Einstein's Field Equations, 253

Riemann, Bernard. *See* Riemannian geometry; extension of geometry to n-dimensional space, 218, 227; generalization of geometry, 46, 59-60; limitations on IE16, 59; on spherical geometry, 59-60, 188-92.; student of Gauss, 227

Riemann curvature tensor: calculation of Gaussian curvature for a sphereical surface, 251-3; in terms of Christoffel symbols, 251; relation to Gaussian curvature, 253; Ricci tensor and Ricci scalars formed from, 253; value of 0 for flat space, 251

Riemannian geometry, 288-9, 294, 304-5, 307. *See* parallel vector displacement (transport) around a closed curve; extends meaning of non-Euclidean geometry beyond hyperbolic and elliptic geometries, 227; n-dimensional geometry with metric tensor, 227-9, 248; parallel vectors, 245-6; vector magnitude and angle between vectors, 238-9

S

Saccheri, Gerolamo, 42,45-53; asymptotic and diverging parallels, 51-3; failure to discover a new geometry, 45-6, 52-3; Hypotheses of Acute (HAA), Obtuse (HOA), and Right Angle (HRA),45-6, 49; Proposition S1, 47; Proposition S2, 47-8; Proposition S3, 49-50; rejection of the HOA, 51; Saccheri's quadrilateral, 46

Saccheri-Legendre Theorem: incompatibility of Euclid's first four postulates with HOA, 58, 188-9; proof of 57-8

Schwarzschild, Karl, 316. *See* General Relativity predictions and verification

space-time 4-vector: force and law of motion, 287; increase of mass with velocity, 285-6; Lorentz transformation of, 277-9, 283; metric tensor in Special Relativity, 283; momentum, 285; momentum invariant, 287; proper time, 283-4; relation of energy to mass, 286-7; relation of momentum invariant to wave nature of light and particles in Quantum Mechanics (fn. cxviii), 287; velocity, 283-4; velocity invariant, 285

Special Relativity, 267, 288, 310, 313, 315, 317, 323, 328, 344. *See* physics prior to Special Relativity, *See* space-time 4-vector; consequences of Lorentz transformation, time dilation, length contraction, and loss of simultaneity, 279; constancy of velocity of light leads to space-time, 280; Michelson-Morley experiments fail to detect changes in the velocity of light required by the hypothesis of an ether, 276; Minkowski's geometric interpretation of space-time, 280-1; Newton's Laws recovered at low velocities compared to that of light, 287; principles, 276

spherical geometry: angle form of Spherical Law of Cosines, 194; comparison with HOA, 60-3; distance of great and small circles on the earth's surface, 195-7; Euclidean Law of Cosines as limit for large radii, 197; geodesic as great circle, 221-3; Spherical Law of Cosines, 192-4 spherical surfaces. *See* geodesics; algebraic description, 167; calculaion of integral curvature using auxiliary sphere, 185-6; calculation of tangent and curvature vector of great circle, 175-6; differential path length ds, 169; geodesic curvature vector of small circle, 176-9; great circle, 168, 175, 221-3; integrated path length s, 169-70; normal curvature vector and magnitude, 177; parameterization of curves on, 168; position vector for spherical surface in Cartesian components with spherical cordinates, 169; small circle, 168; spherical coordinates, 167-8, 195-6

square root of 2: appoximation as a decimal, 106-8; impact of irrationality on Greek mathematics, 13; proof of irrationality cited by Aristotle, 353-4

surfaces. See spherical surfaces; coordinates on a general surface, 170-1; curves on general surface, 168; differential area in terms of the First Form, 173; differential path length ds along a curve, 169-71; Gauss's First Form, 171; integrated area in terms of the First Form, 173-4; integrated curve length s using First Form, 171; position vector with two parameters, 166-8; tangent plane and normal to, 171-2; tangents to curves on a general surface, 175

symbolic algebra: comparison of Renaissance notation of Nicolas Chuquet with modern notation, 87-8; development of various notations in the Renaissance, 87-90; explicit designation of unknowns by Luca Pacioli, 90; importance of, 83, 122; major contribution to its invention by Viète, 89-90; notation for exponentiation, 87, 89-90; primitive approach of Diophantus, 84-5; summation notation introduced by Euler, 142

T

tangent vector, 162-6

Tartaglia, Nicholas: solutions to some cubic equations, 86

Taurinus, Franz A.: interpreting the HOA on the surface of a sphere of imaginary radius, 198, 199; letter from Gauss, 65-6

tensors, 153 239-48. *See* covariant derivative, *See* metric tensor, *See* Ricci tensor and scalar, *See* Riemann curvature tensor; allow geometric properties to be expressed independently of coordinate systems, 227-9; contravariant base vectors, 234-5; contravariant transformation, 232-4; covariant base vector, 236; covariant transformation, 235; critical to formulation of General Relativity, 227-8, 304-6; divergence of covariant and contravariant vectors, 298-300; Einstein convention, 228-9; general transformation for covariant and contravariant forms, 237-8; gradient of scalar as covariant vector, 235; orthogonality of covariant and contravariant base vectors, 236; raising and lowering of indices, 237; rank (order) of, 230; tangent as contravariant vector, 233; vector as tensor of order one with covariant and contravariant forms, 238

Thales: deductive system, 10-1; first mathematician, physicist, philosopher, 9-11; Propositions, 10; propositions included by Euclid, 10, 30

thermodynamics: First Law, 273-4

trigonometric functions: algebraic expressions and identities, 94-5; Cartesian coordinate expression of cosine function, 97; cosine function in inner product,

156, 238, expressed as exponentials and infinite series, 123-4; from similar triangles, 39-40; Law of Cosines, 93, 95-7; Law of Sines, 96; tangent as slope, 137

truth: Bertrand Russell's views on inductive reasoning, 347-8; loss of mathematics as a source of absolute truth, 346-7; mathematical, 346-7, 348-51; non visual nature of, 349-51 personal, 347; scientific, 312, 347-51

U

Universal Law of Gravitation, 306, 312-5, 317; equation of, 264, 295-6, 302; identified as universal inverse square centripetal force, 261-2; illustration of relation to Kepler's planetary laws, 262-4; Newton's lack of an explanation for its cause, 265

universe, 1, 11, 14, 214-5, 228, 248, 288, 314, 327-8, 337-44, 350; age of, 337; Alan Guth - theory of accelerated expansion of early universe (1980), 332-3; dark matter and dark (vacuum) energy, 343; Hubble's comments on relation of his observations to curvature of space, 331; Hubble identifies extragalactic object (1924), 330; Hubble observes receding nebulas implying an expanding universe (1929), 331; Hubble's Law, 331, 336; measurements of cosmic background radiation imply early isotropic expansion (1967), 332 ; Penzias and Wilson discover cosmic background radiation supporting Big Bang Theory (1965), 331-2; scientific view of, up to the late 19th century, 330; stability of stars and black holes, 329-30

Upper Paleolithic Period, 22, 345

V

vectors. *See* curves on a plane, *See surfaces, See* tensors; Cartesian components of vector (cross) product, 157-8; component definition of angle between, 155; definition of Cartesian vector space and operations of triples of real numbers, 154-5; definition of inner product, 156; derivative of Cartesian vector, 163; divergence in Cartesian coordinates, 299-300; divergence definition, 298; Divergence Theorem, 298; early concepts, 153; equation of line parallel to, 157 equation of plane perpendicular to, 157; examples from physics, 153; geometric interpretation of addition and scalar multiplication , 153-4; properties of inner product, 156; properties of vector (cross) product, 159 transformation of Cartesian components under coordinate rotation, 230-2; vector (cross) product definition, 156; vector (cross) product of Cartesian base (unit) vectors , 158; volume of parallelpiped, 160

Viète, Francois. *See* symbolic algebra

W

Weierstrass, Karl: definition of limit, 108

Wessel, Caspar, 120

About the Author

Robert G. Bill was a researcher in fire and fire protection for twenty–five years at FM Global, a major industrial and commercial mutual property insurer which operates the world's largest full–scale fire research facility. There, he was Assistant Vice President and Director of Research for Fire Hazards and Protection, overseeing research in areas of flammability, fire spread, material reactivity, and fire protection systems. Previous to joining FM Global, he was an Assistant Professor in the Department of Mechanical Engineering at Columbia University conducting research in turbulent combustion. He holds BS, MS and PhD degrees in Mechanical Engineering from Cornell University. During his doctoral program, he minored in theoretical physics and took courses from Nobel Laureates, providing a perspective that has led to a continuing avocational interest in physics and this book. Dr. Bill's publications include research in the areas of fluid mechanics, micro–meteorology, combustion, and fire protection. In 1994 and 2003, he received the National Fire Protection Association's (NFPA) Bigglestone Award for communication of scientific concepts in fire protection. From 2002 to 2008 he served on the executive committee of the International Association for Fire Safety Science and in 2009 was elected as a lifetime honorary member equivalent to the grade of Fellow of the Society of Fire Protection Engineering. In 2019, he was a co-winner of NFPA's Philip J. DiNenno Prize for groundbreaking innovations that have had a significant impact in the building, fire and electrical safety fields. Now retired, Dr. Bill enjoys time with family, playing the violin, community volunteering, and walking his fox terrier.

www.ingramcontent.com/pod-product-compliance
Lightning Source LLC
Chambersburg PA
CBHW021347210526
45463CB00001B/15